Cell Cycle Control

Frontiers in Molecular Biology

SERIES EDITORS

B. D. Hames

Department of Biochemistry and Molecular Biology
University of Leeds, Leeds LS2 9JT, UK

AND

D. M. Glover

Department of Anatomy and Physiology
University of Dundee, Dundee DD1 4HN, UK

OTHER TITLES IN THE SERIES

Human Retroviruses
Bryan R. Cullen

Steroid Hormone Action
Malcolm G. Parker

Mechanisms of Protein Folding
Roger H. Pain

Molecular Glycobiology
Minoru Fukuda and Ole Hindsgaul

RNA–Protein Interactions
Kyoshi Nagai and Iain Mattaj

Protein Kinases
Jim Woodgett

DNA–Protein: Structural Interactions
David Lilley

Mobile Genetic Elements
David J. Sherratt

Chromatin Structure and Gene Expression
Sarah C. R. Elgin

Cell Cycle Control

EDITED BY

Christopher Hutchison
Department of Biological Sciences,
University of Dundee

and

David M. Glover
Department of Anatomy and Physiology
University of Dundee

at
OXFORD UNIVERSITY PRESS
Oxford New York Tokyo

Oxford University Press, Walton Street, Oxford OX2 6DP

Oxford New York
Athens Auckland Bangkok Bombay
Calcutta Cape Town Dar es Salaam Delhi
Florence Hong Kong Istanbul Karachi
Kuala Lumpur Madras Madrid Melbourne
Mexico City Nairobi Paris Singapore
Taipei Tokyo Toronto

and associated companies in
Berlin Ibadan

Oxford is a trade mark of Oxford University Press

Published in the United States
by Oxford University Press Inc., New York

A catalogue record for this book is available from the British Library

Library of Congress Cataloging in Publication Data
Cell cycle control / edited by Christopher Hutchison, David M. Glover.
(Frontiers in molecular biology)
Includes bibliographical references and index.
1. Cell cycle. 2. Cellular control mechanisms. I. Hutchison, Christopher. II. Glover, David M. III. Series.
QH605.C4255 1995 574.87'623–dc20 94–47374

ISBN 0 19 963411 4 (Hbk)
ISBN 0 19 963410 6 (Pbk)

Typeset by
Footnote Graphics, Warminster, Wilts.
Printed in Great Britain by
The Bath Press, Avon

Preface

The last decade has seen a unification of hypotheses concerning cycle cycle control, with the realization that many of the regulatory proteins have been highly conserved throughout evolutionary history. For a time it seemed that the unifying hypothesis was also a simplifying one. Now, however, this federation of ideas has begun to fragment, with the investigation of how different organisms have specialized in regulating particular aspects of cell cycle control, resulting in the emergence of a great plethora of proteins that govern cell cycle progression. In this book we have tried to trace the origins of this highly active research field in the early 1970s from the experiments of Masui and Markert that first described maturation promoting factor (MPF) in frogs; the cell fusion experiments of Rao and Johnson that revealed the block to the reinitiation of DNA replication and the dominance of the mototic state; and the work of first Hartwell and then Nurse and their colleagues to define the cell division cycle mutants of the budding and fission yeasts. The unification came, of course, with the realization in the late 1980s that work on this disparate group of organisms was all pointing to a common protein kinase cdc2 regulating the entry into mitosis. We have tried to put these aspects into context within two introductory chapters: the first is by Lee Hartwell who introduces the principles of cell cycle regulation, and the second is by Steve Reed, Chris Hutchison, and Stuart MacNeill who concentrate more upon the experimental organisms that are used in cell cycle research.

Our book is rather focused on the roles of the cyclin dependent protein kinases in cell cycle regulation, a group of enzymes of which cdc2 is the founder member. This is not to say that other regulatory pathways do not play an equally important role, but rather it reflects the fact that most of our knowledge lies in understanding the regulation of this group of enzymes, and that space within our book is limited. In this framework, the chapters by Steve Reed and by Stuart MacNeill and Peter Fantes describe the role of *CDC28* and *cdc2* in *Saccharomyces cerevisiae* and *Schizosacharomyces pombe* respectively. Although these proteins are required for both the G1–S and G2–M transitions, their chapters concentrate on START and the G1–S transition in the budding yeast, and the G2–M transition in fission yeast, reflecting the relative extends to which these transitions have been studied in the two organisms. Such studies in fission yeast first identified the major regulator of the G2–M transition as cdc2 kinase, the vertebrate homologue of which is described in the chapter by Gabriele Basi and Guilio Draetta. In mitosis the kinase is part of a complex with one of two cyclin sub-units, cyclins A and B. Such was our state of knowledge in the late 1980s. In the past five years, it has been realized that there exist large families both of related kinases and their regulatory sub-untis. This 'embarrassment of riches' is described by Jonathon Pines and Tony Hunter in Chapter 6. The regulation of S phase is the topic of Julian Blow's chapter. He looks at the use of cell free systems from *Xenopus* that have led to the 'Licensing Factor model' that explains why chromosomal DNA is only replicated once per cell cycle; and also examines the application of yeast genetics and the SV40 *in vitro* DNA replication system to this problem. He can only cover a small part of these stories, but the interested reader should watch out for the book that he is editing in this series that will develop these themes. The kinetics of cell cycle progression and cell growth in mammalian cells is the topic of the chapter by Anders Zetterberg and Olle Larsson. They consider many aspects of G1 regulation that form an important basis for the chapter by Emma Lees and Ed Harlow on Cancer and the Cell Cycle. This chapter once again looks at the roles of the cyclin dependent kinases, but specifically in relation to progression through G1. Finally Helen White-Cooper and David Glover consider the role of cell cycle regulation in *Drosophila* development. Originally, it was conceived that this chapter would cover the developmental regulation of the cell cycle in a wider variety of organisms. However, circumstances dictated the present format of the chapter, and although we acknowledge that other organisms have been neglected, *Drosophila* is the best characterized multi-cellular organism in this context.

Compiling this volume has been quite a painful process. The field has been advancing so rapidly, it has competed with the speed with which authors can produce chapters. We are very grateful to all the authors who have contributed. We acknowledge especially the patience of those who produced their chapters first, for making last minute revisions, and patiently waiting for the more delinquent rate-limiting numbers. We are conscious, as we go to press, of the important recent findings that we have been unable to cover as the field moves on, but nevertheless we hope that this book gives a timely overall perspective on current cell cycle research.

D.G. and C.J.H.

Dundee
February 1995

Contents

Contributors

GABRIELE BASI
Department of Genetics, Harvard Medical School, 200 Longwood Avenue, Boston, MA 02115, USA.

J. JULIAN BLOW
ICRF Clare Hall Laboratories, Blanche Lane, South Mimms, Potters Bar EN6 3LD, UK.

GUILIO DRAETTA
Mitotix Ltd., One Kendall Square, Building 600, Cambridge, MA 02139, USA.

PETER A. FANTES
Institute of Cell and Molecular Biology, University of Edinburgh, Darwin Building, Kings Buildings, Mayfield Road, Edinburgh EH9 3JR, UK.

DAVID GLOVER
CRC Laboratories, Department of Anatomy and Physiology, University of Dundee, Dundee DD1 4HN.

ED HARLOW
MGH Cancer Center, Building 149 13th Street, Charlestown, MA 02129, USA.

LELAND HARTWELL
Department of Genetics, University of Washington, Seattle, WA 98195, USA.

TONY HUNTER
Salk Institute, Tumor Virus Laboratory, PO Box 85800, La Jolla, CA 92138, USA.

CHRISTOPHER J. HUTCHISON
Department of Biological Sciences, University of Dundee, Dundee DD1 4HN, UK.

OLLE LARSSON
Department of Tumour Pathology, Karolinska Hospital, S-104 01, Stockholm, Sweden.

EMMA M. LEES
DNAX Research Institute, Molecular Biology, 901 California Avenue, Palo Alto, CA 94304, USA.

STUART MacNEILL
Institute of Cell and Molecular Biology, University of Edinburgh, Darwin Building, Kings Buildings, Mayfield Road, Edinburgh EH9 3JR, UK.

JONATHON PINES
Wellcome CRC Institute, Tennis Court Road, Cambridge, UK.

STEVEN I. REED
Scripps Clinic and Research Foundation, Molecular Biology, MB7, 10666 N Torrey Pines Road, La Jolla, CA 92037, USA.

HELEN WHITE-COOPER
Department of Anatomy and Physiology, University of Dundee, Dundee DD1 4HN, UK.

ANDERS ZETTERBERG
Department of Tumour Pathology, Karolinska Hospital, 5-104 01, Stockholm, Sweden.

1 | Introduction to cell cycle controls

LELAND HARTWELL

1. Introduction

A theory for eukaryotic cell cycle control has emerged in the last few years that has achieved a consensus in the research community. Although many surprises undoubtedly await the detailed testing of this theory, it is likely that its major tenets will stand the test of time. In this introductory chapter I will present a broad outline of our current understanding of how the cell cycle is controlled and raise a number of questions that are presently motivating cell cycle research. Most of the topics discussed in this chapter have been extensively reviewed recently; consequently, where possible, I will refer the reader to these reviews for documentation of the primary literature.

2. Many events must be coordinated

During the cell cycle many different macromolecular processes require coordination. Each of these processes is itself a complicated cycle requiring the synthesis of components, their assembly, and their function. For example, the chromosome cycle consists of DNA replication, condensation, and segregation with the attendant synthesis and assembly of the proteins that comprise the chromosome. The spindle pole cycle requires the duplication, separation, and segregation of the poles. A spindle cycle involves polymerization and depolymerization of microtubules, changes in their dynamic behaviour, and changes in their capacity to support oriented movement of organelles including the chromosomes. The nuclear envelope is disassembled and reassembled. Actin filaments are assembled, contracted, and disassembled to permit cytokinesis. Changes in many other cellular events are also orchestrated including secretion patterns, organelle reproduction and migration, cytoskeletal morphogenesis, and in some organisms cell wall growth and division. Moreover, the continuous processes of the cell, the accumulation of protein, RNA, and other components, must also be integrated with these discontinuous events during the cell cycle in order to maintain a more or less constant cell size.

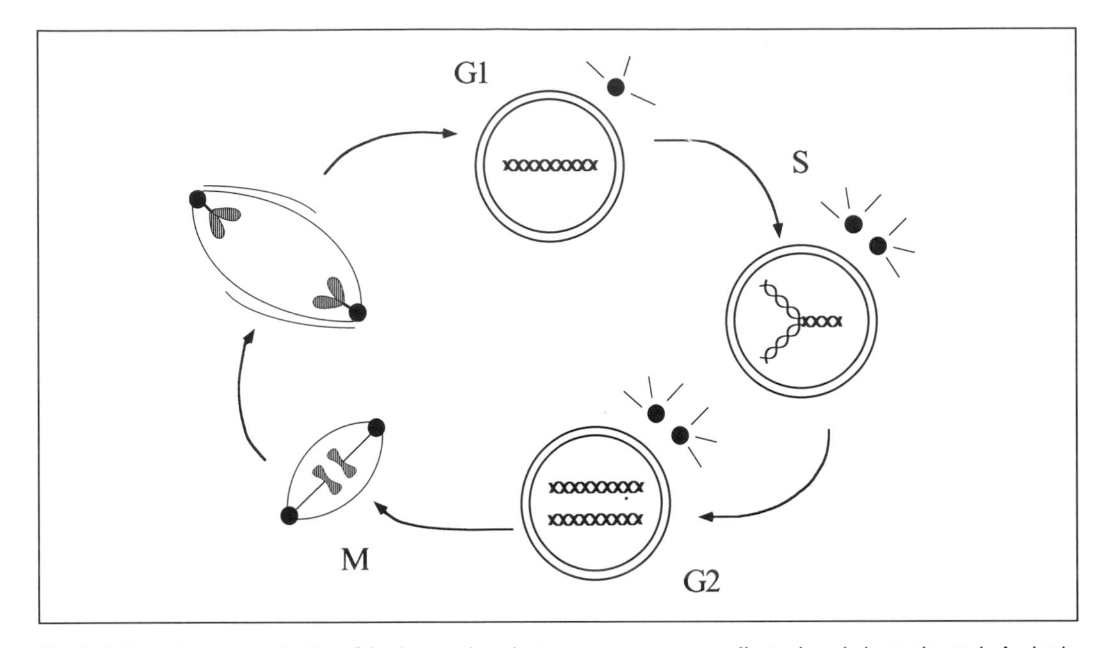

Fig. 1 Cell cycle events. In the G1 phase, the chromosomes are unreplicated and decondensed. A single spindle pole (centrosome) is present and microtubules do not extend from the pole into the nucleus. In most cells extracellular signals that control proliferation act to prevent the transition from G1 to S. The duplication of the spindle poles occurs in late G1 or at early S phase. In the S phase DNA is replicated and the spindle poles migrate apart. In G2 the chromosomes are completely replicated. The poles migrate apart at the beginning of mitosis (prophase) in most organisms; in *S. cerevisiae* the poles migrate earlier in the cell cycle. At M phase (mitosis) the nuclear envelope breaks down and the spindle poles organize an intranuclear spindle composed of microtubules; the chromosomes condense, attach to microtubules, organize themselves in the centre of the spindle, and then at anaphase sister chromatids separate to the spindle poles as a result of chromosome migration along the microtubules and movement of the poles apart. At telophase the chromosomes decondense, the intranuclear spindle is disassembled, the nuclear membrane reforms around the chromosomes, cytokinesis separates the two daughter cells, and the cells re-enter G1.

Thus the major problem that the cell faces is how to coordinate a large number of processes in both space and time. For example, the duplicated and separated spindle poles, and the fully replicated and condensed chromosomes must be available at the time that the mitotic spindle is organized. We might have found that each of these events was under a separate control and that their integration involved a very complex network of interconnections. However, recent findings suggest that the timing of all the events that transpire during the cell cycle may be controlled in a similar way, through protein phosphorylation. This hypothesis has produced the unified theory of the cell cycle.

3. The unified theory

What has been called the unified theory of the cell cycle proposes that the cell as a whole proceeds through a series of states during the cell cycle (1–3). In each state critical proteins are phosphorylated or not. Whether or not they are phosphorylated may regulate their activity, assembly, and/or cellular localization. Thus the events that occur together at any one time in the cell cycle are presumed to have been activated by the phosphorylation of critical proteins that mediate each of the events. A particular class of protein kinases, the cyclin-dependent kinases (cdk), is responsible for these phosphorylations and thus for orchestrating these changes of state.

It has been suggested that the simplest cell cycle may consist of only two changes in state, the S state (during DNA replication) and the M state (during mitosis) with critical proteins existing in two forms, phosphorylated during S or phosphorylated during M (1). These simplified cell cycles occur during the early embryonic division of organisms with very large eggs, like *Xenopus* and *Drosophila*. In these eggs, all the components needed for many divisions have been synthesized and stored in the cytoplasm during the oogenesis. Consequently, divisions can occur extremely rapidly, and the pauses, G1 and G2, that usually occur before and after DNA synthesis are absent. Whether or not the two-phase model for this embryonic system turns out to be correct, the concept is easily expanded to incorporate the existence of additional states.

4. The role of the *S. pombe* cdk in the G2/M transition

Cyclin-dependent protein kinases are now thought to be the primary agents mediating these changes in state. Research on a variety of organisms converged to identify cdk as the central players. A primary, cell cycle-controlling gene or activity had been identified through genetic or biochemical experiments in several highly divergent species (*Saccharomyces cerevisiae, Schizosaccharomyces pombe, Xenopus laevis, Aspergillus nidulans*, and vertebrate cells) and in each case the identity of this component was ultimately revealed to be a cdk (3–8).

The best understood example of a cdk is the cdc2 protein kinase responsible for the initiation and completion of mitosis in *S. pombe*, and it serves as our paradigm (refs 3, 9, and Chapter 2). This is because a large amount of genetic data has been accumulated on the genes controlling the expression and activity of this enzyme, thereby identifying the components that act *in vivo*. This genetic analysis has been correlated with biochemical observations analysing the activity and form of the enzyme in cell extracts made from synchronous cultures at various times in the cell cycle and from various mutants.

The kinase activity referred to as cdc2 kinase is composed of three proteins, Cdc2p, containing the catalytic site, Suc1p, a small protein of unknown function, and Cdc13p, a cyclin. Cyclins were first implicated in mitosis and derive their name by their periodic synthesis and destruction in synchrony with the cell cycle of invertebrate marine embryos (8). cdc2 kinase activity is usually assayed in cell extracts by its ability to phosphorylate histone H1. cdc2 kinase activity is low in extracts prepared from G1 cells, accumulates during the cell cycle prior to mitosis, and then disappears abruptly as mitosis occurs. Moreover, Cdc2p moves from the

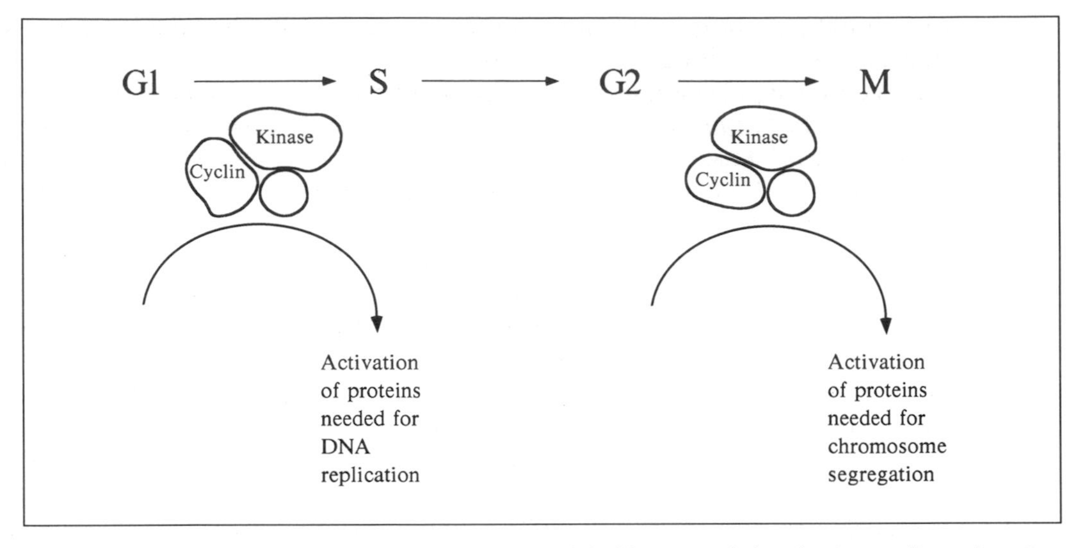

Fig. 2 cdk function in the cell cycle. The cdk composed of the kinase catalytic subunit, a cyclin, and another small protein act at the G1–S transition to phosphorylate proteins that are necessary for DNA replication and at the G2–M transition to phosphorylate proteins that are necessary for mitosis. A different catalytic subunit of the cdk and a different cyclin are utilized in vertebrates at the G1–S and the G2–M transitions while in the yeasts the same catalytic kinase subunit but different cyclins are utilized. It is possible that other, as yet undiscovered, transitions are mediated by other forms of the cdk.

cytoplasm to the nucleus as cells approach mitosis and abruptly leaves the nucleus as mitosis occurs. These changes in activity and location correlate well with genetic evidence showing a requirement for Cdc2p to activate mitosis (*cdc2* mutants arrest in G2) and are largely accounted for by changes in phosphorylation and subunit composition.

Prior to the activation of cdc2 kinase in G2, Cdc2p is found to be phosphorylated on three sites. One of these sites of phosphorylation, Tyr15, inhibits kinase activity and it is the removal of this phosphate that activates mitosis. Tyr15 phosphorylation is catalysed by two partially redundant kinases encoded by the *wee1* and *mik1* genes and its dephosphorylation is catalysed by the *cdc25* product (10).

cdc2 kinase activity must be inactivated for the cell to complete mitosis. This inactivation is accomplished not by rephosphorylation of Tyr15 but through some activity of the Suc1p and by degradation of the cyclin, Cdc13p.

The identification of the cdc2 kinase and the proteins that control it (Wee1p, Mik1p, Cdc25p) provide a remarkable example of the utility of genetic analysis. We are confident of the roles of these components in the cell because mutations in these genes produce characteristic phenotypes. A deficiency for a protein that inhibits cdc2 function (Wee1p) results in premature mitosis and overexpression of this function prevents or delays mitosis. A deficiency of Cdc25p, the function that activates Cdc2p, prevents or delays mitosis and overexpression of Cdc25p accelerates mitosis. Double mutants have been used to explain the order of action of different components. For example, a cdc2 protein containing a mutation changing the inhibiting phosphorylation site from tyrosine to phenylalanine is no longer dependent upon Cdc25p for its activation.

As complicated as it is, the story is only partially revealed. The activation of cdc2 kinase is thought to be autocatalytic and additional kinases and phosphatases have been reported that affect the timing of mitosis. These may act upstream of the wee1 or mik1 kinases and the cdc25 phosphatase implicating cascades of phosphorylation in the control of cdc2 kinase.

It is likely that most of the details worked out for the control of cdc2 kinase from *S. pombe* apply as well to that of other organisms (11, 12). Not only can the homologous kinases from *S. cerevisiae* and vertebrates substitute for Cdc2p in *S. pombe* cells but site-specific mutagenesis of the sites of phosphorylation suggest that the *S. cerevisiae* and vertebrate enzymes are controlled in *S. pombe* much as the *S. pombe* enzyme is.

5. The role of the *S. cerevisiae* cdk in the G1/S transition

The role of cdk in the G1–S transition is best understood in *S. cerevisiae* (ref. 12 of Chapter 1). The story here seems quite similar to that of the G2–M transition in *S. pombe*. A cdk (the *CDC28* product) works together with cyclins and a small protein of unknown function (Csk1), the *S. cerevisiae* G1 analogue of the *S. pombe* Suc1 protein. In *S. cerevisiae*, cells that are nutritionally deprived and cells that have responded to mating pheromones in preparation for conjugation are arrested in G1. Upon release from either of these conditions, cells first pass START, a step that commits them to a mitotic cell cycle and that requires the function of the cdk. Both temperature-sensitive mutants in the *CDC28* gene and deficiencies for G1 cyclins cause arrest at START. Conversely, overexpression of cyclins causes premature execution of START, evidenced by budding at a smaller than normal size and/or by less sensitivity to arrest by mating pheromones. The G1 cyclins are controlled by periodic transcription and by protein stability as is the G2 cyclin of *S. pombe*. Cyclin protein and mRNA are in low abundance in starved or pheromone-arrested cells, appear early in G1, and disappear as the cells pass START. Mutations that stabilized G1 cyclins from degradation also result in premature execution of START.

The sense of unity in cell cycle control was further strengthened when it was realized that the cdc2 kinase of *S. pombe* and the *CDC28* kinase of *S. cerevisiae* were not restricted to activating the G2–M transition or the G1–S transition, respectively. It has become clear that both cdk perform essential functions at least twice in the cell cycle, at the G1–S transition and at the G2–M transition. However, there must be some differences between the G1–S forms and the G2–M forms because they activate different functions and phosphorylate different proteins. The major known distinction is that different cyclins associate with the catalytic subunit in G1 and G2 and it is presumed that the cyclin dictates substrate specificity. One amusing experiment suggests that the state of the *cdc2* gene in *S. pombe* might also be important in keeping track of where the cell is in the cell cycle (13).

How many states of phosphorylation are there? The only known cdk of *S. cerevisiae* (*CDC28*) and *S. pombe* (*cdc2*) are required at least twice in the cell cycle, once in G1 and once in G2. Two forms of the kinase, if activated independently, could specify as many as four cell cycle states. Furthermore, in *S. cerevisiae* there are at least eight cyclins (14–16). Any single cyclin can be deleted without killing the cell; hence none is essential, but combinations of deletions are lethal. These cyclins are expressed in three sequential waves during the cell cycle and certain deletion combinations result either in arrest in G1, the initiation of S, mid-S phase or G2; these results suggest that there may be four phosphorylation states in the *S. cerevisiae* cell cycle. Two different cdks have been shown to have roles at different phases of the *Xenopus* embryonic cell cycle, cdk2 is required for S phase and cdc2 for mitosis (see Chapters 6 and 7). At least a dozen different cdks (17) and numerous cyclins have been cloned from human cells. These facts allow for the possibility of many combinations of cdks and cyclins and many potential phosphorylation states.

6. Synthesis, assembly, and location of proteins by phosphorylation

Many proteins important for cell cycle progress undergo changes in phosphorylation state during the cell cycle. It is easy to detect electrophoretic changes in protein migration due to phosphorylation, and this technique together with the generation of antibodies specific for cell cycle proteins has resulted in a wealth of knowledge about changes in phosphorylation during the cell cycle. It is also relatively straightforward to establish the significance of this phosphorylation by peptide mapping of the sites of phosphorylation, followed by site-specific mutagenesis of these serine, threonine, or tyrosine residues. However, identifying which of these phosphoproteins is actually the substrate of the cdks has been difficult (2, 11). One can demonstrate that the protein is phosphorylated *in vivo*, that the phosphorylation coincides temporally with the activation of the cdk, and, if mutants are available in the cdk or its associated cyclin, that phosphorylation of the suspected substrate is dependent upon active cdk *in vivo*. However, even with this evidence the suspected substrate could be a substrate of a kinase that is itself activated by the cdk. Since cascades of kinase activation by phosphorylation are well known, and may be the rule rather than the exception, this is quite likely. Other evidence that can be amassed is that the purified cdk will phosphorylate the suspected substrate *in vitro* and that the sites of phosphorylation *in vitro* are the same as those *in vivo*.

A number of potential cdk substrates have been examined in this way, including vimentin, the retinoblastoma gene product, and nuclear lamins (11). The nuclear lamins are probably the best studied and provide the clearest example of a likely substrate where the consequences of phosphorylation are dramatic. Nuclear lamins form a network of insoluble fibres immediately underlying the inner nuclear envelope and this meshwork is thought to provide important functional domains, not only for the maintenance of nuclear size and shape, but also for chromatin attachment sites. At mitosis the nuclear envelope is dissolved into vesicles, and the lamin network is solubilized. Concomitant with these changes the nuclear lamin proteins are phosphorylated. The phosphorylation can be stimulated by the mitotic form of the cdk kinase in cell extracts and the sites of phosphorylation include many of the same sites that are phosphorylated *in vivo*. Mutation of these sites results in lamins that cannot be solubilized and nuclear envelopes that do not break down at mitosis (see Chapter 5).

7. Cell cycle checkpoints

In most cells, the events in the nuclear cycle (spindle pole duplication and separation, DNA replication, spindle formation, chromosome segregation) are organized into a dependent sequence of steps (18, 19). For example, if DNA replication is inhibited with a temperature-sensitive mutation in DNA polymerase, mitosis and cell division are arrested as well. Why in this circumstance does the cdk not undertake the appropriate phosphorylation events to assemble the spindle and segregate chromosomes? There is nothing in the simple form of the unified theory that would allow the cell to monitor progress and be aware of its interruption.

There are two ways to explain this dependence of cell cycle events. In one model the activations and inactivations of the cdk would be steps within the dependent sequence. For example, the completion of DNA replication might be necessary for activation of the cdk that is, in turn, responsible for activating mitosis. In a second model activation of the cdk that activates mitosis might be inhibited by a separate control mechanism only if DNA replication is inhibited. In the first model the changes in state of the cdk are part of the dependent series of cell cycle events. In the second model the changes in state of the cdk occur independently of cell cycle events unless there is a perturbation to the course of these events.

The second model predicts the possibility of mutations that relieve the normal dependencies (20). If inhibition of DNA replication in turn inhibits activation of cdk then there must be components that mediate the inhibitory signal to the cdk; their elimination by mutation would result in a cell where inhibition of DNA replication no longer inhibited mitosis. Such mutations have been found.

A number of observations of this type have led to a second theory. We might call it the surveillance theory of the cell cycle (20–22). The cell contains a number of systems that are responsible for monitoring the proper completion of events. When such a system detects the failure of some event it signals the inhibition of downstream events. The surveillance systems have been called *checkpoints* because they define stages of the cell cycle at which the cell checks to see if it is okay to pass on. We think of them as intracellular signal transduction systems. They must be able to detect the failure of an early event, generate a signal, amplify and transmit that signal to a destination, and inhibit the machinery of the later events.

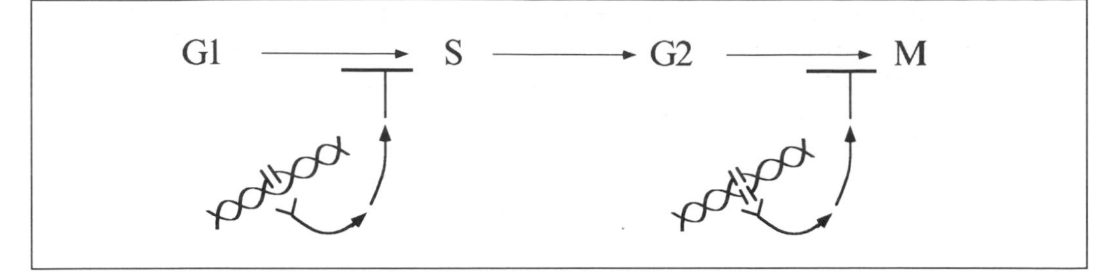

Fig. 3 Checkpoints in the cell cycle. Damage to DNA, shown here as single strand nicks, delays the transition from G1 to S phase, presumably to allow time for DNA repair prior to replication. DNA damage, shown here as double strand breaks, also inhibits the transition from G2 to M. The G2–M transition is also inhibited by unreplicated DNA and by unassembled microtubules. It is likely that other checkpoints, as yet undiscovered, exist in the cell cycle.

Like other signal transduction systems, there is evidence that some of these systems exhibit adaptation. That is, if the signal is not removed due to repair of the stalled event, the signal is in some cases eventually attenuated, allowing the cell to continue anyway.

These checkpoints are sensitive to extrinsically imposed perturbations to the cell cycle as well (see Fig. 3). DNA damage introduced by UV or X-ray irradiation arrests cells in G1, so that damaged DNA can be repaired before the DNA is replicated, and in G2 in order to repair chromosome breaks before chromosomes are segregated at mitosis.

The criterion for identifying a checkpoint control is the demonstration first that a dependent relationship exists between two events and second that this dependence can be relieved by mutation. These criteria have only been satisfied in a few instances. Entry into S is controlled by DNA damage; entry into mitosis is controlled by DNA damage, DNA replication, and microtubule polymerization. These examples document only two arrest stages, one at the G1/S boundary and one at the G2/M boundary. However, there is reason to suspect that there might be more than two checkpoint-mediated arrest stages..Temperature-sensitive *cdc* mutants of *S. cerevisiae* define at least six cell cycle stages where an arrest occurs that prevents the execution of late events when early events are inhibited. Whether all of these represent checkpoint-mediated arrests is at present uncertain since mutations have not yet been found that relieve the dependence of late events on each of the early events. However, the fact that the cell can pause for essentially a full cell cycle at intermediate states of progress suggests that additional controls to prevent uncoordinated progress may be present.

8. The cell cycle as an evolutionary paradigm

Research on organisms displaying cell cycles with rather different modes of control has revealed an underlying unity that is a lesson in evolution. We are beginning to see how the same cellular components can be used with subtle variations to achieve fundamentally different outcomes. The *S. cerevisiae* cell cycle exhibits a primary control point in G1 where growth is integrated with division and where mating pheromones can interrupt cell cycle progress. In these respects the *S. cerevisiae* cell cycle is like that of vertebrate somatic cells. In *S. pombe* growth is integrated with division in G2 for rapidly growing cells. Although we do not understand the precise reasons for these differences, the current expectation is that they will turn out to represent minor details in the way the cdks are controlled. In both organisms a single cdk functions to activate an early G1 event and the G2–M transition. The different biology could be explained by the hypothesis that growth is tied to the accumulation of a G1 cyclin in *S. cerevisiae* and to a G2 cyclin in *S. pombe*.

Another difference between the cdks of the two organisms is that phosphorylation on Tyr15 is more critical for control of the *S. pombe* cell cycle than it is for *S. cerevisiae*. If this tyrosine is converted to the amino acid phenylalanine, which cannot be phosphorylated, *S. pombe* loses the dependence of mitosis upon prior DNA replication but *S. cerevisiae* does not (23). Furthermore, this difference lies in components of the cell other than the cdk because if *S. cerevisiae* is running its cell cycle using the *S. pombe* cdc2 protein kinase it is insensitive to the phenylalanine replacement, whereas if *S. pombe* is utilizing the *S. cerevisiae CDC28* gene product it is sensitive to the phenylalanine replacement.

9. The cell cycle in development

Many variations in the cell cycle occur during development. In some organisms the early embryonic divisions are very rapid and, in contrast to the later divisions, occur without accompanying growth. In some cell cycles DNA replication occurs without nuclear division, producing polyploid or polytene nuclei, and in others, nuclear division occurs without cell division, producing multinucleate cells. During development cell differentiation is often coordinated with the lineage of cell divisions. In germ cells a special division, meiosis, occurs that is dramatically different from mitosis. How are these changes on the basic mitotic plan accomplished and how are they related to the activity of cdk and cell cycle checkpoints?

In *Drosophila* embryos the cells divide synchronously up to division 13 without net growth (Chapter 10). After this time transcription from the embryonic genome becomes prominent, the cell cycles slow down, and division is no longer synchronous in the embryo as a whole. However, even at this stage domains of contiguous cells remain synchronous and these domains are related to the developmental fate of the cells that compose them. Regulation at the level of the cdk is important in these transitions (24). During the first 13 divisions, the cell cycle is being driven by a store of maternal proteins and, although it is not known which protein is rate limiting in the early *Drosophila* embryo, in *Xenopus* embryos at this early stage cyclin synthesis is known to be the limiting component (25, 26). However, after division 13 the rate-limiting component in *Drosophila* becomes the cdc25 protein (27), the phosphatase that activates the mitotic form of the cdk.

Checkpoints are also different in some organisms in the early divisions compared to the later ones. For example, the early embryonic cell cycles of *Drosophila* and *Xenopus* differ from the later somatic divisions in that, in the embryo, late events are in general not dependent upon the completion of early events (20, 24). For example, if one blocks DNA replication with mutants or inhibitors, the cells of the early *Xenopus* embryo attempt mitosis but cells later in development do not (28). These observations would suggest that the early *Xenopus* embryo cell cycle is fundamentally different, perhaps lacking checkpoint controls. However, if one injects the one-celled embryo with extra DNA, then a dependence of mitosis upon prior DNA replication can be generated prematurely. Apparently, the signal transduction system that links mitosis to completion of DNA replication is present from the outset but it does not come into play until the amount of DNA reaches a sufficient concentration at about the 800 cell stage. When this checkpoint becomes active it acts to inhibit the mitotic cdk activity by increasing its phosphorylation on tyrosine (29).

These observations raise an interesting dilemma for the early embryo. Presumably, the reason that some embryos do dispense with checkpoints early in development is because their developmental strategy requires rapid and synchronous divisions. Checkpoints would delay the cell cycle in some cells, those that have experienced stochastic errors, resulting in asynchrony among cells. I have argued that checkpoints exist to ensure the fidelity of the mitotic programme. If the embryo dispenses with checkpoints, then its mitoses should have a greater risk of incurring errors. However, we know that mitotic errors contribute to developmental abnormalities and cancer so this would seem to be a risk that the embryo would not want to take. One way out of this dilemma would be for the embryo to have a method for assessing the quality of cells and disposing of those that are abnormal. This might seem like a difficult task, but we know that humans discard nearly all aneuploid fetuses at a very early stage of development and that organisms are capable of counting their X chromosomes so that males and females have equal expression of X-linked genes. Thus monitoring aneuploidy or other karyotypic abnormalities during development might not be impossible. In fact, there is evidence that *Drosophila* embryos do discard some aberrant nuclei at the blastoderm stage. Nuclei that fail to divide in *pal* mutant embryos (30), or fail to completely segregate their chromosomes if they have extra long chromosomes (31), are removed from the blastoderm surface, and segregated to the yolk mass; the yolk nuclei do not contribute to the tissues of the fly.

The nematode, *Caenorhabditis elegans*, provides a dramatic illustration of the relationship between the lineage of cell divisions and cell differentiation. This worm consists of only about 1000 somatic cells. The complete lineage of divisions has been mapped and is invariant from worm to worm (32). The fate of any particular cell is determined by its division history. Clearly, the control of gene expression must be integrated with the cell division programme. Although we know little about the nature of this integration in the nematode, an interesting model has been worked out in the yeast, *S. cerevisiae* (33). In this yeast there are three possible cell fates: haploid *MATa*, haploid *MATalpha*, and diploid *MATa/alpha*. In homothallic strains, a haploid *MATa* spore will produce two *MATa* cells at the first division, one being the mother cell and the other being the daughter cell. At the next division the mother will switch cell type and produce two *MATalpha* cells and the daughter will produce two *MATa* cells; the *MATa* and *MATalpha* cells then mate to produce two diploid *MATa/alpha* cells. This precise lineage raises several questions. We need to understand how mother cells switch mating type, why both of their progeny are switched, and why daughters do not switch. While all of these questions have not been fully answered, we know that switching occurs as the result of a site-specific double strand break introduced at the *MAT* gene by a specific endonuclease, HO. This endonuclease is turned on only in mother cells and only in G1. Its activation requires the expression of the cdk, *CDC28*, in G1 (34). Some of the genes responsible for controlling the expression of the *HO* gene have been identified, and how their activity is integrated with expression of the cdk in G1 is an active area of investigation (35).

Meiosis consists of two successive divisions (see Chapter 7). In the first meiotic division, DNA is replicated, homologous chromosomes pair, high levels of recombination occur, and at anaphase, homologous chromosomes separate rather than sister chromatids. At the second meiotic division there is no DNA replication and sister chromatids separate at anaphase. How are these changes from the mitotic programme achieved? We do not know the answer to this question. However, most of the genes that are essential for mitosis in *S. cerevisiae* are also essential for meiosis (36), indicating that the two programmes utilize most of the same machinery, including the cdk. However, temperature-sensitive mutations in the *S. cerevisiae* cdk, *CDC28*, arrest after DNA replication in the first meiotic division, whereas in mitosis they arrest prior to DNA replication (37). Control over the cdk of *Xenopus* oocytes is also used to hold mature oocytes arrested for long periods at the first meiotic metaphase. This arrest is achieved by an inhibitor of the cdk, cytostatic factor.

10. The cell cycle in disease

Any hyper- or hypoproliferation is potentially a cell cycle disease. The most obvious such disease is cancer (see Chapter 9). Cancer is a genetic disease. Mutations alter normal genes to produce cells that no longer obey the signals that usually control somatic proliferation. Many oncogenes and tumour suppressor genes have been identified because genetic alterations at these genes are reproducibly associated with certain types of cancer and by the fact that transfection of non-tumour cells with these altered genes can convert them to tumour cells or to cells that are more likely to produce tumours. We would expect mutations that produce cancer to be located in genes that are important for controlling cell division, for example, genes that produce and interpret signals telling cells to divide or not to divide. Growth factors, receptors that bind growth factors, and signal transduction components that mediate transmission of the signal from the receptor to its site of action indeed constitute the majority of the identified cancer genes (38). Furthermore, mutations in cyclin genes, inappropriate expression of cyclins (39), and inappropriate expression of the meiotic cytostatic factor *Mos* (40) have been implicated in tumourigenesis. These results suggest that the ultimate target of the signals that control cell proliferation to prevent cancer may be a cdk.

Genes that normally ensure the fidelity of mitosis also play a role in cancer (41). Tumour cells typically have grossly rearranged karyotypes, displaying chromosomal aneuploidy, and subchromosomal deletion, amplification, and translocation. There are probably two reasons why karyotypic rearrangements contribute to cancer. One is that a translocation or amplification can alter the regulation of an important gene. Another is that these changes might eliminate the one wild-type copy of an important gene leaving the cell with a single copy that may contain or acquire a recessive tumour-promoting mutation.

Many changes occur in cells as they become tumourigenic, including release from growth factor regulation, loss of contact inhibition, increased proteolysis, avoidance of immune surveillance, acquisition of immortality, promotion of angiogenesis and metastasis. It is unlikely that mutation at a single gene is responsible for all of these changes. Studies on colon cancer suggest that a series of mutations involving as many as six genes may be necessary (42). It was suggested many years ago that the karyotypic instability of tumour cells might be an expression of a higher mutation rate than normal cells (41). When we consider the number of phenotypic changes that a normal cell must undergo to become a metastatic cancer cell it might not surprise us that an elevated mutation rate would be necessary. Direct measurements of mutation rates in tumour cells compared to their normal counterparts have revealed only modest changes, if any, in the rate of point mutation (43). However, all tumour cells that have been examined exhibit greatly increased rates of gene amplification and this is likely to signal an increased rate of all types of chromosomal rearrangement (44).

What changes are responsible for the karyotypic instability of tumour cells? Recent results suggest that the most common tumour suppressor gene, p53, as well as the genes that are defective in a human genetic disorder which confers a proclivity to cancer, ataxia telangiectasia, is responsible in normal cells for a checkpoint control (45). Normal cells that experience DNA damage in G1 do not enter S phase until they have repaired their DNA, while cells mutant in p53 or in the ataxia genes enter replication with damage. This behaviour could lead to genomic instability; replication over damaged DNA would, for example, convert DNA single strand breaks to double strand breaks.

11. The future

It should be clear from this brief introduction that the unified theory of the cell cycle has raised many questions that will require years to answer. This is, of course, the measure of a good theory. As the subsequent chapters will show, much of the current work in cell cycle control is being driven by the expected role of cdk under the unified theory and by attempts to further test the role of surveillance mechanisms in the cell cycle.

Acknowledgements

I would like to thank Mandy Paulovich and Wendy Raymond for critical comments on the manuscript, and acknowledge the support of the American Cancer Society and the National Institutes of Health.

References

1. Murray, A. W. and Kirschner, M. W. (1989) Dominoes and clocks: the union of two views of cell cycle regulation. *Science*, **246,** 614.
2. Lewin, B. (1990) Driving the cell cycle: M phase kinase, its partners, and substrates. *Cell*, **61,** 743.
3. Nurse, P. (1990) Universal control mechanism regulating onset of M-phase. *Nature*, **344,** 503.
4. Draetta, G. (1990) Cell cycle control in eukaryotes: molecular mechanisms of cdc2 activation. *Trends Biochem. Sci.*, **15,** 10.
5. Maller, J. L. (1991) Mitotic control. *Curr. Opin. Cell Biol.*, **3,** 269.
6. Morris, N. R. and Enos, A. P. (1992) Mitotic gold in a mold. *Trends Biochem. Sci.*, **8,** 32.
7. Kirschner, M. (1992) The cell cycle then and now. *Trends Biochem. Sci.*, **17,** 281.
8. Minshull, J., Pines, J., Golsteyn, R., Standart, N., Mackie, S., Colman, A., Blow, J., Ruderman, J. V., Wu, M., and Hunt, T. (1989) The role of cyclin synthesis, modification and destruction in the control of cell division. *J. Cell Sci., Suppl.*, **12,** 77.
9. Forsburg, S. L. and Nurse, P. (1991) Cell cycle regulation in the yeasts *Saccharomyces cerevisiae* and *Schizosaccharomyces pombe. Annu. Rev. Cell Biol.*, **7,** 227.
10. Millar, J. B. A. and Russell, P. (1992) The cdc25 M-phase inducer: an unconventional protein phosphatase. *Cell*, **68,** 407.
11. Norbury, C. and Nurse, P. (1992) Animal cell cycles and their control. *Annu. Rev. Biochem.*, **61,** 441.
12. Reed, S. I. (1992) The role of p34 kinases in the G1 to S-phase transition. *Annu. Rev. Cell Biol.*, **8,** 529.
13. Broek, D., Bartlett, R., Crawford, K., and Nurse, P. (1991) Involvement of p34cdc2p in establishing the dependency of S phase on mitosis. *Nature*, **349,** 388.
14. Reed, S. I. (1991) G1-specific cyclins: in search of an S-phase promoting factor. *Trends Genet.*, **7,** 95.
15. Fitch, I., Dahmann, C., Surana, U., Amon, A., Nasmyth, K., Goetsch, L., Byers, B., and Futcher, B. (1992) Characterization of four B-type cyclin genes of the budding yeast *Saccharomyces cerevisiae. Mol. Biol. Cell*, **3,** 805.
16. Richardson, H., Lew, D. J., Henze, M., Sugimoto, K., and Reed, S. I. (1992) Cyclin-B homologs in *Saccharomyces cerevisiae* function in S-phase and in G2. *Genes Dev.*, **6,** 2021.
17. Myerson, M., Ender, G. H., Wu, C. L., Su, L. K., Gorka, C., Nelson, C., Harlow, E., and Tsai, L. H. (1992) 10 kinase genes structurally related to p34cdc2; the 7 novel genes are broadly expressed with each displaying some tissue specificity. The cdk3 gene can complement cdc28 mutants. *EMBO J.*, **11,** 2909.

18. Hartwell, L. H., Culotti, J., Pringle, J. R., and Reid, B. J. (1974) Genetic control of the cell division cycle in yeast. *Science*, **183,** 46.
19. Nurse, P., Thuriaux, P., and Nasmyth, K. (1976) Genetic control of the cell division cycle in the fission yeast *Schizosaccharomyces pombe. Mol. Gen. Genet.*, **146,** 167.
20. Hartwell, L. H. and Weinert, T. A. (1989) Checkpoints: controls that ensure the order of cell cycle events. *Science*, **246,** 629.
21. Enoch, T. and Nurse, P. (1991) Coupling M phase and S phase: controls maintaining the dependence of mitosis on chromosome replication. *Cell*, **65,** 921.
22. Murray, A. W. (1992) Creative blocks: cell-cycle checkpoints and feedback controls. *Nature*, **359,** 599.
23. Sorger, P. K. and Murray, A. W. (1991) S-phase feedback control in budding yeast independent of tyrosine phosphorylation of p34cdc28. *Nature*, **355,** 365.
24. Glover, D. M. (1991) Mitosis in the *Drosophila* embryo—in and out of control. *Trends Genet.*, **7,** 125.
25. Murray, A. W. and Kirschner, M. W. (1989) Cyclin synthesis drives the early embryonic cell cycle. *Nature*, **339,** 275.
26. Minshull, J., Blow, J. J., and Hunt, T. (1989) Translation of cyclin mRNA is necessary for extracts of activated *Xenopus* eggs to enter mitosis. *Cell*, **56,** 947.
27. Edgar, B. A. and O'Farrell, P. H. (1990) The three postblastoderm cell cycles of *Drosophila* embryogenesis are regulated in G2 by *string*. *Cell*, **62,** 469.
28. Dasso, M. and Newport, J. W. (1990) Completion of DNA replication is monitored by a feedback system that controls the initiation of mitosis *in vitro*: studies in *Xenopus*. *Cell*, **61,** 811.
29. Smythe, C. and Newport, J. W. (1992) Coupling of mitosis to the completion of S phase in *Xenopus* occurs via modulation of the tyrosine kinase that phosphorylates p34cdc2. *Cell*, **68,** 787.
30. Sullivan, W., Minden, J., and Alberts, B. (1990) Daughterless-abo-like, a *Drosophila* maternal-effect mutation that exhibits abnormal centrosome separation during the late blastoderm divisions. *Development*, **110,** 311.
31. Sullivan, W., Daily, D., Fogarty, P., Yook, K., and Pimpinelli, S. (1993) Delays in anaphase initiation occur in individual nuclei of the syncytial *Drosophila* embryo *Mol. Biol. Cell*, **4,** 885.
32. Sulston, J. (1988) Cell Lineage. In *The nematode Caenorhabditis elegans*. Wood, W. B. (ed.). Cold Spring Harbor Press, New York, p. 123.
33. Horvitz, H. R. and Herskowitz, I. (1992) Mechanisms of asymmetric cell division: two Bs or not two Bs, that is the question. *Cell*, **68,** 237.
34. Nasmyth, K. (1985) A repetitive DNA sequence that confers cell-cycle START (CDC28)-dependent transcription of the HO gene in yeast. *Cell*, **42,** 225.
35. Andrews, B. J. and Herskowitz, I. (1990) Regulation of cell cycle-dependent gene expression in yeast. *J. Biol. Chem.*, **265,** 14057.
36. Simchen, G. (1974) Are mitotic functions required in meiosis? *Genetics*, **76,** 745.
37. Shuster, E. O. and Byers, B. (1989) Pachytene arrest and other meiotic effects of the start mutations in *Saccharomyces cerevisiae. Genetics*, **123,** 29.
38. Bishop, J. M. (1991) Molecular themes in oncogenesis. *Cell*, **64,** 235.
39. Hunter, T. and Pines, J. (1991) Cyclins and Cancer. *Cell*, **66,** 1071.
40. Sagata, N., Watanabe, N., Van de Woude, G. F., and Ikawa, Y. (1989) The c-mos proto-oncogene product is a cytostatic factor responsible for meiotic arrest in vertebrates eggs. *Nature*, **342,** 512.

41. Nowell, P. C. (1976) The clonal evolution of tumor cell populations. *Science*, **194,** 23.
42. Vogelstein, B., Fearon, E. R., Scott, E. K., Hamilton, S. R., Preisinger, A. C., Nakamura, Y., and White, R. (1989) Allelotype of colorectal carcinomas. *Science*, **244,** 207.
43. Barrett, J. C., Tsutsui, T., Tlsty, T., and Oshimura, M. (1990) Role of genetic instability in carcinogenesis. In *Genetic mechanisms in carcinogenesis and tumor progression*. Harris, C. C. and Liotta, L. A. (ed.). Wiley-Liss, p. 97.
44. Tlsty, T. D., Margolin, B. H., and Lum, K. (1989) Three rat epithelial cell lines of tumorigenicity 0, 11–50, and 100% all gave amplification to PALA resistance but the tumorigenic clones were about 10 and 30× higher frequency. *Proc. Natl. Acad. Sci. USA*, **89,** 9441.
45. Hartwell, L. (1992) Defects in a cell cycle checkpoint may be responsible for the genomic instability of cancer cells. *Cell*, **71,** 543.

2 | A brief history of the cell cycle

STEVEN I. REED, CHRISTOPHER J. HUTCHISON,
and STUART MACNEILL

1. Introduction

A particular feature of research into the eukaryotic cell cycle is the diverse range of experimental organisms in use in laboratories around the world, ranging from unicellular yeasts, filamentous fungi (*Aspergillus*), marine invertebrates (e.g. clams and starfish), and flies, to vertebrates such as *Xenopus*, mice, and humans. In this chapter we briefly review the history of the eukaryotic cell cycle in terms of the organisms and experimental approaches used, and show how diverse species have contributed, and will continue to contribute, to our current understanding of the workings of the cell cycle.

2. Cell cycle analysis in yeasts

Cell cycle analysis in yeast has its origins in the 1950s with the pioneering work of Murdoch Mitchison in Edinburgh on the fission yeast *Schizosaccharomyces pombe* (1). Mitchison was attracted to the high degree of regularity in shape and size of this organism, as well as its symmetrical mode of division, allowing for relatively straightforward determinations of cell growth based on microscopic observation of individual cells (2). Mitchison's studies established relationships between a number of physiological parameters associated with growth and division and in doing so laid the foundation for later genetic studies. Genetic analysis of fission yeast also began around this time (3), although it was not until the 1970s that Nurse and co-workers first applied genetic methods to the cell cycle with the isolation of the first fission yeast cdc (cell division cycle) mutants (4). Genetic analysis of the budding yeast *Saccharomyces cerevisiae* also began in the 1950s (5, 6), but in contrast to the situation in *S. pombe*, physiological studies on the *S. cerevisiae* cell cycle did not significantly pre-date cell cycle genetics (7–9), with the result that the cell cycle problem in budding yeast tends to be framed in the context of cell cycle genetics, with many physiological avenues of investigation being opened up in response to the discovery and analysis of cdc mutants.

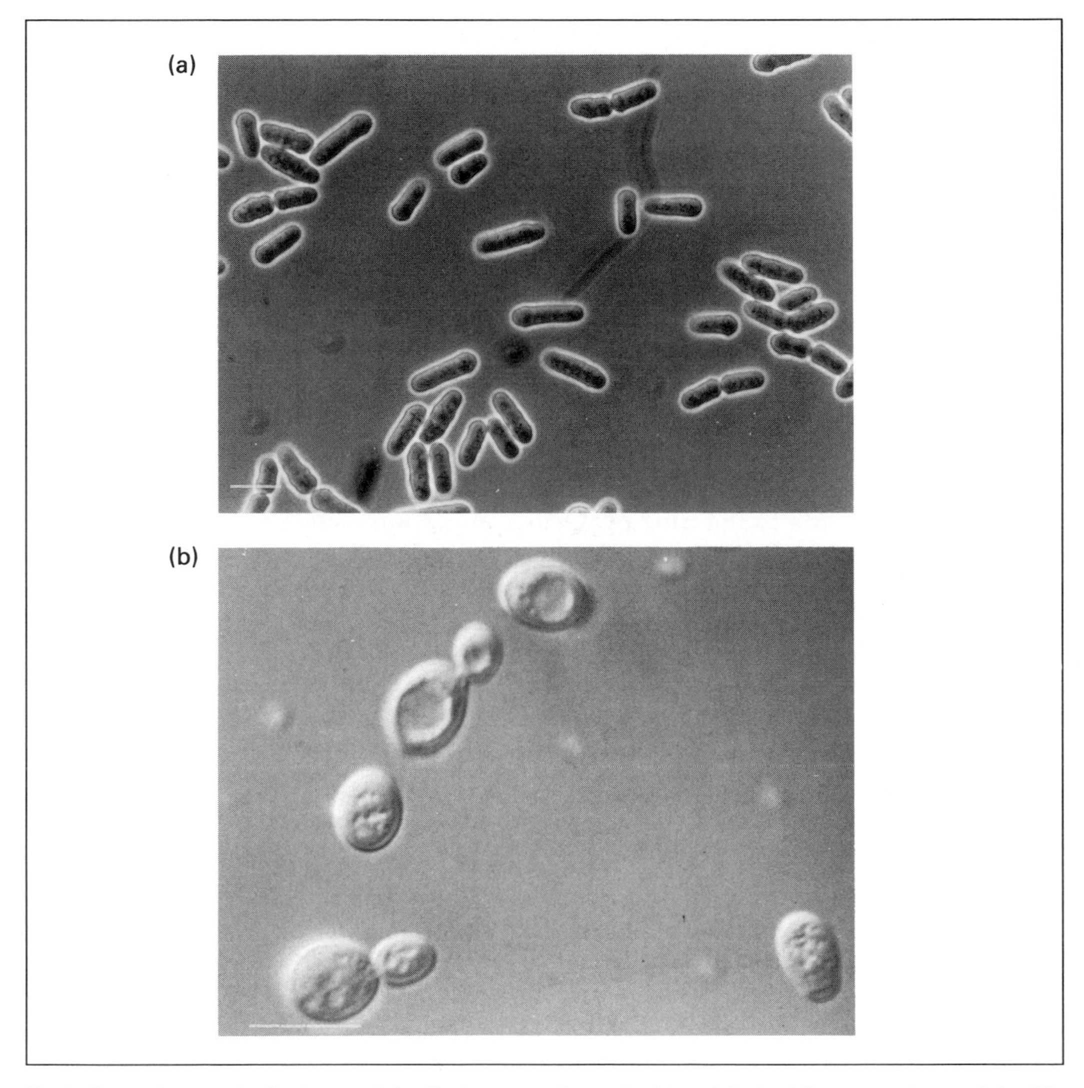

Fig. 1 Photomicrograph of cultures of the fission yeast *S. pombe* (a) and the budding yeast *S. cerevisiae* (b). Scale bar = 10 μm.

In spite of their different historical contexts, the development of the two organisms converged in the mid-1970s and has subsequently proceeded in parallel. The most notable advance was the application of recombinant DNA methodology to cell cycle genetics in both systems in the late 1970s and early 1980s (10, 11). With the ability to clone, characterize, and manipulate cell cycle genes, using relatively routine methodologies, the field has advanced rapidly to its present status, where both organisms serve as corner-stones for the aggregate of effort on the cell cycle problem.

3. Biology of the yeast cell cycle

3.1 *S. cerevisiae* versus *S. pombe*

Despite the fact that the two are frequently considered together, the yeasts *S. cerevisiae* and *S. pombe* are not particularly closely related in evolutionary terms. Indeed they appear to be as divergent from each other as they are from human cells (12). This lack of relatedness is manifest in a broad range of cellular processes, including the cell cycle; a fundamental difference between budding and fission yeast cells is that a short intranuclear spindle begins to form in budding yeast cells early in the cell cycle, at a point that corresponds approximately to the beginning of the G2 interval, whereas in *S. pombe* the spindle is present only during mitosis (Fig. 2 and ref. 3). Although the structure of the *S. cerevisiae* spindle changes significantly at mitosis, the existence of the pre-mitotic spindle appears to be a unique characteristic of budding yeasts. The assembly of this pre-mitotic spindle corresponds to the time of nuclear migration, when the nucleus approaches and then enters the canal between the mother cell and the bud, and the pre-mitotic spindle may be involved in some aspect of nuclear migration and/or positioning during this process (Fig. 2). In addition, budding yeast chromosomes do not become obviously condensed during the mitotic process, while those of fission yeast and higher eukaryotes do.

These phenomenological distinctions of budding yeast have been used to argue that this organism's cell cycle is structured differently from that of the other eukaryotes. Indeed, it has been suggested that fission yeast, with its more straightforward cell cycle (the fission yeast cell cycle has discrete G1, S, G2 and M phases), may be a better model for cell cycle studies (see ref. 14). However, more recent evidence, notably the elucidation of many of the molecular mechanisms that underlie cell cycle progression, suggests that these distinctions are likely to be superficial only and that the fundamental organization of the eukaryotic cell cycle is conserved in budding yeast at the physiological level.

3.2 Differences between the cell cycles of yeast and higher cells

Although the cell cycles of budding and fission yeast are dissimilar in many ways, they share some characteristics that render them distinct from higher cells. The most notable of these distinctions is the 'closed' mitosis characteristic of all fungi. Because the nuclear envelope remains intact during mitosis, the process of mitosis in fungi is often referred to as nuclear division. The topological necessities of such a mitosis dictate that the microtubule organizing centres of the mitotic spindle, known as the spindle pole bodies, exist as distinct organelles embedded in the nuclear envelope (see Fig. 2 and ref. 15). The mitotic spindles of both organisms are intranuclear but go through similar mechanics, particularly elongation during anaphase (anaphase b), as do the spindles of higher cells, although there is no

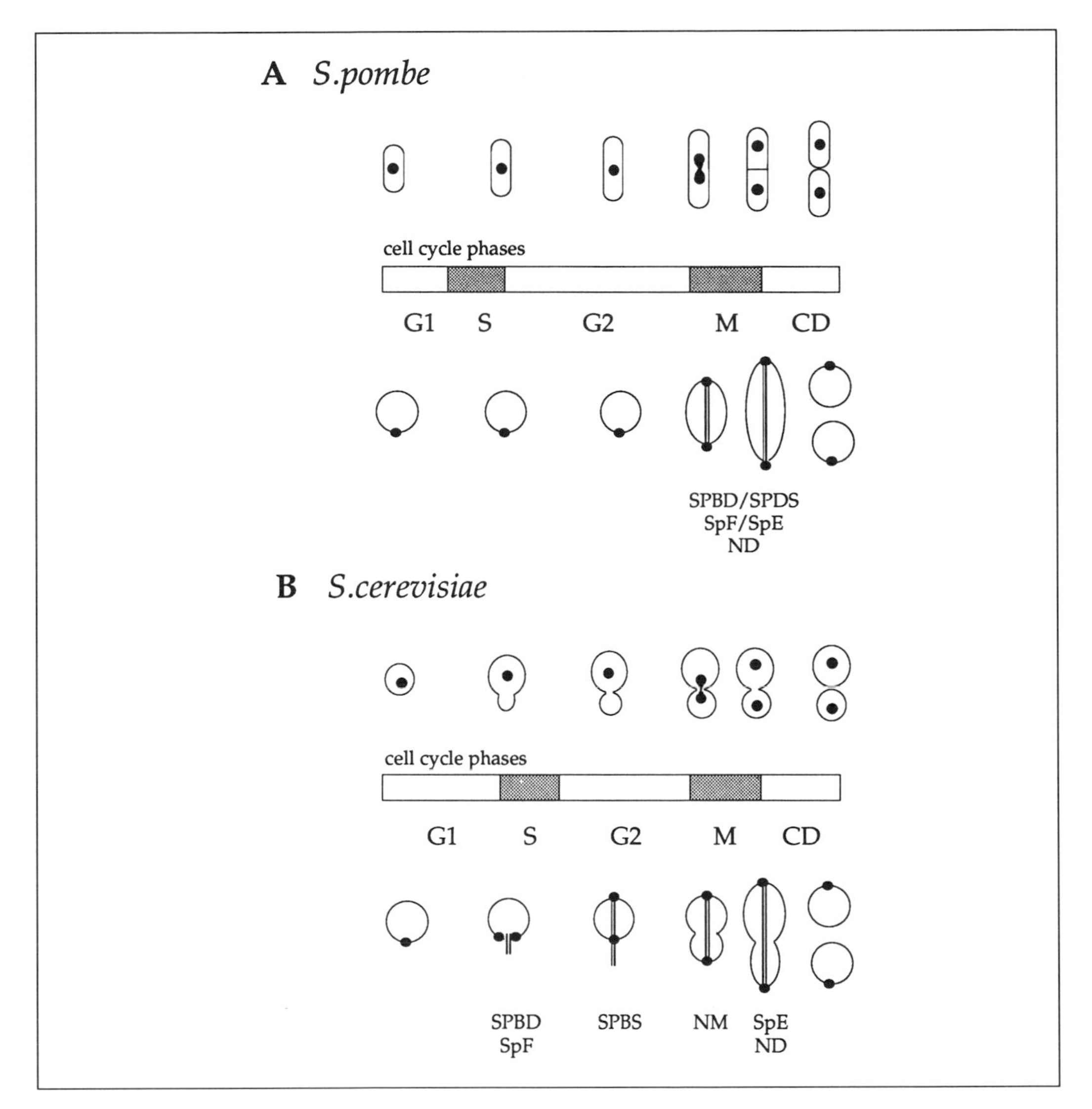

Fig. 2 Schematic representation of the organization of the cell cycles of *S. pombe* (A) and *S. cerevisiae* (B), showing cell morphology (upper part), the relative duration of the cell cycle phases (centre) and the behaviour of the spindle pole bodies and spindle (lower part). Abbreviations: SPBD (spindle pole body duplication), SPBS (spindle pole body separation), SpF (spindle formation), SpE (spindle elongation), NM (nuclear migration into bud neck), and ND (nuclear division).

evidence for polar chromosomal movement along spindle microtubules (anaphase a). Clearly, in both yeasts, the nucleus stretches and deforms significantly from a spherical shape during mitosis, suggesting the possibility of an alteration in the internal nuclear architecture, perhaps akin to disassembly of nuclear lamins in higher cells. However, the existence of molecules related to or analogous to lamins has not been conclusively demonstrated in either organism. It has been reported that in both yeasts there is not a cessation of macromolecular synthesis and secretory processes during mitosis, as has been demonstrated for animal cells. However, because it has been difficult to obtain a synchronous population of yeast cells that can be considered truly mitotic, it has been difficult to address this issue.

4. Cell cycle genetics

4.1 Origins

The ascendence of yeast as a cell cycle model resulted to a large degree from a chance observation and an inspired interpretation of it. In a screen for random temperature-sensitive (ts) lethal mutations in the budding yeast *S. cerevisiae*, targeted to identify elements essential for macromolecular synthesis, Hartwell *et al.* (16) noticed that a class of ts mutants exhibited an abnormal but homogeneous morphology at the restrictive temperature. In addition, whereas these mutant cells continued to enlarge for some time, they usually ceased division within one population doubling time. This led Hartwell *et al.* to conclude that this class represented mutants that were not defective in growth but rather in cell cycle progression.

From a genetic perspective, the implications of this work were far-reaching: single gene products might be responsible for single functions within the cell duplication process. The most sophisticated genetic analysis at that point in history directed towards a morphogenetic process was bacteriophage morphogenesis and the potential parallel with these putative yeast cdc mutants was not lost on early investigators in the field. In essence, much of the initial genetic analysis was framed in the context of cell division being a self-assembly process conceptually analogous to the processes of phage morphogenesis (17).

4.2 Essential concepts

Conditional lethal cdc mutations are defined by two parameters. The first of these, known as the terminal phenotype (16, 18), corresponds to the final morphological and physiological state of the mutant cell after division has ceased. The second parameter associated with cdc mutations is the execution point (16, 18). The execution point or transition point, simply defined, is the point in the cell cycle where the mutationally conferred defect becomes critical, ultimately leading to arrest. These concepts will be expanded below.

4.2.1 Terminal phenotype

The most obvious notable aspect of cdc mutants, the characteristic feature that initially caught the eye of Hartwell *et al.* (16), is the homogeneous appearance of cells in an arrested culture. This intrinsic property of a cdc mutation is referred to as the terminal phenotype. There are several implications of the existence of homogeneous terminal phenotypes. The first is that the loss or alteration of the function of a single gene can lead cells to a distinct morphological and, presumably, physiological state. Second, the same loss or alteration of gene function does not affect progression of a mutant cell through the majority of the cell cycle. In effect, homogeneous arrest from an asynchronous culture suggests that the cell cycle machinery utilizes some individual gene products in a highly specialized and process-specific fashion. This revelation may not seem extraordinary or even surprising in the current context of advances in the cell cycle field. However, the notion that single gene products would be devoted to highly specialized individual cell division functions was not intuitive at the time that cdc mutants were initially described. There is one common misconception concerning the terminal phenotype. It is often assumed that the terminal phenotype corresponds literally to a point in the cell cycle where cells have become temporally frozen due to inability to progress. In fact, terminally arrested mutant cells do not reflect any state normally encountered in the course of cell cycle progression. The terminal phenotype most likely reflects a departure from the normal cell cycle, presumably from the point where the cell cycle is blocked. Why this progression to an aberrant state occurs will be discussed below in the context of dependency of cell cycle events. Nevertheless, this fact must be taken into account in any assessment of possible function of a gene product marked by a mutation relative to the terminal phenotype produced. In essence, it is necessary to extrapolate back to the point of departure from the cell cycle in order to consider possible targets of the mutational lesion.

4.2.2 Execution point

The execution point (or transition point) defines the last point in the cell division cycle where a gene product is required. If a cell is at an early point in the cell cycle at the time at which it is deprived of a gene product, then it will proceed to the execution point and not complete the current cell cycle. If, however, the cell has already passed the execution point, it will undergo one division and then arrest in the subsequent cell cycle. The framing of the execution point as a demarcation between arrest prior to division and arrest subsequent to division allows the establishment of the execution point of any gene product for which a loss-of-function mutation is available. It is assumed that most of the ts cdc mutations are of this type. Simply stated, if an execution point is early in the cell cycle, most cells in a population at any time will already have passed it. These will undergo one division before arresting. However, if an execution point is late, few cells will be able to divide before arresting. In theory, therefore, one need only to shift an asynchronous ts mutant population to the restrictive temperature and measure the increase in cell number to establish the execution point of the corresponding gene product.

5. Cell cycle mutants and the nature of dependency

The fact that cdc mutants can produce distinct terminal phenotypes definable by single execution points reflects a fundamental aspect of cell cycle organization in yeast, that is that cell cycle progression can be halted in a highly characteristic manner by impairment of the functions of individual gene products. This implies that dependency relationships exist between cell cycle events so that later events cannot be initiated if earlier ones are blocked. That terminally arrested mutant cells do not precisely resemble normal cycling cells at any point during cell cycle transit however, suggests that not all cell cycle events are on the same dependency pathway. Although several methods have evolved for establishing dependency between events, none of these addresses their molecular basis.

Two models have been proposed to explain the dependency relationships described in the preceding section. The first might be called the 'self-assembly' model and is based largely on bacteriophage morphogenetics. It was shown that the process of phage assembly could be broken down genetically into dependent pathways. If an event in one of these pathways was blocked by mutation, precursors in that pathway would accumulate and mature phage would not be produced. The obvious explanation for the dependency relationships exhibited in such self-assembling systems is that early events produce substrates for later events. Thus, if an early event is blocked, a later event, dependent for its substrate on prior processes, cannot be initiated or completed. The 'self-assembly' model interprets the dependency relationships revealed in the course of genetic analysis of the yeast cell cycle within the same framework. This model is attractive for its logic and simplicity, as one need only assume precursor–product relationships between successive events in the course of assembly of a new cell. It is certainly likely that relationships of this type exist in the cell cycle, but none has yet been demonstrated.

The alternative model to explain dependency has been termed the 'checkpoint model' (19, 20). Within this framework, dependency relationships are not intrinsic to successive cell cycle events. Instead, they are superimposed as regulatory networks in order to maintain an orderly progression through the cell cycle. In effect, the completion of an earlier event would produce a signal required for a subsequent event to begin. In contrast to the paucity of evidence for the 'self-assembly' model, there is ample evidence that regulatory 'checkpoints' govern much of the dependency observed in the yeast cell cycle. First, in primitive cell cycles such as those observed in early cleavage embryos, the same events occur as in the yeast cell cycle, yet much of the dependency is absent. It has been argued that in such cell cycles rapidity is more important than fidelity so that many checkpoint controls are removed. The most compelling evidence for checkpoint regulatory mechanisms, however, is the existence of mutations that apparently remove them.

The best characterized of the checkpoints in the *S. cerevisiae* and *S. pombe* cell cycles is that which couples the completion of DNA replication and the integrity of the replicated DNA to the initiation of mitosis. Under normal circumstances, cells that have not completed DNA replication or which have experienced DNA damage will become arrested or delayed in G2—mitosis only ensues when replication is completed and damage has been repaired. In budding yeast *RAD9* is one of several genes that can be mutated to undermine this regulation (19). In a *rad9* mutant, cells that have not completed replication (e.g. *cdc9* mutants defective in DNA ligase at the restrictive temperature) or which have suffered radiation damage to their DNA have lost the capacity to arrest in G2. Instead, they proceed into mitosis with lethal consequences. In fission yeast similar phenotypes are displayed by mutants in the *rad1*$^{+}$ gene amongst others (refs 21, 22, and see Chapter 4).

The molecular basis for *RAD9*- or *rad1*-mediated checkpoint controls has not yet been determined. In either organism, however, the similar cdc phenotypes conferred by mutations that impair DNA replication can now be explained by the existence of a replication checkpoint as can the dependency of mitotic events upon the replication events defined by these mutations. In both organisms screens to identify checkpoint mutations are being employed that utilize sensitivity to agents that block DNA replication as a criterion for loss of checkpoint control. Such checkpoint mutants, as is the case with *rad9* in *S. cerevisiae* or *rad1* in *S. pombe*, are expected to commit suicide when replication is blocked or delayed.

Recently, other checkpoints have been defined in budding yeast and fission yeast. One of these controls the link between metaphase and anaphase (23–25). Mutations defining elements in this system have been identified based on sensitivity to benomyl, an anti-microtubule drug that impairs spindle formation. As with the case of agents inhibiting DNA replication and DNA checkpoint mutations, metaphase checkpoint mutations appear to die as a result of undergoing an abortive mitosis in the presence of benomyl. Another checkpoint identified in *S. cerevisiae* controls the link between bud formation and mitosis. When budding is blocked mutationally or as a result of environmental conditions, mitosis is normally delayed (26). It has been possible to isolate checkpoint mutations that override this delay.

Perhaps the best studied and understood checkpoint occurs at the beginning of the cell cycle in G1 and is known as START. In *S. cerevisiae*, this is the point in the cycle where nutritional, hormonal, and cell size control are all integrated. Checkpoint mutations exist for START control, causing the cell cycle to be initiated prematurely or under inappropriate conditions. Since this aspect of cell cycle control is discussed in detail in Chapter 3, we merely present it here as another example of the checkpoint paradigm and to underscore the pervasiveness of this mode of regulation in the organization of the cell cycle. It should be noted, however, that not all critical cell cycle defects lead to well-defined terminal phenotypes. One explanation for this is that some of these lesions bypass checkpoint controls altogether and since it is primarily the checkpoint controls that lead to distinct cdc phenotypes, none is observed in these cases. Thus, atypical phenotypes associated with essential cell division cycle functions may be indications that these functions operate outside of the checkpoint control systems. An alternative possibility is that certain genes may be required for both completion of a specific cell cycle event (such as DNA replication) as well as for establishing conditions in which a checkpoint can operate. An example of this phenomenon is provided by the *cdc18*$^{+}$ and *cut5*$^{+}$ genes in *S. pombe*: cells lacking either *cdc18*$^{+}$ or *cut5*$^{+}$ function, which are unable to replicate their DNA, proceed into mitosis, indicating that the pre-mitotic checkpoint is inoperative in these circumstances (refs 27, 28, and see Chapter 4).

6. Genetic and molecular methods of yeast cell cycle analysis

6.1 Genetics analysis

Mutants are an essential part of cell cycle analysis in yeast. Three general types of mutant that have proved especially useful in cell cycle studies over the last twenty years (recessive loss-of-function, dominant gain-of-function, and dominant negative mutants) are discussed below and their effects illustrated schematically in Fig. 3.

The first type of cell cycle mutant to be identified and exploited (as discussed in Section 3.1) was the recessive conditional lethal. Heat-sensitive (usually referred to as temperature-sensitive) mutations have been the most widely used of this type, although cold-sensitive (cs) mutants and mutants that are unusually sensitive to drugs that block DNA replication of mitosis have been put to good use in both organisms. It is often assumed (incorrectly) that such mutations completely eliminate the function of the mutant protein under restrictive conditions, but in the vast majority of cases this seems unlikely to be the case. Almost all of the temperature-sensitive alleles of the fission yeast *cdc2* gene, for example, despite causing cell cycle arrest at the restrictive temperature, retain sufficient function so that when overexpressed, cell cycle progression can continue. In addition, several cases have been identified where the phenotype of cells carrying a chromosomal deletion of a gene (a true null allele) behave markedly differently from those in which a temperature-sensitive protein has been inactivated.

Dominant gain-of-function mutations have also played a key role in helping to unravel the complexities of cell cycle gene functions. Particularly informative are

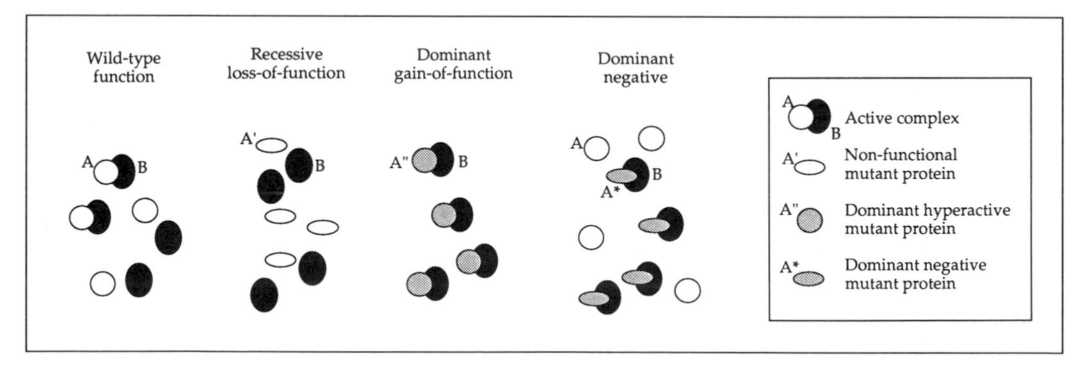

Fig. 3 How mutations affect protein function: an example. Two proteins (labelled A and B) form an active complex required for progression through the cell cycle, and whose activity determines the timing of a key cell cycle step. (1) In the example shown, the effect of recessive loss-of-function mutations in protein A (labelled A′) is to cause dissociation and inactivation of the complex—cell cycle progression is therefore halted. (2) Dominant gain-of-function mutations in protein A (A″) cause formation of a more active A–B complex than would otherwise exist with the result that the timing of the A–B-dependent cell cycle step is advanced. (3) Dominant negative mutations in protein A (A*) sequester the functional B into inactive complexes. Since the amount of B is limiting, the wild-type A protein cannot form a functional complex and cell cycle progression is halted.

those that bring about the acceleration of key cell cycle steps. In budding yeast, two dominant mutant alleles of the *CLN3* gene (specifically *CLN3–1* and *CLN3–2*) were isolated by virtue of their ability to bring about passage of START at a smaller cell size than in wild-type cells, indicating that *CLN3* function is rate-limiting at this point in the cycle. Subsequent analysis of these mutant alleles has shown that they encode hyperstable truncated cyclin proteins (see Chapter 3). In fission yeast, dominant (wee) mutations in the *cdc2*$^+$ gene have been isolated that cause premature entry into mitosis and cell division at a reduced cell size. In these cases the mutant proteins are altered in their interactions with their regulators, the cdc25 and wee1 proteins (see Chapter 4).

A second type of dominant mutation (Fig. 3) which is proving increasingly useful is the dominant negative (also called dominant lethal). In this case the mutant proteins exert their effects on the cell by interfering with the function of the corresponding wild-type protein (for example, by titrating out the activity of an essential activator) to cause cell cycle arrest. Mutants of this type have typically been identified by screening banks of randomly mutagenized plasmids carrying the wild-type gene (see Section 5.2) (29), though in some cases targeted mutagenesis of the wild-type gene has thrown up dominant alleles by chance (30).

6.2 Molecular analysis

The approaches described above for the most part reveal the relationships that exist between various aspects of cell cycle progression but provide little basis for these relationships at the molecular level. Many of the principal breakthroughs in characterizing the molecular basis for cell cycle progression and control came about through the application of molecular biological (recombinant DNA) technologies to yeast cell cycle genetics. Central to this was the ability to isolate cell cycle genes by complementation, which was itself dependent upon the development of methods for the introduction of plasmid DNA into yeast cells at reasonable efficiency (10, 11) and of plasmid vectors capable of being maintained in both yeast and *E. coli* (31–34). In budding yeast, these methods allowed the isolation of cell cycle genes by marker rescue of conditional-lethal cdc mutants (32, 33). One of the first *S. cerevisiae* genes to be cloned was *CDC28* (discussed at length in Chapter 3). Transformation methodologies for *S. pombe* followed shortly after (35) and in this organism the *cdc2*$^+$ gene was the first cell cycle gene to be cloned (ref. 36, and see Chapter 4).

Such has been the effectiveness of this approach that it is now safe to conclude that most, if not all, of the cell cycle genes defined mutationally in the two yeasts have been cloned, sequenced, and scanned against databases of known protein primary structures. In a number of cases pre-existing chromosomal mutant alleles have also been cloned and sequenced, allowing location of mutated sites in the primary sequence. Cloned sequences and recombinant DNA techniques have allowed the preparation of antibodies prior to knowing anything about gene product properties or functions. The result is that in a significant proportion of cases, rapid progress has been made towards a molecular understanding of cell cycle events. For the majority of genes, at least some useful and ultimately interpretable information has been gained. Only in a minority of instances has the molecular genetic strategy proven totally uninformative.

6.2.1 Manipulating cloned genes

The molecular cloning of a cell cycle gene opens the way towards a range of methodologies unrivalled in any other eukaryotic system (summarized in Fig. 4). The ability to create a null allele of any cloned gene in a routine fashion has been one of the most powerful genetic tools in both yeast systems. Double-stranded DNA fragments corresponding to chromosomal sequences serve as efficient substrates for gene conversion in both *S. cerevisiae* and *S. pombe* (37). Operationally,

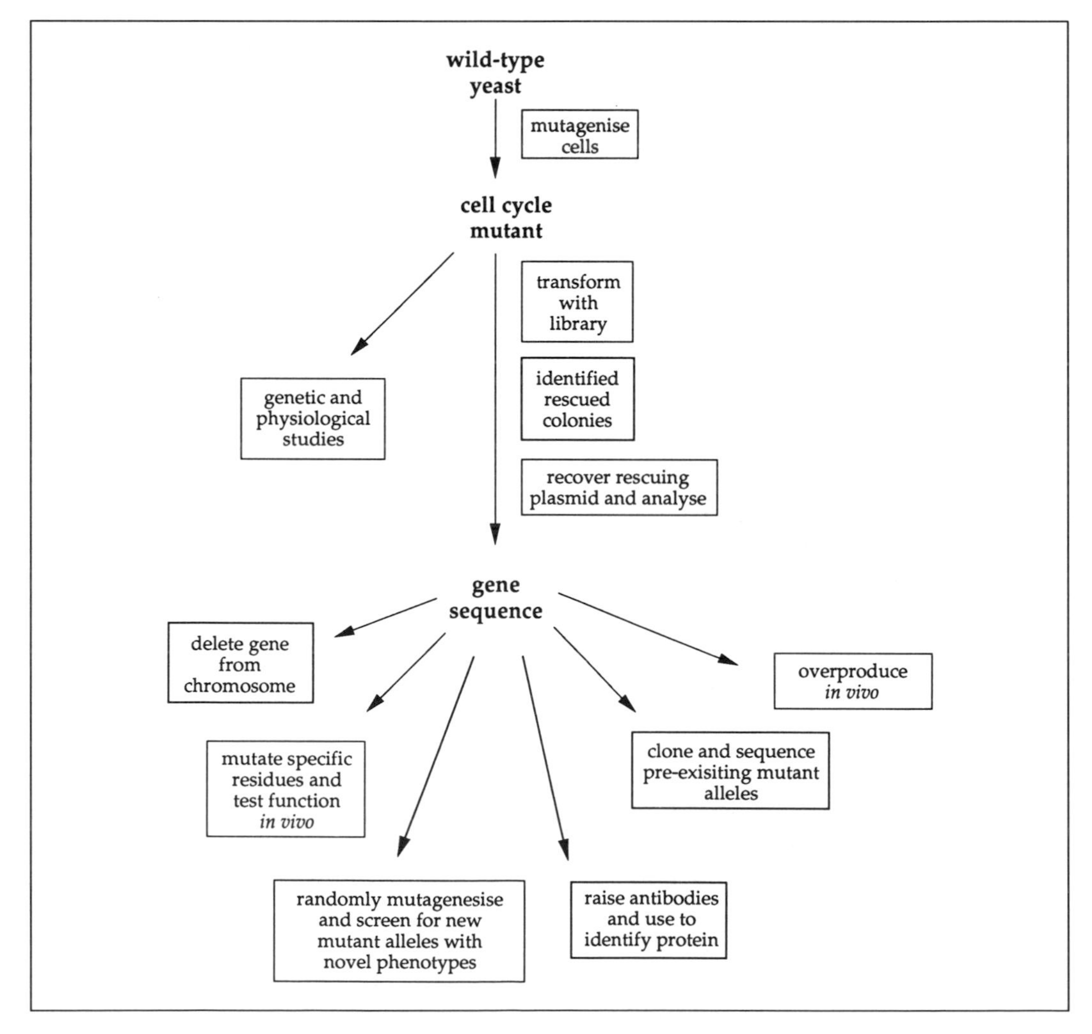

Fig. 4 The awesome power of yeast genetics.

this means that fragments corresponding to null alleles, usually insertions or substitutions of selectable markers, can be introduced by transformation procedures, and gene convertants containing null alleles can be selected. These manipulations are frequently carried out in diploid cells (only one copy of the target gene is disrupted) which can then be sporulated and the phenotype of haploid cells carrying null alleles analysed. Multiple gene disruptions are also possible (see for example ref. 38). Analogous procedures in most other organisms, though possible, remain laborious. In addition to eliminating gene function altogether by deletion or disruption, a number of systems are available that facilitate high-level constitutive or regulatable expression of cloned genes *in vivo*. Thus genes can be free from the normal restraints of regulation.

The ability to create novel mutant alleles *in vitro* by site-directed or random mutagenesis and subsequently to test their function *in vivo* (in the presence or absence of a functional copy of the endogenous gene) is another powerful approach that has been put to excellent use in a number of important studies over recent years (see for example ref. 39). In particular, this method allows the functional dissection of proteins and the identification of key amino acid residues and/or domains of structure. New alleles constructed *in vitro*, such as the dominant negatives described above, can also provide the basis of new genetic screens (see Section 5.3).

6.2.2 Crossing phylogenetic boundaries

One novel approach that has proven highly effective is to utilize mutants in one yeast to clone functionally homologous genes from the other. Initially, *S. cerevisiae* genomic DNA libraries (constructed using budding yeast vectors) were used to rescue *S. pombe* mutations because such plasmids can generally be propagated in *S. pombe* and because *S. cerevisiae* genes are typically free of introns. This approach has been applied to a variety of different types of genes, including cdc genes (36). More recently, with the preparation of cDNA libraries in *S. cerevisiae* and *S. pombe* expression vectors, it has also been possible to identify the homologues of yeast cdc genes from higher eukaryotes (40–42).

6.3 Widening the net; identifying interacting gene functions

Perhaps the greatest strength of yeast genetics is the ability, after identifying one component of a genetic system, to then identify other interacting elements. This has proven central and essential for cell cycle genetics, particularly as it is becoming clear that a large fraction of genes involved in cell cycle progression or control have not been identified as cdc genes based on screening for recessive mutations. The reasons for this most likely include functional duplications of genes, structural constraints on the encoded gene products rendering conditional (ts or cs) alleles unlikely, and mutant phenotypes (non-cdc) that are difficult to score. However, it has been demonstrated that such elements can be identified as suppressors of a primary cdc mutation. Two suppression strategies are commonly employed:

mutational suppression and dosage suppression, but the respective rationales are similar. Rescue is demanded of one defective element in an interactive genetic system by a compensatory change in another element. In mutational suppression, a compensatory mutational change re-establishes function. In dosage suppression, increased concentration of an interacting component might stabilize a crucial complex or promote a reaction simply by mass action. Both of these approaches have been employed successfully and have led to the identification of a large number of genes that would most likely not have been found otherwise (see refs 38, 43, and 44). More recently, strategies have been put forward to identify genetic components that interact with suppressor loci themselves. As a result a sparse landscape of essential functions defined by a relatively small number of cdc genes is being developed into a more complete inventory of components and relationships. In some areas, this mode of genetic analysis pursued in an interactive context with biochemical approaches has led to a high degree of understanding at the molecular level. The greatest contribution of the yeast system, however, is that genetics, in the end, is oblivious to the ability to set up a biochemical assay. Many, if not most, of the genetic components identified encode products that do not lend themselves easily to biochemical analysis or assay, at least at a crude level. However, as various aspects of cell division get reduced and refined to problems of biochemistry, the genetic components that could only be identified in a system such as yeast will become critical pieces of the puzzle. The demonstration of strong evolutionary conservation in the cell cycle is reassuring in that it suggests that this puzzle is universal.

7. The use of oocytes and early cleavage embryos in cell cycle analyses

7.1 Ideal tools for cell biology

The preceding discussion has highlighted the value of the marriage between yeast genetics and molecular biology in identifying genes involved in cell cycle genetics. Of equal importance in the recent advances in our understanding of the control of cell division has been the availability of simple experimental systems which are tractable to biochemical analyses. Of these systems, cleavage embryos from a variety of organisms have provided invaluable insights into the mechanisms of cell cycle control (reviewed in ref. 45). Although many fascinating experiments have been performed over a number of years, here we will discuss only those experiments which best illustrate the peculiarities and usefulness of these experimental systems.

The large size of oocytes and eggs of amphibians and echinoderms makes them ideal for experimental manipulations such as microinjection (see for example ref. 46). In addition, the early cleavage stage of development is characterized by rapid and synchronous cell divisions (47) which occur in the absence of transcription (48) and with only a minimal requirement for translation (reviewed in ref. 49). This is

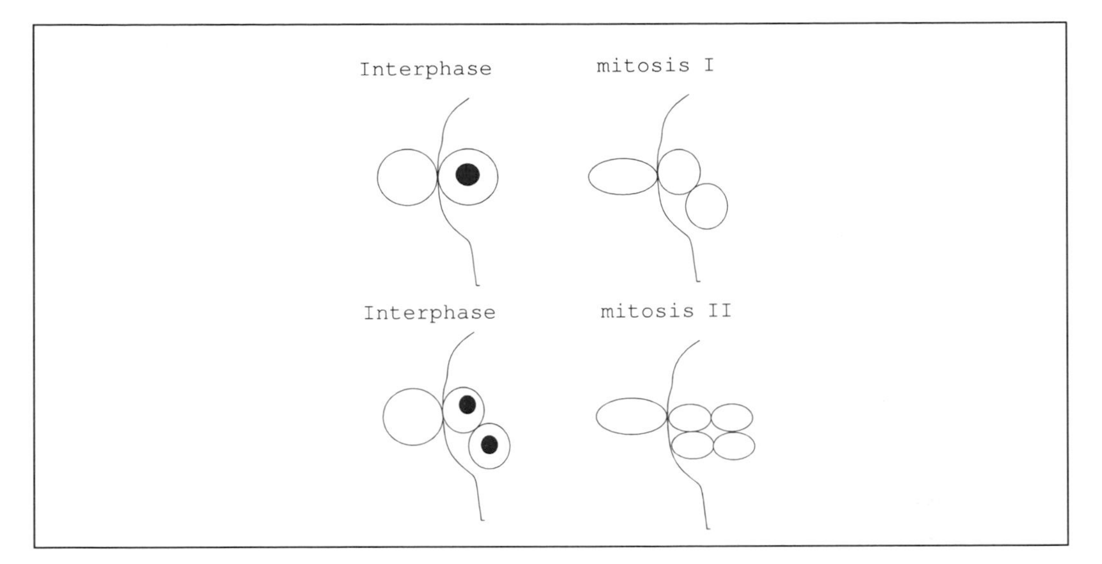

Fig. 5 Schematic representation of the ligation experiment performed by Hara *et al.* (51). A fertilized *Xenopus* egg is ligated using a pulled loop formed by a human hair giving rise to one nucleated half and one anucleate half. At mitosis I, the nucleate half divides (and the nucleus breaks down) but the anucleate half contracts. As the embryo re-enters interphase, nuclei re-form in the newly formed blastomeres and the surface of the anucleate half of the embryo relaxes. At mitosis II, further divisions in the nucleated half of the embryo occur and again the anucleate half of the embryo contracts.

achieved because, throughout oogenesis, stockpiles of proteins and mRNA are accumulated which provide a store of material which is sufficient to provide all of the material needed to replicate the equivalent of 12 000 genomes (50). In the amphibian *Xenopus laevis*, each cleavage division is completed in just 20 min. Moreover, the cell cycle is very simple and comprises S phase and mitosis but does not contain G1 and G2 periods (47). The simplicity of this basic cell cycle led Kirschner and co-workers (51) to perform a very straightforward but elegant study, the results of which have influenced much of our current thinking about the control of cell division. The experiment is summarized in Fig. 5. Using a loop of fine human hair a newly fertilized egg was split in two with a constriction. Thus only one half contained the zygotic nucleus. The half that contained the zygotic nucleus continued to divide on cue, while the anucleate half was prevented from dividing. However, time-lapse cinematography revealed periodic changes in the height of the anucleate (non-dividing) half of the embryo which resulted from changes in stiffness of the cell cortex. The most fascinating feature of these so-called surface contraction waves was that they occurred in close synchrony with the divisions of the nucleate half of the embryo.

7.2 Cell cycle oscillators and MPF

The observations described above led Kirschner to propose that the simple cell cycle in early amphibian embryos was driven by a cytoplasmic oscillator which periodically drives the cell into mitosis thus preventing chromosomes from responding to other cytoplasmic signals which initiate DNA synthesis (discussed in more detail below). While Kirschner and his colleagues certainly formalized this theory, evidence for the existence of oscillatory properties in amphibian eggs had been obtained independently by Masui and co-workers by microinjecting cytoplasmic extracts of eggs of the frog *Rana pipiens* into mature recipient oocytes of the same species.

Earlier Masui and Makert (52) had devised an elegant assay which is depicted in Fig. 6. To become fertilizable, fully grown amphibian oocytes undergo characteristic changes in their nuclear and cytoplasmic activities in a process termed oocyte maturation. The most striking feature is the completion of the first meiotic division which involves breakdown of the oocyte nucleus, termed the germinal vesicle, extrusion of the first polar body, and arrest of the second meiotic division at metaphase (reviewed in ref. 53). At its arrest point, the unfertilized egg is characterized by a large white spot at the surface of the dark pigmented animal hemisphere. Usually, oocyte maturation is promoted by hormone stimulation, but Masui and Makert (52) discovered that cytoplasm taken from unfertilized eggs could induce maturation when microinjected into recipient oocytes. They proposed that after hormone stimulation oocytes develop a cytoplasmic activity which they termed 'maturation promoting factor' or 'MPF' which was capable of inducing maturation in unstimulated oocytes. Later studies revealed that in extracts prepared from crushed *Rana* eggs, MPF was also present and its activity oscillated with storage (54). A modification of these microinjection experiments led to the identification of a further cytoplasmic activity. The arrest of unfertilized egg at metaphase of second meiosis implied an unusual stability of the meiotic state in these cells. By preparing extracts of these cells in the presence of the calcium chelator EGTA, the stability of the egg cytoplasm could be maintained. Moreover, when such extracts were injected into one blastomere of a two-cell embyro, that blastomere became arrested in metaphase of the second mitotic division, while the neighbouring blastomere continued to divide. The metaphase arrest in the injected blastomere was explained by the presence of an additional calcium-sensitive activity in unfertilized eggs, termed the 'cytostatic factor' or CSF (52, 55).

While it was difficult to explain the oscillation of MPF in cytoplasmic extracts, the observations made by Masui and co-workers encouraged Kirschner and co-workers to undertake a series of exciting but technically demanding experiments. By modifying the microinjection procedure to allow transfer of cytoplasm directly from dividing embryos into immature oocytes and also into CSF-arrested blastomeres, several important characteristics of embryonic cell cycles were identified:

(a) MPF activity oscillated during meiotic and mitotic cell division, peaking at metaphase and disappearing just after anaphase (56).

(b) When MPF activity was low or undetectable in cells, an antagonistic activity termed anti-MPF could be recovered (56).

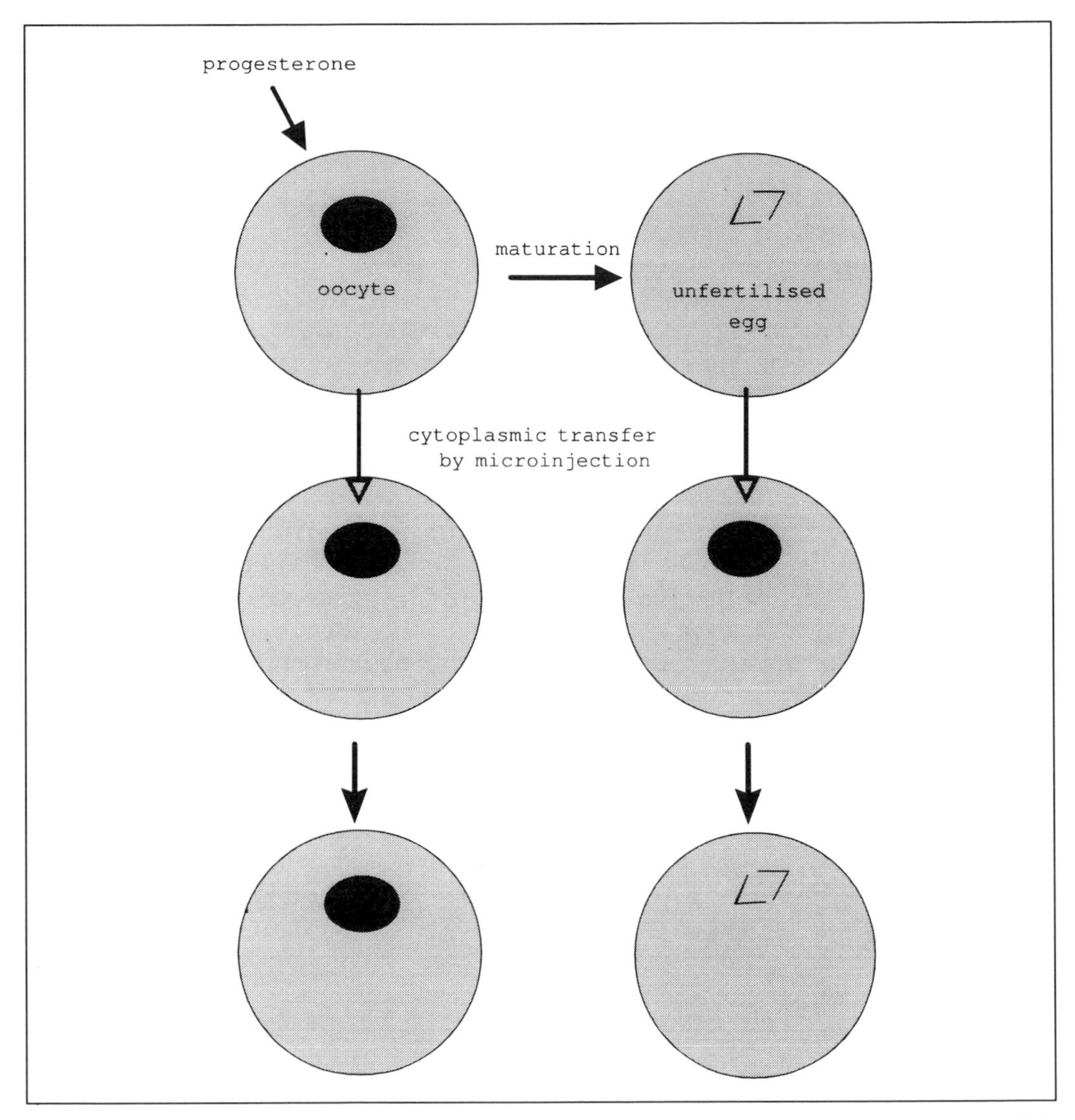

Fig. 6 Schematic representation of the microinjection experiments performed by Masui and Makert (52). Stage VI oocytes can be induced to complete first meiosis by hormone stimulation and arrest a metaphase of second meiosis as unfertilized eggs. Cytoplasm transferred from unfertilized eggs into stage VI oocytes, but not cytoplasm transferred by one oocyte to another, also induces oocyte maturation.

(c) The generation of MPF activity was dependent upon continuous protein synthesis (56, 57).

(d) The generation of MPF activity was not dependent upon either DNA replication or spindle assembly (56, 58).

Although the generation of MPF activity is independent of prior DNA replication, the initiation of DNA replication in fertilized *Xenopus* eggs is strictly entrained to cell cycle progression. Using density substitution analysis, Harland and Laskey (59) demonstrated that microinjected DNA templates could undergo only a single round of DNA replication in fertilized eggs which were prevented from dividing by treatment with protein synthesis inhibitors. In contrast, in dividing embryos, microinjected DNA templates could be re-replicated. Thus as with cells growing in culture, re-replication of DNA templates was dependent upon passage through mitosis (ref. 60, and see Chapter 7). Nevertheless, although re-replication was prevented in cells which were unable to divide, the cells retained the capacity to initiate *de novo* replication in virgin DNA templates (59).

The experiments described above suggested that the cleavage divisions in early embryos incorporated two parallel cycles, termed the chromosome cycle and the MPF cycle. The MPF cycle is driven by protein synthesis and is independent of the events of the chromosome cycle (DNA replication and chromosome segregation). DNA replication is initiated by signals which are constitutive in the extract, but chromosomes are periodically withdrawn from the influence of these signals as a result of condensation and/or nuclear envelope breakdown in response to MPF. However, normally the chromosome cycle is entrained to the MPF cycle because the block to re-replication is only lifted when the cell passes through mitosis (Fig. 7, and ref. 58). While this basic cell cycle appears to differ considerably from the dependent series of events which typify somatic cell divisions and

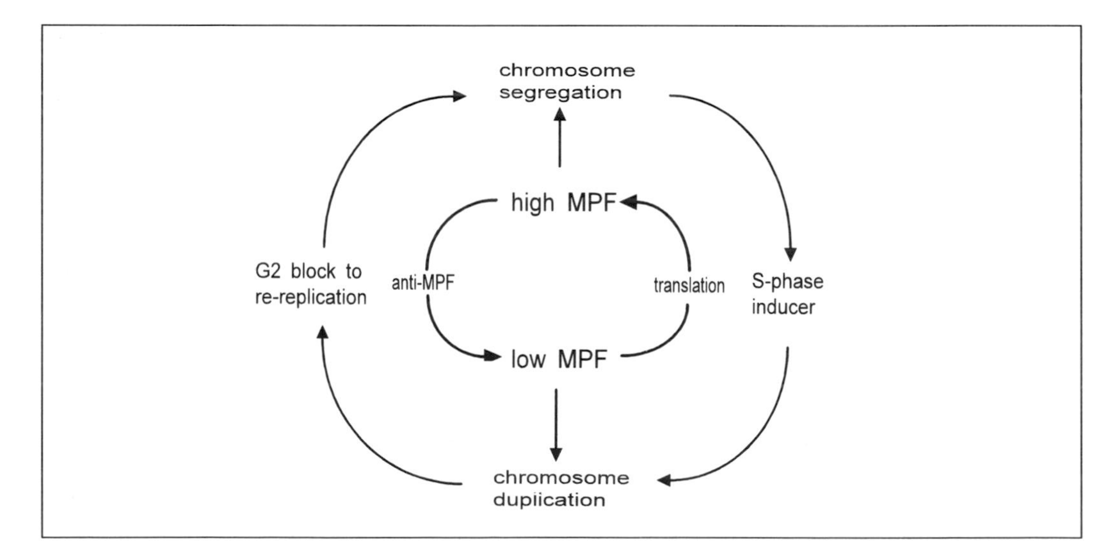

Fig. 7 The MPF and chromosome cycles in cleavage embryos. In cleavage embryos each blastomere oscillates between a mitotic and an interphase state by cyclic generation and destruction of MPF. MPF generation is dependent upon protein synthesis and its destruction is dependent upon the presence in anaphase cells of anti-MPF. High levels of MPF **induce** nuclear envelope breakdown, chromosome condensation, and segregation. Low levels of MPF **permit** chromosome decondensation, nuclear envelope assembly, and DNA replication. The permissive action of low levels of MPF is explained because S phase-inducing signals are constitutive in early embryos. The entrainment of the chromosome cycle to the MPF cycle is explained because chromosomes which have undergone DNA replication are prevented from re-replicating until they have passed through mitosis.

yeast cell cycles (see Section 2.1), as discussed above, the basic conservation of mechanisms which underpin the MPF cycle and the chromosome cycle can easily be modified by superimposition of checkpoint controls which then link each cycle and introduce G1 and G2 periods (see Chapter 1; Section 5 this chapter and ref. 61).

Moreover, by formalizing a basic theory for the cell cycle achievable goals in terms, the immediate aims of cell cycle research were established:

(a) The identification and characterization of MPF.

(b) Elucidation of the mechanism of MPF activation/inactivation.

(c) Defining the nature of the G2 block to re-initiation.

(d) Identifying the activities which initiate DNA replication.

It is probably true to say that progress in each of these goals has been largely achieved through the use of cell-free extracts which recapitulate events of the cell cycle *in vitro*.

8. Cell-free extracts

8.1 Purification of MPF

While useful for its identification, microinjection proved to be limited as an assay system for the purification of MPF (62). However, the generation of cell-free nuclear assembly extracts from fertilized and unfertilized amphibian eggs provided a simple tool with which to achieve this goal. Cell-free extracts were first produced from *Rana* eggs by the simple procedure of crushing large numbers of eggs by centrifugation and then treating the resulting cytosolic fraction with drugs which prevent microfilament assembly (63). The resulting extracts were able to assemble nuclear structures around exogenously added demembranated sperm heads and could respond to the addition of MPF by inducing nuclear envelope breakdown and chromosome condensation (64). Similar extracts were later prepared from *Xenopus* eggs and these displayed the same properties (65, 66). Because extracts were easy to store they were ideal as an assay system to test the activity generated during the purification of MPF. Thus the goal of purifying and characterizing MPF was achieved in 1988 (67). Highly purified fractions of MPF contained two polypeptides of M_r ~32 kDa and 45 kDa. The 32 kDa protein reacted with antibodies against the conserved region of cdc2 (see Chapter 5 and ref. 68) and was inhibited by suc1 protein (69), identifying it as the *Xenopus* homologue of cdc2/CDC28. In active MPF the 32 kDa subunit was always complexed with the 45 kDa polypeptide (or an equivalent polypeptide) which was later identified as cyclin B. Moreover, this protein complex contained a highly active histone H1 kinase activity (see for example ref. 70). Thus, purified MPF was found to consist of a complex between a protein kinase subunit, cdc2, and cyclin B. The formation and function of this complex is comprehensively discussed in Chapter 5.

8.2 Period extracts and the role of cyclins

Cyclins were discovered after a series of fortuitous experiments performed in echinoderm embryos. In order to investigate the utilization of maternally derived mRNA, Joan Ruderman, Tim Hunt, and co-workers (71, 72) performed a series of pulse chase experiments to identify species which were translated at fertilization. Their results were unexpected. The most abundant species to be translated after fertilization were destroyed abruptly as the embryos entered M phase only to be synthesized again as the embryos re-entered interphase (71, 72). Moreover, the periodicity of cyclin synthesis and destruction coincided with the activation and inactivation of MPF (see Fig. 8), the peak of MPF activity coinciding with the highest levels of cyclin accumulation. While suggestive of the fact that cyclins were involved in MPF activation, these observations did not provide formal proof. However, such proof quickly followed. By modifying the procedure of Lohka and Maller (66), it was possible to recapitulate the basic embryonic cell cycle in a cell-free extract of *Xenopus* eggs. All of the properties of the embryonic cell cycle were displayed by these extracts, including periodic DNA replication in which re-replication was dependent upon passage through mitosis, which in turn was dependent upon protein synthesis (73, 74). Using anti-sense oligonucleotides to cyclin B, Minshull *et al.* (75) were able to specifically deplete such extracts of cyclin B mRNA with the result that nuclei formed in the extracts arrested in a G2-like state. In a complementary series of experiments, Murray and Kirschner (76) used pancreatic RNase to remove all mRNA from the extract which resulted in a G2 arrest. However, by re-adding cyclin B mRNA in the presence of an RNase inhibitor, cyclin B became the only translated product and the synthesis of this protein

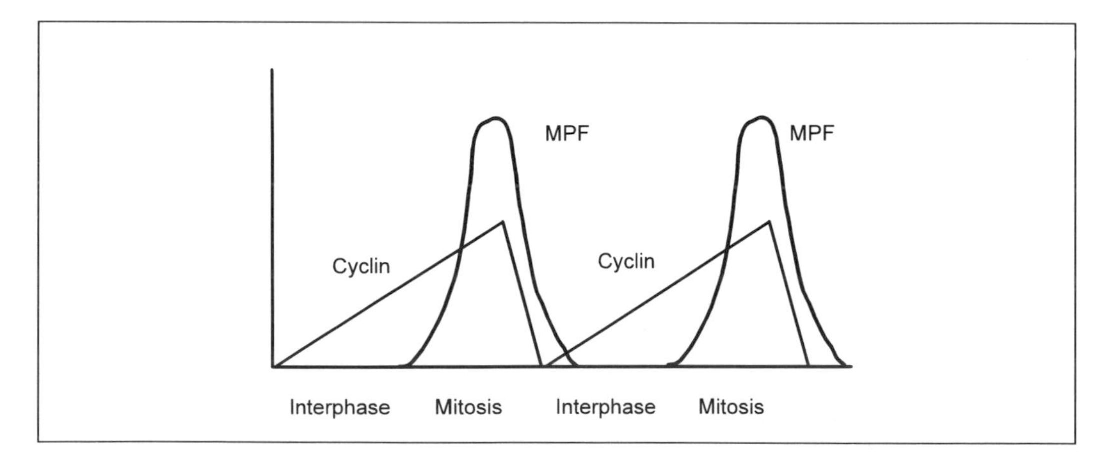

Fig. 8 Cyclin synthesis in cleavage embryos is required for MPF activation. During interphase of each cell cycle, cyclin is synthesized at a constant rate, eventually reaching a threshold level at which point MPF activity increases abruptly and the cell enters mitosis. At the peak of MPF activity, cyclin is rapidly degraded to undetectable levels. At the same time or shortly after this event MPF activity disappears and the cell re-enters interphase. The process is then repeated.

was sufficient to drive the extract into mitosis. Thus it was established that cyclin B was both necessary and sufficient to drive the embryonic cell cycle *in vitro*.

8.3 Summary

Clearly, regulation of entry into mitosis does not solely rely on synthesis of cyclin B and its subsequent association with cdc2. A more thorough review of cdc2/cyclin B activation will be found in Chapter 5. In addition, we should not ignore the contribution made to our understanding of cell cycle progression by cell-fusion experiments (60) and kinetic studies in cell culture systems (77). However, these topics are dealt with in depth in Chapters 7 and 8. Moreover, aspects of S phase regulation, referred to in Section 7.2, have received much attention recently but since these topics are also reviewed thoroughly in Chapters 6 and 7, we have not discussed them here.

We have chosen to highlight the experiments discussed above because, from a historical perspective, we cannot overemphasize their importance to the development of contemporary cell cycle theory. Much of the explosion of knowledge which will be described in the succeeding chapters has its roots in these very simple experimental systems.

References

1. Mitchison, J. M. (1957) The growth of single cells. I. *Schizosaccharomyces pombe. Exp. Cell Res.*, **13**, 244.
2. Mitchison, J. M., Passano, L. M., and Smith, F. H. (1956) An integration method for the interference microscope. *Q. J. Microsc. Sci.*, **97**, 287.
3. Leupold, U. (1950) Die Verebung von Homothallie und Heterothallie bei *Schizosaccharomyces pombe. C. R. Trav. Lab. Carlsberg, Ser. Physiol.*, **24**, 381.
4. Nurse, P., Thuriaux, P., and Nasmyth, K. (1976) Genetic control of the cell division cycle in fission yeast *Schizosaccharomyces pombe. Mol. Gen. Genet.*, **146**, 167.
5. Winge, O. and Lautsen, O. (1937) On two types of spore germination, and on genetic segregations in *Saccharomyces*, demonstrated through single-spore cultures. *C. R. Trav. Lab. Carlsberg, Ser. Physiol.*, **22**, 99.
6. Lindegren, C. C. and Lindegren, G. (1943) Segregation, mutation and copulation in *Saccharomyces cerevisiae. Ann. Mo. Bot. Gard.*, **30**, 453.
7. Hartwell, L. H., Culotti, J., and Reid, B. (1970) Genetic control of the cell division cycle in yeast I. Detection of mutants. *Proc. Natl. Acad. Sci. USA*, **66**, 352.
8. Williamson, D. H. (1965) The timing of deoxyribonucleotide acid synthesis in the cell cycle of *Saccharomyces cerevisiae. J. Cell Biol.*, **25**, 517.
9. Hartwell, L. H. (1967) Macromolecule synthesis in temperature-sensitive mutants of yeast. *J. Bacteriol.*, **93**, 1662.
10. Beggs, J. D. (1978) Transformation of yeast by replicating hybrid plasmid. *Nature*, **275**, 104.
11. Hinnen, A. A., Hicks, J. B., and Fink, G. R. (1979) Transformation of yeast. *Proc. Natl. Acad. Sci. USA*, **75**, 1929.

12. Sipiczki, M. (1989) Taxonomy and phylogenesis. In *Molecular biology of the fission yeast*. Nasim, A., Young, P., and Johnston, B. F. (ed.). Academic Press, London, UK, p. 431.
13. Hagan, I. M. and Hyams, J. S. (1988) The use of cell cycle division mutants to investigate the control of microtubule distribution in the fission yeast *Schizosaccharomyces pombe*. *J. Cell Sci.*, **89,** 343.
14. Nurse, P. (1985) Cell cycle control genes in yeast. *Trends Genet.*, **1,** 51.
15. Byers, B. and Goetsch, L. (1975) Behaviours of spindles and spindle plaques in the cell cycle and conjugation in *Saccharomyces cerevisiae*. *J. Bacteriol.*, **124,** 511.
16. Hartwell, L. H., Culotti, J., Pringle, J. R., and Reid, B. J. (1974) Genetic control of the cell division cycle in yeast. *Science*, **183,** 46.
17. Epstein, R. H., Bolle, A., Steinberg, C. M., Kellenberg, E., Boy de la Tour, E., Chevallay, R., Edgar, R. S., Susman, M., Denhardt, G. H., and Lielausis, A. (1964) Physiological studies of conditional mutants of bacteriophage T4D. *Cold Spring Harbor Symp. Quant. Biol.*, **28,** 375.
18. Hartwell, L. H., Mortimer, R. K., Culotti, J., and Culotti, M. (1973) Genetic control of the cell division cycle in yeast. V. Genetic analysis of yeast mutants. *Genetics*, **74,** 267.
19. Weinert, T. A. and Hartwell, L. H. (1988) The *RAD9* gene controls the cell cycle response to DNA damage in *S. cerevisiae*. *Science*, **241,** 317.
20. Hartwell, L. H. and Hartwell, T. A. (1989) Checkpoints: controls that ensure the order of cell cycle events. *Science*, **246,** 629.
21. Al-Khodairy, F. and Carr, A. M. (1992) DNA repair mutants defining G2 checkpoint pathways in *Schizosaccharomyces pombe*. *EMBO J.*, **11,** 1343.
22. Rowley, R., Subramani, S., and Young, P. G. (1992) Checkpoint controls in *Schizosaccharomyces pombe*: rad1. *EMBO J.*, **11,** 1335.
23. Hoyt, M. A., Totis, L., and Roberts, B. T. (1991) *S. cerevisiae* genes required for cell cycle arrest in response to loss of microtubule function. *Cell*, **66,** 507.
24. Li, R. and Murray, A. W. (1991) Feedback control of mitosis in budding yeast. *Cell*, **66,** 519.
25. Frankhauser, C., Marks, J., Reymond, A., and Simanis, V. (1993) The *S. pombe* cdc16 gene is required for both maintenance of p34 cdc2 kinase activity and regulation of septum formation: a link between mitosis and cytokinesis? *EMBO J.*, **12,** 2697.
26. Lew, D. J. and Reed, S. I. (1995) Cell cycle checkpoint monitors cell morphogenesis in budding yeast *J. Cell Biol*, in press.
27. Kelly, T. J., Martin, G. S., Forsburg, S. L., Stephen, R. J., Russo, A., and Nurse, P. (1993) The fission yeast *cdc18*$^{+}$ gene product couples S-phase to START and mitosis. *Cell*, **74,** 1.
28. Saka, Y. and Yanagida, M. (1993) Fission yeast *cut5*$^{+}$, required for S-phase onset and M-phase restraint, is identical to the radiation damage gene *rad4*$^{+}$. *Cell*, **74,** 383.
29. Mendenhall, M. D., Richardson, H. E., and Reed, S. I. (1988) Dominant negative protein kinase mutations that confer a G1 arrest phenotype. *Proc. Natl. Acad. Sci. USA*, **85,** 4426.
30. MacNeill, S. A. and Nurse, P. (1993) Mutational analysis of the fission yeast p34 cdc2 protein kinase gene. *Mol. Gen. Genet.*, **236,** 415.
31. Broach, J. R., Strathern, J. M., and Hicks, J. B. (1979) Transformation in yeast: development of a hybrid cloning vector and isolation of the *CAN1* gene. *Gene*, **8,** 212.
32. Nasmyth, K. A. and Reed, S. I. (1980) The isolation of genes by complementation in yeast: the molecular cloning of a cell cycle gene. *Proc. Natl. Acad. Sci. USA*, **77,** 2119.
33. Botstein, D. and Davis, R. W. (1982) Principles and practice of recombinant DNA

research in yeast. In *The molecular biology of the yeast Saccharomyces: metabolism and gene expression*. Stranthern, J. N., Jones E. W., and Broach, J. R. (ed.). Cold Spring Harbor Laboratory, New York, p. 607.
34. Clarke, L. and Carbon, J. (1980) Isolation of yeast centromere and construction of small circular chromosomes. *Nature*, **287,** 504.
35. Beach, D. and Nurse, P. (1981) High frequency transformation of the fission yeast *Schizosaccharomyces pombe. Nature*, **260,** 140.
36. Beach, D., Durkacz, B., and Nurse, P. (1982) Functionally homologous cell cycle control genes in budding and fission yeast. *Nature*, **300,** 706.
37. Rothstein, R. J. (1983) One-step gene disruption in yeast. *Methods Enzymol.*, **101,** 202.
38. Hadwinger, J. A., Wittenberg, C., Richardson, H. E., de Barros Lopes, M., and Reed, S. I. (1989) A family of cyclin homologs that control the G1 phase in yeast. *Proc. Natl. Acad. Sci. USA*, **86,** 6255.
39. Gould, K. L. and Nurse, P. (1989) Tyrosine phosphorylation of the fission yeast *cdc2*$^{+}$ protein kinase regulates entry into mitosis. *Nature*, **342,** 39.
40. Lew, D. J., Dulic, V., and Reed, S. I. (1991) Isolation of three novel human cyclins by rescue of G1 cyclin (Cln) function in yeast. *Cell*, **66,** 1197.
41. Xiong, Y., Connolly, T., Futcher, B., and Beach, D. (1991) Human D-type cyclin. *Cell*, **65,** 691.
42. Lee, M. G. and Nurse, P. (1987) Complementation used to clone a human homologue of the fission yeast cell cycle control gene *cdc2*$^{+}$. *Nature*, **327,** 31.
43. Hayles, J., Aves, S., and Nurse, P. (1986) *suc1* is an essential gene involved in both cell cycle and growth in fission yeast. *EMBO J.*, **5,** 3373.
44. Booher, R. and Beach, D. (1988) Involvement of *cdc13*$^{+}$ in mitotic control in *Schizosaccharomyces pombe*: possible interaction of the gene product with microtubules. *EMBO J.*, **7,** 2321.
45. Murray, A. W. and Kirschner, M. W. (1991) What controls the cell cycle? *Scientific American*, **264,** 56.
46. Gurdon, J. B. (1990) Nuclear transplantation in *Xenopus*. In *Methods in Cell Biology*. Vol. **36**. *Xenopus laevis: practical uses in cell and molecular biology*. Kay, B. K. and Peng, H. B. (ed.). Academic Press, New York, p. 299.
47. Graham, C. F. and Morgan, R. W. (1966) Changes in the cell cycle during early amphibian development. *Dev. Biol.*, **14,** 439.
48. Newport, J. W. and Kirschner, M. W. (1982) A major developmental transition in early *Xenopus* embryos. I. Characterization and timing of cellular changes at the mid blastula stage. *Cell*, **30,** 675.
49. Ford, C. C. (1985) Maturation promoting factor and cell cycle regulation. *J. Embryol. Exp. Morphol.*, **89** (Suppl.), 271.
50. Newport, J. and Spann, T. (1987) Disassembly of the nucleus in mitotic extracts: membrane vesicularisation, lamin disassembly and chromosome condensation are independent processes. *Cell*, **48,** 219.
51. Hara, K., Tydeman, P., and Kirschner, M. A. (1980) A cytoplasmic clock with the same period as the division cycle in *Xenopus* eggs. *Proc. Natl. Acad. Sci. USA*, **77,** 462.
52. Masui, Y. and Makert, C. L. (1971) Cytoplasmic control of nuclear behaviour during meiotic maturation of frog oocytes. *J. Exp. Zool.*, **177,** 129.
53. Masui, Y. and Clarke, H. J. (1979) Oocyte maturation. *Int. Rev. Cytol.*, **57,** 185.
54. Masui, Y. (1982) Oscillatory activity of maturation promoting factor (MPF) in extracts of *Rana pipiens* eggs. *J. Exp. Zool.*, **224,** 389.

55. Meyerhof, P. G. and Masui, Y. (1977) Ca and Mg control of cytoplasmic factors from *Rana pipiens* oocytes cause metaphase and cleavage arrest. *Dev. Biol.*, **61,** 214.
56. Gerhart, J., Wu, M., and Kirschner, M. W. (1984) Cell cycle dynamics of an M-phase-specific cytoplasmic factor in *Xenopus laevis* oocytes and eggs. *J. Cell Biol.*, **98,** 1247.
57. Miake-Lye, R., Newport, J., and Kirschner, M. (1983) Maturation-promoting factor induces nuclear envelope breakdown in cycloheximide-arrested embryos of *Xenopus laevis. J. Cell Biol.*, **97,** 81.
58. Newport, J. W. and Kirschner, M. W. (1984) Regulation of the cell cycle during early *Xenopus* development. *Cell*, **37,** 731.
59. Harland, R. M. and Laskey, R. A. (1980) Regulated replication on DNA microinjected into eggs of *Xenopus laevis. Cell*, **21,** 761.
60. Rao, P. and Johnston, R. T. (1970) Mammalian cell fusion: studies on the regulation of DNA synthesis and mitosis. *Nature*, **225,** 159.
61. Dasso, M. and Newport, J. W. (1990) Completion of DNA replication is monitored by a feedback system that controls the initiation of mitosis *in vitro*: studies in *Xenopus. Cell*, **61,** 811.
62. Wu, M. and Gerhart, J. C. (1980) Partial purification and characterisation of the maturation-promoting factor from eggs of *Xenopus laevis. Dev. Biol.*, **79,** 465.
63. Lohka, M. L. and Masui, Y. (1983) Formation *in vitro* of sperm pronuclei and mitotic chromosomes induced by amphibian ooplasmic components. *Science*, **220,** 719.
64. Lohka, M. L. and Masui, Y. (1984) Roles of cytosol and cytoplasmic particles in nuclear envelope assembly and sperm pronuclear formation in cell-free preparations from amphibian eggs. *J. Cell Biol.*, **98,** 1222.
65. Miake-Lye, R. and Kirschner, M. W. (1985) Induction of early mitotic events in a cell-free system. *Cell*, **41,** 165.
66. Lohka, M. L. and Maller, J. (1985) Induction of nuclear envelope breakdown, chromosome condensation and spindle formation in cell-free extracts. *J. Cell Biol.*, **101,** 518.
67. Lohka, M. J., Hayes, M. K., and Maller, J. L. (1988) Purification of maturation-promoting factor, an intracellular regulator of early mitotic events. *Proc. Natl. Acad. Sci. USA*, **85,** 3009.
68. Gautier, J., Norbury, C., Lohka, M., Nurse, P., and Maller, J. (1988) Purified maturation-promoting factor contains the product of a *Xenopus* homolog of the fission yeast cell cycle control gene *cdc2*$^{+}$. *Cell*, **54,** 433.
69. Dunphy, W. G. and Newport, J. W. (1989) Fission yeast p13 blocks mitotic activation and tyrosine dephosphorylation of the *Xenopus* cdc2 protein kinase. *Cell*, **58,** 181.
70. Meijer, L., Arion, D., Golsteyn, R., Pines, J., Brizuela, L., Hunt, T., and Beach, D. (1989) Cyclin is a component of the sea urchin egg M-phase specific histone H1 kinase. *EMBO J.*, **8,** 2275.
71. Rosenthal, E. T., Hunt, T., and Ruderman, J. V. (1980) Selective translation of mRNA controls the pattern of protein synthesis during early development of the surf clam, *Spisula solidissima. Cell*, **20,** 487.
72. Standart, N., Minshull, J., Pines, J., and Hunt, T. (1987) Cyclin synthesis, modification and destruction during meiotic maturation of the starfish oocyte. *Dev. Biol.*, **124,** 248.
73. Hutchison, C. J., Cox, R., Drepaul, R.-S., Gomperts, M., and Ford, C. C. (1987) Periodic DNA synthesis in cell-free extracts of *Xenopus* eggs. *EMBO J*, **6,** 2003.
74. Hutchison, C. J., Cox, R., and Ford, C. C. (1988) The control of DNA replication in a cell-free extract that recapitulates a basic cell cycle *in vitro*. *Development*, **103,** 553.

75. Minshull, J., Blow, J. J., and Hunt, T. (1989) Translation of endogenous B-type cyclin mRNA is necessary for extracts of activated *Xenopus* eggs to enter mitosis. *Cell*, **56,** 947.
76. Murray, A. W. and Kirschner, M. W. (1989) Cyclin synthesis drives the early embryonic cell cycle. *Nature*, **339,** 275.
77. Pardee, A. B., Coppock, D. L., and Yang, H. C. (1986) Regulation of cell proliferation at the onset of DNA synthesis. *J. Cell Sci.* (*Suppl.*), **4,** 171.

3 | START and the G1–S phase transition in budding yeast

STEVEN I. REED

1. Introduction: the concept of START

Mechanisms to ensure that division occurs in a coordinated and regulated fashion are prerequisite for fitness and ultimately for viability of all cells. This means that the division process must be responsive to a host of internal and external signals that mediate, among other things, integration of division with the extracellular milieu, coordination of division with cellular growth and the occurrence of cell cycle events in the proper order. In the budding yeast, *Saccharomyces cerevisiae*, both the regulation of division by external signals and the coordination of growth and division occur in the G1 interval of the cell cycle (1). Operationally, there are two types of signal that severely restrain division in G1 in this organism: nutritional limitations (2–13) and the action of mating hormones (4, 14–22). Although the regulatory intents of these signals are quite disparate, each causes a cell cycle arrest characterized by completion of the ongoing cell cycle and arrest in the subsequent G1 interval (1). Additionally, changes in the cellular growth rate are coordinated with the division process to ensure a relatively constant cell size by altering specifically the length of the G1 interval: slow growth is accommodated by extending the length of G1 while rapid growth shortens G1 (5, 23–38). These observations led to the hypothesis of a universal G1 control mechanism in yeast (1). In its simplest formulation, this hypothesis states that all signals governing passage through G1 and entry in S phase do so by regulating the completion of an essential G1 function or event designated START; completion of START commits a cell to S phase and division; delay of completion of START delays entry into S phase; failure to complete START prevents initiation of S phase and imposes alternative fates upon cells.

1.1 START as the master regulator of alternative developmental pathways

S. cerevisiae cells can undergo stable vegetative (mitotic) growth either as haploids or diploids. The transition from the diplophase to haplophase occurs via meiosis

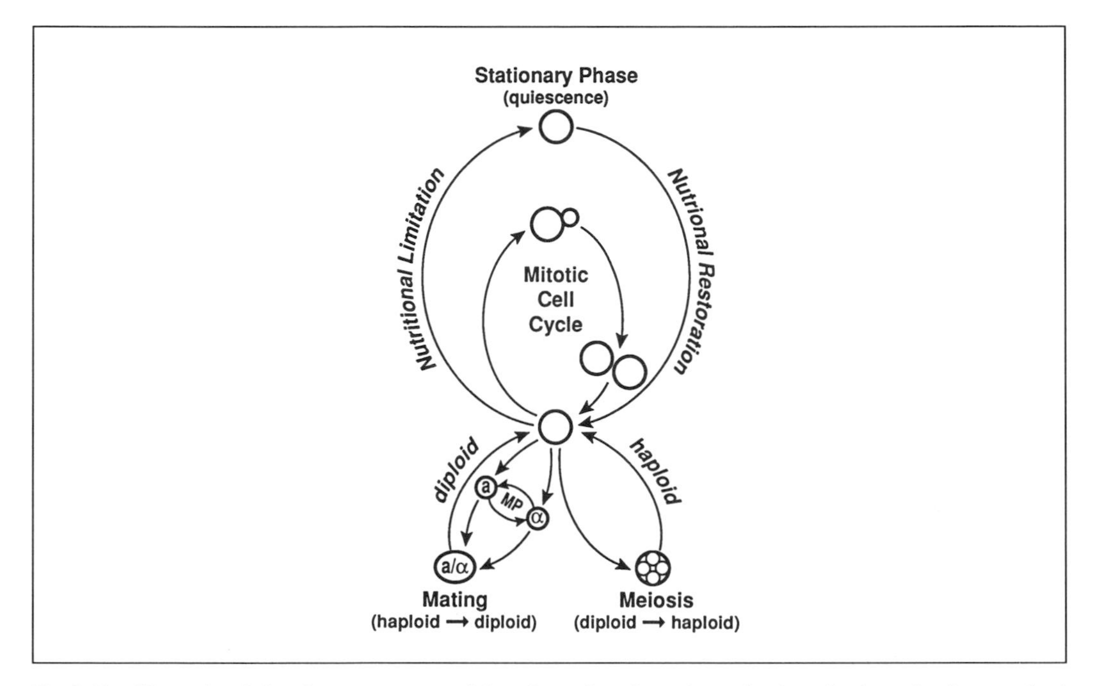

Fig. 1 The life cycle of *Saccharomyces cerevisiae*. Central to the scheme is the mitotic cycle characterized by the budding mode of division. Other developmental pathways diverging from the mitotic cycle are entry into stationary phase in response to nutrient limitation, conjugation during which haploid cells fuse to form diploid cells, and meiosis during which diploid cells give rise to haploid ascospores. In budding yeast both haploid and diploid cells are capable of mitotic growth.

and sporulation whereas restoration of the diplophase occurs via conjugation of haploids (Fig. 1). Cells can also exit from the vegetative division cycle into a stable quiescent state when nutrients become exhausted (Fig. 1) (2–13). The decisions that either commit a cell to the mitotic cell cycle or to one of these alternative developmental fates are made in the G1 phase of the cell cycle. To the degree of precision that is possible with genetic experiments, it appears that cells exit the cell cycle to quiescence, mating, or meiosis from the same point, consistent with the idea of a master regulatory event or START (Fig. 2). When cells are subjected to starvation, they arrest in G1, prior to developing the physiological repertoire associated with quiescence. Likewise, haploid cells responding to mating pheromones arrest in G1 as part of the mating response, presumably to synchronize mating partners prior to zygote formation. By most criteria, cells responding to starvation or mating pheromones arrest at the same point in the cell cycle, that is prior to all developmental commitments to the mitotic cycle: cells arrest without buds and with unduplicated microtubule organizing centres or spindle pole bodies (4). More significantly, cells that are arrested by starvation can still be blocked by mating pheromones if nutrients are restored and vice versa (39), suggesting that these stimuli send signals that impinge on the same targets. Meiosis, a specialized

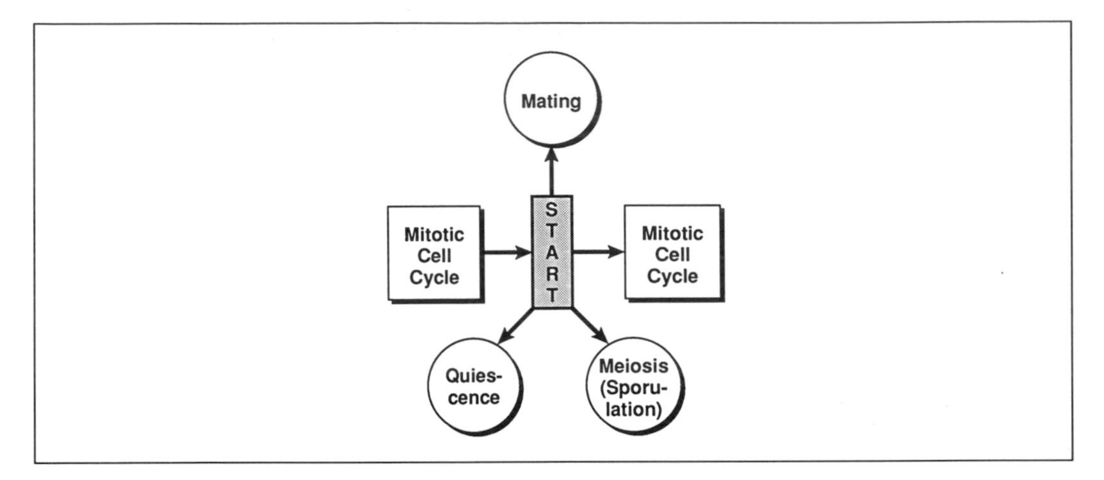

Fig. 2 All developmental options diverge from START. In *Saccharomyces*, the START event poses a gate between concurrent mitotic cell division cycles. START is also the point of divergence for other developmental pathways including mating (conjugation), sporulation (meiosis), and quiescence in response to nutrient limitation.

variation on the theme of starvation, occurs when diploid cells are severely limited for certain nutrients (see ref. 40 for review). As with haploid cells, the initial response is G1 arrest, followed by commitment to the meiotic pathway that ultimately concludes with the formation of quiescent haploid spores (40). Based on these observations it would appear that the START event serves as a point of embarkation for all possible cellular fates (Fig. 2). Plentiful nutrient supply promotes execution of START and commitment to the mitotic cycle. Nutrient starvation or the response to mating pheromones blocks the execution of START, thus preventing commitment to the mitotic cycle while promoting commitment to other developmental pathways.

1.2 START governs commitment to the vegetative cell cycle

START serves as a crossroads for alternative developmental fates in yeast. To some degree, this regulation is negative in that failure to complete START is a prerequisite for embarkation on other pathways. Conversely, completion of START commits a cell to an entire round of division. Thus, starvation and mating pheromones have little impact on the cell cycle once START has been traversed and only have regulatory consequences in the subsequent G1 interval. Consistent with this idea, conditions that slow growth, either by nutrient limitation or by inhibiting protein synthesis, expand the pre-START component of G1 but leave the post-START component of the cell cycle largely invariant (5, 23–38). One can conclude, therefore, that completion of START commits cells to a rigid deterministic temporal programme of division. The timing of START is governed in part, as indicated above, by hormonal and nutritional circumstances. In addition, under conditions of unrestrained division, START is tightly co-ordinated with cell growth. Thus, cells become committed to the cell cycle only after achieving a requisite growth requirement. This ensures that populations maintain a constant optimal average cell size.

1.3 Genetic analysis of START: START mutants

The idea of a G1 gating event was bolstered by the isolation of G1 arrest mutations (1, 41–44). These conditional mutations conferred arrest in G1 with phenotypes similar to those exhibited by cells that have been treated with mating pheromones. It was therefore suggested that these mutations might occur in genes encoding molecular components of the START regulatory machinery and that such proteins might be the targets of signalling pathways that regulate cell cycle progression via modulating START. Although some START mutations were found to confer G1 arrest by ectopically activating the pheromone response pathway (42, 44), at least one complementation group, *CDC28* (1, 41 42), appeared to define a central element in cell cycle control. Mutations in the gene *CDC28* have been a major tool in unravelling the molecular basis for START control. Mutations in a second gene, *CDC37*, although less well characterized, confer a similar phenotype (42).

Cloning and sequence analysis of the *CDC28* gene indicated that it encoded a protein kinase, suggesting a role for protein phosphorylation in the execution of START (45, 46). The discovery, shortly thereafter, that the *cdc2* protein kinase of fission yeast (47, 48) and other organisms, implicated in mitotic induction (49, 50), was highly homologous was initially puzzling in the context of the genetically determined G1 role for *CDC28* (see Chapters 4 and 5). However, as discussed below, this apparent discrepancy is now easily interpretable in the context of universal cell cycle regulatory roles for members of this protein kinase family. Sequence analysis of the *CDC37* gene was not revealing in terms of its role or biochemical activity. However, recently a *Drosophila* homologue has been identified in a screen for signalling elements downstream of the *Sevenless* cell surface receptor (51), suggesting a role in linking extracellular and/or intracellular signals to the cell cycle progression machinery.

2. START and the Cdc28 protein kinase

2.1 Execution of START correlates with activation of the Cdc28 kinase

As stated above, the sequence of the *CDC28* gene implicated protein phosphorylation as having a role in the execution of START (45). However, the preferred *in vitro* substrate, bovine histone H1, shed little light on the *in vivo* function of the Cdc28 protein kinase. The resemblance of cells arrested at START by the action of mating pheromones or starvation to those arrested by *cdc28* mutations led to the proposal that START control might be mediated at the level of controlling the activity of the Cdc28 kinase. Direct assay of the Cdc28 kinase confirmed that mating pheromones and starvation both arrest cells with the kinase in an inactive state (52). Furthermore, assay of Cdc28 kinase activity as a function of cell cycle progression in synchronized populations revealed a lack of kinase activity during the G1 interval, followed by an increase correlating temporally with the execution of START (53). None of these observations, however, unequivocally place the Cdc28 kinase at the centre of the START event. This has been accomplished largely through the study of G1 cyclins.

2.2 G1 cyclins are G1-specific activators of the Cdc28 kinase

Gel filtration analysis of Cdc28 kinase in yeast lysates suggested that the regulation of activity observed might be mediated at the level of interaction with other polypeptides. Whereas active protein kinase complexes from cycling cells chromatographed at molecular weights significantly higher than monomeric Cdc28 polypeptide, starved cells or cells responding to mating pheromones contained none of these active high molecular weight complexes (52). Furthermore, monomeric Cdc28 polypeptide was found to have no associated protein kinase activity under any circumstances (52). Understanding the molecular basis of Cdc28 activation, however, was largely a function of genetic analysis and parallel studies on other organisms. G1 cyclins, or Clns, were discovered by three different genetic screens. High dosage suppressors of temperature-sensitive (ts) *cdc28* mutations were sought with the aim of identifying regulatory elements that interacted with the Cdc28 kinase (54). A second screen was based on small cell size (55) and a third on resistance to cell cycle arrest by mating pheromones (56). The first screen yielded two highly homologous genes, *CLN1* and *CLN2* (54), while the second and third yielded a gene now designated *CLN3* (originally designated *WHI1* and *DAF1*, respectively) (55–57). The translation products of these three genes were found to have low but significant homology to a class of proteins known as cyclins (Fig. 3) (54, 57). The discovery that cyclins were the activators of the related cdc2 kinase of *Xenopus* and a number of marine invertebrates for mitotic functions (see refs 58–64, for review) led to the hypothesis that Clns were G1-specific activators

```
CLN3  DKYSSRFIIKSYNYQLLSLTALWISSKFWDSKNR
CLN1  DRYCSKRCCLKDQAKLVVGTCLWLAAKIWGGCNHIINNVSIPTGGRFYGPNPRA
CYCA  DRFLSKISVLRGKLQLVGAASMFLAAKYEE                   IYPPD

CLN3   MATLKVLQNLCCN Q YSIKQFITMEMHLFKSLDWSICQSAYFDSYIDIF
CLN1  RIPRLSELVHYCGGSDLFDESMFIQMERHILDTLNWDVYEPMIND YILNV
CYCA        VKEFA YITD DTYTSQQVLRMEHLILKVLTFDVAVPTINW FCEDF
```

Fig. 3 Similarity between G1 cyclins and cyclin A. The 'cyclin box' region of Cln1 is compared with the homologous region of Cln3 (above) and cyclin A from clam (below). Virtually all similarity between G1 cyclins and other cyclins, as well as between G1 cyclin classes defined by Cln1 and Cln3, respectively, is limited to the region shown (less than 30% of each complete polypeptide).

of the Cdc28 kinase. This idea has received both genetic and biochemical confirmation. First, mutational elimination of Clns 1, 2, and 3 confers G1 arrest (65), as would be expected for a functionally overlapping set of activators for an essential G1 protein kinase activity. Second, the Cln polypeptides could be co-immunoprecipitated with Cdc28 and the complexes were shown to have Cdc28-specific protein kinase activity (66, 67). Finally, Cln1 and Cln2 protein and associated protein kinase activity were found to accumulate with dramatic periodicity through the cell cycle, peaking near the G1/S phase boundary and reaching a minimum late in the cell cycle (G2/M) (53, 66). Although the regulation of Cln3 is not yet clear, it does not appear to exhibit this high degree of periodicity at the protein level (67).

Given that the Clns are activators of the Cdc28 protein kinase, the most convincing case for G1 activation of Cdc28 being synonymous with execution of START is based on dominant *CLN* mutations. As pointed out above, *CLN3* was identified through dominant alleles *WHI1–1* (55) and *DAF1–1* (56). Mutants harbouring these alleles are abnormally small because they are prematurely advanced into S phase and are resistant to cell cycle arrest in response to mating pheromones. Similar mutations have been described for *CLN1* and *CLN2* (54, 68). These defects are directly attributable to an inability to coordinate the execution of START with cell growth and with the failure to restrain START in response to mating pheromones and, in the case of *CLN2*, in response to starvation. In other words, all of these mutations advance the execution of START in cycling cells and prevent or impair the capacity of regulatory signals to block the execution of START. These phenotypes are easily interpretable within the context of the role of Clns as G1 activators of the Cdc28 kinase and the molecular anatomy of the mutant *CLN* alleles. All of the dominant mutations of this type eliminate, by truncation, carboxy-terminal sequences associated with protein instability (69). Thus, by mutationally increasing the stability of Cln proteins, the kinetics of activation of the Cdc28 kinase and its accessibility to regulatory signals are altered. Consistent with this interpretation, constitutive overexpression of Cln proteins also advances START (56, 57, 65). The fact that mutational stabilization and chronic overexpression of Cln proteins have dramatic ramifications for the execution of START provides the strongest evidence that execution of START and G1 activation of the Cdc28 kinase are one and the same.

2.3 G1 vs. G2 functions of the Cdc28 kinase

The discussion so far has focused on the role of the Cdc28 kinase in the G1 to S phase transition in *S. cerevisiae*. However, it has been pointed out that the Cdc28 protein kinase is highly homologous to the cdc2 kinase of vertebrates, thought to be specialized for mitotic functions (see Chapter 5). In fission yeast, however, cdc2 was shown by genetic analysis to have essential functions both in G1 and G2, establishing a precedent for duality of function (ref. 50, and see Chapter 4). This model appears to be applicable to *S. cerevisiae* as well. Whereas most temperature

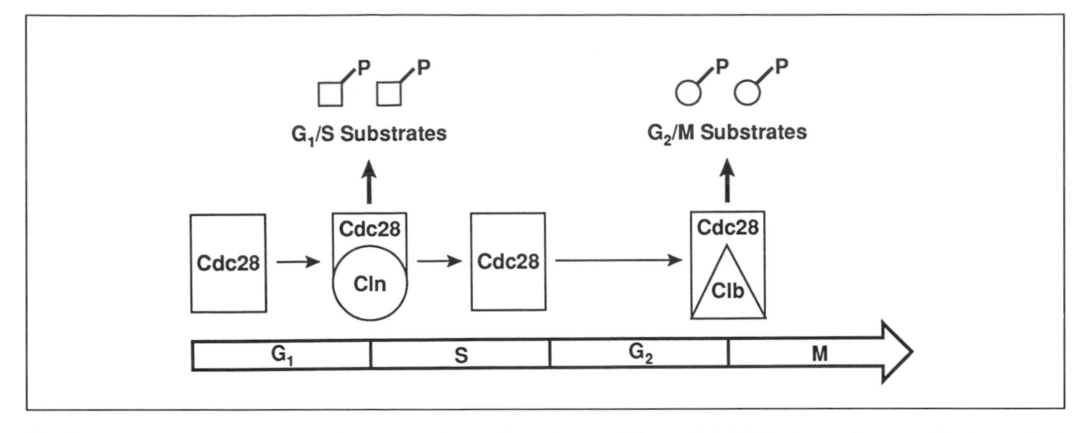

Fig. 4 Different cyclins control the two major cell cycle transitions. Cdc28 is the unique cyclin-dependent kinase of *Saccharomyces*. However G1 cyclins (Clns) activate the kinase for G1/S functions, presumably targeting the activity to a specific class of G1/S substrates. B-type cyclin (Clbs) activate the kinase for mitotic functions, presumably by targeting the activity to G2/M substrates.

sensitive *cdc28* mutations confer a START-arrest phenotype (1, 42), this is likely to be largely a function of the dynamics of thermolabile decay of Cdc28 activity coupled with a greater sensitivity of G1 function to activity loss. The experimental design can be manipulated, however, to confer homogeneous G2 arrest for many alleles previously described as START specific (70). Furthermore, an atypical allele, *cdc28–1N*, although not conferring a homogeneous G2 arrest, does confer a post-START arrest containing G2 cells (see below) (71). The G2 functions of the Cdc28 kinase have been clarified most effectively, however, by identification of B-type cyclins or Clbs, in *S. cerevisiae* (72–75). Whereas Clns show only weak homology to metazoan cyclins, Clbs show a relatively high level of homology to the primary class of mitotic cyclins, known as B-type cyclins. This system has been somewhat refractory to analysis due to a high degree of functional redundancy: at least four yeast B-type cyclins have been shown to collaborate in performing essential mitotic functions (74, 75). Thus, as is the case for fission yeast and metazoan organisms, the Cdc28 protein kinase has a primary role in mitotic induction mediated by mitotic cyclins. This function is separable from the G1 function of the kinase because each uses a different set of positive regulatory subunits: Clns for G1 and Clbs for G2/M (Fig. 4).

3. Control of and by G1 cyclins

In budding yeast, as in many organisms, the G1 to S phase transition serves as the primary regulatory target for extracellular signals relating to growth and differentiation (1). In addition, many cells, including budding yeast, use this same point in the cell cycle to coordinate growth with cell division (23). If G1 cyclins are essential as well as rate-limiting for the G1 to S phase transition, as the evidence suggests, then it is likely that the regulatory signals mentioned, both internal and external, impact directly on the accumulation or function of Clns.

3.1 Regulation of Cln biosynthesis through the cell cycle

The first convincing evidence that Cln1 and Cln2 were cyclins, in the classical sense that they accumulated periodically through the yeast cell cycle, came when RNA blot analysis was performed as a function of cell cycle position (66). Blots prepared across a synchronous time-course revealed that Cln1 and Cln2 mRNAs accumulated dramatically in late G1, near the G1/S phase boundary but almost disappeared in the subsequent G2 interval. This periodicity at the level of mRNA is reflected in Cln1 and Cln2 protein levels and ultimately in Cln1- and Cln2-associated protein kinase activity (53, 66). Consistent with the parallel relationship of Cln1 and Cln2 protein levels to mRNA levels is the demonstration that these proteins have an extremely short half-life (66). Unlike the mRNAs for Cln1 and Cln2, the Cln3 mRNA appears to be maintained at a constant level through the cell cycle (57, 66). It is not yet clear whether Cln3 protein levels or associated activity are periodic via a post-transcriptional mechanism. Nevertheless, like Cln1 and Cln2, the Cln3 protein has a short half-life (67, 76). Although the mechanism of Cln turnover is not known, all three protein primary structures contain so-called PEST sequences (69), associated with metabolic instability in a large number of proteins. Truncation mutations that remove the PEST-containing domains confer increased stability to the Cln proteins, consistent with this idea (54, 66, 76). Furthermore, it has recently been demonstrated that Clns are targets of ubiquitination, although the significance of this is not yet known (77).

3.1.1 Cell cycle-dependent transcriptional regulation of *CLN1* and *CLN2*

It was observed that *CLN1* and *CLN2* transcripts exhibited a cell cycle periodicity identical to that previously described for the *HO* gene that controls mating-type interconversion in homothallic yeast strains (78). The *cis*-acting element that drives cell cycle periodicity has been well defined and has been designated 'SCB' for 'Swi4 cell cycle box' (see below) (79–81). Both the *CLN1* and *CLN2* genes have been found to contain SCB consensus sequences in their transcriptional regulatory regions, accounting for the observed G1 periodicity of transcription (82). Analysis of *HO* transcription has defined two transcription factors, Swi4 and Swi6, that appear to drive periodic transcription (83–85). At the primary structure level, Swi4 and Swi6 have a significant degree of homology relative to each other (84, 85). However, DNA binding studies suggest that only Swi4 contacts DNA (86, 87). Nevertheless, both Swi4 and Swi6 are present in what are minimally ternary complexes with SCB DNA suggesting an interaction between Swi4 and Swi6 (86–90). Whereas the basis for *trans*-activation by these factors has not yet been elucidated, it is assumed that a Swi4/Swi6 factor activates transcription by sequence-specific binding of Swi4 to the SCB. The G1 specificity of this transcription appears, in part, to be mediated by G1 specific transcription of the *SWI4* gene itself (91). The *SWI6* mRNA does not show a high degree of periodicity (91).

The situation, however, is complicated by the existence of a second *cis*-acting motif conferring G1-specific transcription known as the MCB for 'MluI cell cycle box', so named because the consensus contains an MluI restriction site (92–94). The MCB was initially defined in the context of G1 specific transcription of a large number of genes involved in S phase functions (reviewed in ref. 95) and has been shown to be directly Cdc28-dependent, in that activation of the Cdc28 kinase in the absence of protein synthesis is sufficient for induction of MCB transcription (96). Both *CLN1* and *CLN2* (87) have one or more MCB-related sequences and *SWI4* has several (91), helping to account for the G1 periodicity of all of these transcription units. Interestingly, the *trans*-acting factor for the MCB contains Swi6 (97, 98), as does the SCB factor, but instead of Swi4, it contains a protein of 120 kDa, termed Mbfl, that makes direct contact with MCB DNA (97). The regulation of the Mbfl factor has not yet been elucidated. Thus the cell cycle transcription of *CLN1* and *CLN2* is regulated by two different *cis*-acting elements shown to confer G1 specificity.

3.1.2 Cdc28-dependent positive feedback

It has been known for some time that transcription of *HO* and a variety of MCB-containing genes is dependent on Cdc28 function (81, 96, 99). Transcripts from these genes remain undetectable or at basal levels when ts *cdc28* mutants are shifted to their restrictive temperature. A dependence of SCB and MCB transcription on Cdc28 function would appear to be paradoxical in the context of *CLN1* and *CLN2*, the G1 activators of Cdc28, being dependent on SCB and MCB elements. The demonstration of a regulatory positive feedback loop involving Cdc28 kinase and Cdc28-dependent *CLN1* and *CLN2* transcription in part resolves the paradox (Fig. 5) (89, 90, 100–102). Newborn G1 cells contain no significant level of *CLN1* and *CLN2* transcripts (100–102). In a *cdc28* mutant maintained at the restrictive temperature, a basal level of *CLN1* and *CLN2* transcripts appears when the cells have achieved a defined amount of growth but this level does not increase (100–102). A wild-type cell, on the other hand, accumulates maximal levels of *CLN1* and *CLN2* transcripts rapidly from this same starting point (100, 102). Thus, Cdc28 kinase activity is required for this amplification of *CLN1* and *CLN2* transcription from a Cdc28-independent basal level. The most likely candidate for a target of positive feedback is the Swi6 transcription factor. First, it is part of both the SCB and MCB *trans*-activating complexes (86–90, 97, 98). Second, its primary structure reveals consensus sites for the Cdc28 kinase (87). Finally, it has been shown to be a substrate for Cdc28 kinase *in vitro* (87). Thus, transcriptional activity driven by SCB and MCB elements might be activated by phosphorylation of Swi6. This hypothesis, of course, needs to be explored further at the experimental level.

3.2 Regulation of the G1/S phase transition by Clns

Given that the Cln proteins are essential G1 activators of the Cdc28 protein kinase, the accumulation and functions of these proteins are likely targets of cell cycle regulatory signals. The demonstration that *CLN* mutations are capable of confer-

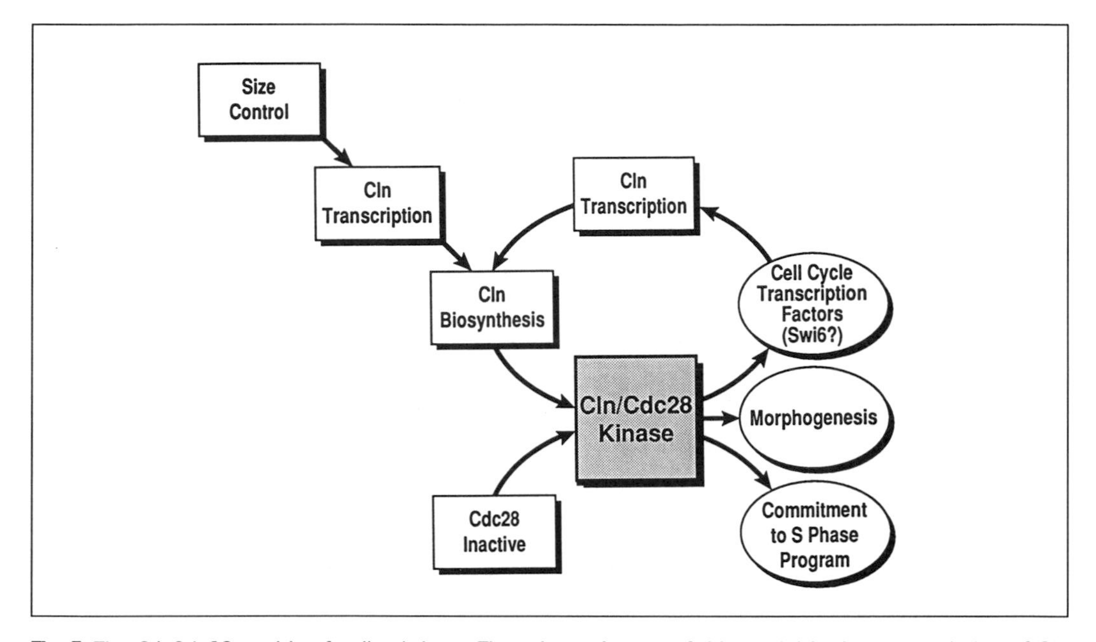

Fig. 5 The G1 Cdc28 positive feedback loop. The primary feature of this model is that transcription of Cln genes is under dual control, Cdc28 kinase-independent growth or size control and Cdc28 kinase-dependent control at the level of activation of transcription factor(s) (most probably Swi6). Cdc28-independent transcription of Cln mRNAs results in a basal level of Cdc28 kinase activity. This then leads to additional Cdc28-dependent transcription of Cln mRNAs resulting in full and decisive activation of the Cdc28 kinase. Full kinase activation induces both the morphogenetic (budding) and nuclear division (DNA replication) programmes associated with cell cycle commitment.

ring resistance to mating pheromones and inability to G1 arrest in response to starvation is consistent with this idea (54, 56, 57). Therefore, it is not surprising that treatment with mating pheromones and starvation both lead to rapid turnover of Clns and loss of Cdc28 kinase activity (52, 66, 103).

3.2.1 The mating pheromone response

The mating pheromone response leads to transcriptional repression of *CLN1* and *CLN2* (66, 100, 104). However, this alone cannot account for the inactivation of the Cdc28 kinase. First, *CLN3*, rather than being repressed is induced (57, 66). Second, removal of *CLN1* and *CLN2* from transcriptional control by mating pheromones still leaves cells responsive (105). Therefore regulation must be primarily at the post-transcriptional level. At least two elements appear to be involved. The *FAR1* gene, identified in a screen for mutations that uncoupled cell cycle arrest from the mating pheromone response, has been shown to be required for post-translational inactivation and turnover of Cln1 and Cln2 (105, 106). Far1 appears to be activated by phosphorylation in response to mating pheromones and functions by binding directly to Cln/Cdc28 complexes (107, 108). A second element involved in pheromone-directed inactivation of Cln/Cdc28 kinase activities is the Fus3 protein kinase, a yeast Erk (or MAP kinase) homologue (104, 109). Fus3, which is activated by the mating pheromone response pathway and is required for transcriptional activation of pheromone-responsive genes, appears to be responsible for activating Far1 (108) as well as inactivating Cln3 in a Far1-independent manner, possibly by direct phosphorylation (104, 109). How Far1 and Fus3 inactivate and destabilize Clns remains to be elucidated.

3.2.2 Starvation

It has been observed that *CLN2*-1 mutants, expressing a truncated form of Cln2 that is stabilized, cannot G1 arrest in response to starvation (54). The implication of this result is that G1 arrest depends on Cln protein turnover. Consistent with this prediction, it has been observed that Cln2 protein decays rapidly in response to nitrogen starvation, well in advance of cell cycle arrest (103). In this instance the response again appears to be largely post-transcriptional, since mRNA levels decline significantly more slowly during the time-course. The behaviour of Cln1 and Cln3 under these conditions has not been established.

3.2.3 Size control

Cln proteins also appear to be involved in the coordination of growth with division. As has been mentioned, dominant *CLN* mutations thought to stabilize Cln proteins confer a small cell-size phenotype due to a loss of this coordination (54–57). Basal levels of *CLN* transcription occur at a characteristic cell size, suggesting a higher level of control linking *CLN* transcription to growth (100, 102). The mechanism of this regulation is not yet known. Although it has been reported that different Cln proteins are critical for size control in mother vs. daughter cells (102), more recent experiments show that the interpretation of data leading to this conclusion was incorrect. It has been observed additionally that the size at which cells bud is modulated by the nutrient environment: in rich medium cells bud at a large size whereas in poor medium cells bud at a small size (26, 29, 30). Since it is likely that this modulation of budding size reflects a modulation of the execution of START relative to cell growth, the regulation of Cln accumulation or action must be involved. Although mutational modulation of cyclic AMP (cAMP) levels mimics some of these nutritional effects (high cAMP leads to budding at a larger size) (110, 111), there is no evidence that cAMP levels change significantly when cells are shifted between most nutrient conditions Therefore the signalling pathways that link nutrient environment to fine-structure control of execution of START remain to be elucidated.

4. The relationship of START to S phase

While the simplest model for the control of the G1/S phase transition has the Cln/Cdc28 kinase activity generated at START directly phosphorylating substrates involved in initiation of DNA replication, a number of observations argue for at least two distinct phases of regulation: one higher order of control executed at START and a second level of regulation occurring temporally and mechanistically much closer to actual initiation events. First, START occurs well before cells initiate DNA replication (15 min under optimal growth conditions), which corresponds to a major segment of the cell cycle (102). Second, a number of stage-specific cdc mutational blocks have been identified between START and the initiation of S phase (1, 41). These data suggest that there are a number of intervening events between START and S phase, even within the short interval of time that separates them. And finally, other classes of cyclins appear to have S phase roles not related to passage through START (75, 112, 113, 141, 142).

4.1 Does START control act directly on initiation of DNA replication?

As pointed out above there is both temporal and genetic evidence that START does not directly control the chemistry of DNA replication. In particular, there are a number of cdc mutants that confer cell cycle arrest prior to replication but subsequent to START (Fig. 6) (1, 41). Mutants of one class, including *cdc4* and *cdc34*, arrest with buds and with duplicated but unseparated spindle pole bodies (the yeast microtubule organizing centres) (1, 4, 41). This is in contrast to cells arrested at START, which are unbudded and have unduplicated spindle pole bodies (1, 4, 41). The second class, typified by *cdc7*, also arrests with buds but with duplicated and separated spindle pole bodies, indicating arrest at a later stage (1, 4, 41). Thus, cells can be blocked at two physiologically distinct phases subsequent to START but before S phase. These simple observations have been confirmed by a number of genetic experiments demonstrating that START functions have already been executed at these later cdc block points (39). This leads to the conclusion that the phosphorylations associated with execution of START are not sufficient to initiate S phase and that once START has been executed, the kinase signal associated with

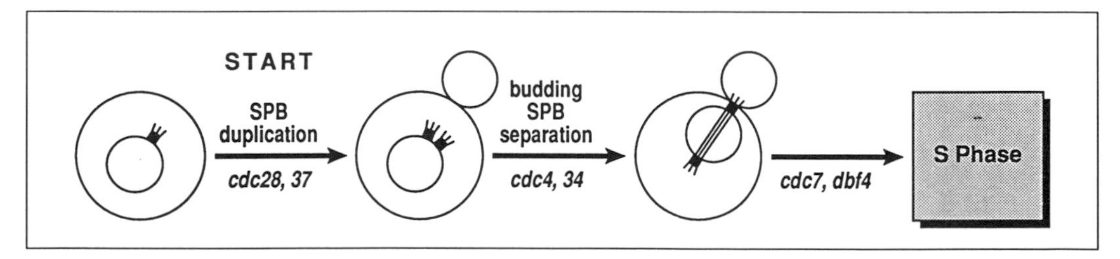

Fig. 6 There are several genetically defined steps between START in S phase. Mutations in genes *cdc28* and *cdc37* confer arrest at START. This is characterized by the absence of buds and unduplicated spindle pole bodies, the microtubule organizing centres of yeast. Mutations in genes *cdc4* and *cdc34* confer arrest prior to S phase with buds and duplicated but unseparated spindle pole bodies. Mutations in genes *cdc7* and *dbf4* also confer arrest prior to S phase but with separated spindle pole bodies. Only after these genetically defined steps have been completed can DNA replication begin. It is possible, however, that these stages correspond to checkpoint blocks triggered by the respective mutational lesions.

START is unlikely to be required directly for initiation of S phase. There are several caveats pertinent to this interpretation. First, it is possible that the cdc blocks described above do not define essential cell cycle functions but instead are indicative of checkpoint controls (ref. 114, and see Chapter 1) designed specifically to prevent initiation of DNA replication in response to specific cellular lesions. Although this scenario is difficult to argue against when the cellular defect is not known, it has been observed that mutants defective in proper spindle pole body assembly do not necessarily undergo a pre-S phase cell cycle block (115). Second, there are clearly a host of Cdc28 activities driven by cyclins other than the Clns that might substitute for START-driven S phase functions at later times. This issue will be addressed below. Finally, a large number of genes that encode proteins important for S phase have been shown to be transcriptionally regulated by MCB elements (see ref. 95, for review). As pointed out above, the MCB element appears to be directly controlled by the Cln/Cdc28 kinase at START. MCB-regulated genes encode enzymes required to maintain deoxyribonucleotide pools (e.g. *CDC21* encoding thymidylate synthase (116) and *RNR1* encoding ribonucleotide reductase large subunit (117)) and enzymes directly involved in DNA replication (e.g. *CDC9* encoding DNA ligase (99, 118) and *CDC17* encoding DNA polymerase α (94, 119)). Of particular interest is the MCB-controlled gene *CDC6* and its product (120). Although the molecular function of *CDC6* is not known, the phenotype of ts *cdc6* mutants is pre-S phase arrest (120, 121). In fission yeast, a functional and structural homologue of *CDC6*, *cdc18*, was isolated as a suppressor of a ts START mutant, *cdc10*, and *cdc10* encodes a component of the MCB transcription factor of that organism (122). The simplest interpretation of the fission yeast data is the MCB-mediated transcription controls initiation of S phase via *cdc18*. However, *cdc18* expression does not suppress deletion of *cdc10* suggesting other important targets. In budding yeast, on the other hand, it is not clear that the activation of MCB transcription achieved at START is required for DNA replication, as most MCB-driven genes show significant levels of basal transcription (92–96). Thus the relevance of Cln/Cdc28 activation of MCB transcription to *CDC6* function, for instance, remains to be determined.

4.2 Are there S cyclins regulating S phase functions?

Perhaps the best evidence for a distinct regulatory transition at the actual initiation of S phase comes from the study of B-type cyclins in budding yeast. B-type cyclins are normally associated with mitosis in higher organisms and at least two of the *S. cerevisiae* cyclins of this family, Clb1 and Clb2, appear to have primarily such a role (72–75, 123). However, simultaneous deletion of the genes encoding these cyclins and those encoding two additional B-type cyclins, Clb3 and Clb4, conferred cell cycle arrest with nuclei containing a half-replicated DNA content (75). Thus, these four cyclins collaborate to perform a mid-S phase function, perhaps controlling the initiation of late-replicating DNA. Although the molecular nature of the mid-S phase B-type cyclin requirement is not known, it establishes a role for cyclin-dependent kinase activities downstream of START in DNA replication. Another class of B-type cyclins, Clb5 and Clb6, however, may have a direct role in S phase initiation. Cells containing deletions in *CLB5* and *CLB6* are severely impaired for initiation of S phase, although they have no START defect (112, 113, 141, 142). Cells deleted for Clbs 3 through 6 are inviable, blocked at the initiation of S phase (113). Thus Clb3 and Clb4 apparently collaborate with a different set of B-type cyclins for a post-START replication initiation function. These overlapping functions are consistent with patterns of synthesis, with Clb5 and Clb6 being synthesized with dynamics identical to Clns, with Clb3 and Clb4 synthesized during S phase and G2, and Clb1 and Clb2 synthesized primarily during G2 (72–75, 112, 113, 123, 141, 142/Cdc28). Taken together, these data support the idea of S phase regulation mediated by Cdc28 activities involving B-type cyclins occurring downstream of START (Fig. 7). The elucidation of the targets of these kinase activities will require elucidation of the mechanism of initiation at chromosomal origins. The inability of Clns to rescue these Clb deficiencies argues against a direct role for Cln/Cdc28 in S phase initiation. However, the requirement of Cln/Cdc28 kinase activity for the transcription of *CLB5* and *CLB6* suggests that execution of START controls S phase indirectly by initiating an S phase transcription programme, including the genes for S phase cyclins. The issue is somewhat confused by the ability of *CLB5* and *CLB6* to rescue *cln*-deficient mutant strains with a high degree of efficiency (112, 113, 141, 142), suggesting that these B-type cyclins may normally contribute to START functions. In many ways the relationship of Clns to S phase Clbs in *S. cerevisiae* resembles the relationship between mammalian G1 cyclins and cyclin A, which appears to have an S phase role (124, 125).

4.3 START also controls morphogenesis

The discussion above has focused on the relationship of START to S phase and chromosomal DNA replication. However, in budding yeast, a complex morpho-

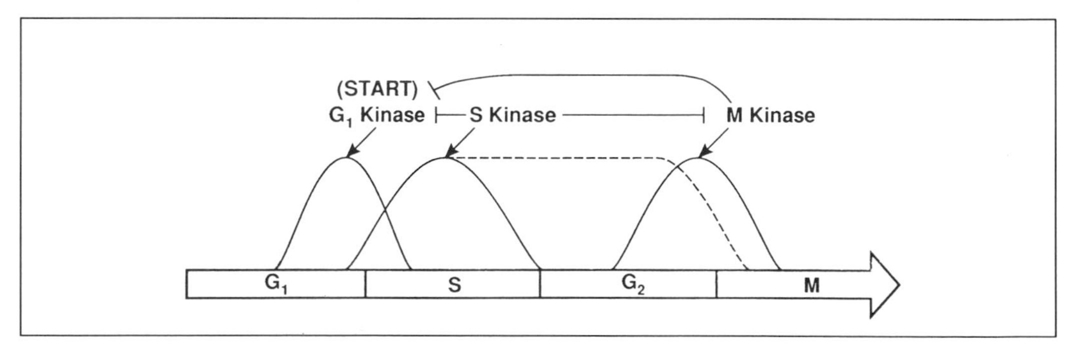

Fig. 7 Cdk functions at START, S phase and at the G2/M transition. Although Clns mediate the START transition and a subset of B-type cyclins (Clbs) are specialized for mitotic induction, some B-type cyclins appear to have roles in S phase activities, all via activation of the Cdc28 kinase. In addition both S phase and M phase cyclins negatively regulate G1 cyclins as the cell cycle progresses. At least one class of B-type cyclins, characterized by Clbs 3 and 4 appear to have sequential S phase and M phase functions. This may be similar to cyclin A in mammalian cells.

genetic programme is also required in order for cell division to be successful. This programme begins with the choice of a bud site and the mobilization of the acton cytoskeleton and secretory apparatus and ends with the formation of a budded cell undergoing mitotic division. Although a detailed discussion of morphogenesis in yeast is not within the scope of this review, it is important to point out that START controls budding as it does DNA replication. Conditions or mutations that prevent execution of START also prevent budding (1, 4, 15, 16, 41–44). In the normal course of events, the cytoskeletal polarization that precedes budding cannot be separated temporally from execution of START and is *CDC28*-dependent (126). Two other observations suggest a direct relationship between the G1 form of the Cdc28 kinase and cytoskeletal polarization. First, cytoskeletal polarization can occur upon activation of the Cdc28 kinase with G1 cyclins in the absence of protein synthesis (126) precluding any intervening level of gene expression. Second, constitutive overexpression of G1 cyclins leads to a chronic hyperpolarization of the cytoskeleton (126), suggesting a causal relationship between the Cln/Cdc28 kinase and polarization. However, as is the case for other critical targets of the Cdc28 kinase at START, the key targets for cytoskeletal polarization remain to be discovered.

5. Conclusions

The concept of a master regulatory point in the cell cycle to serve as an arbiter between different cellular activities and fates has gained wide acceptance for both yeast and mammalian cells. In yeast, a G1 event, given the designation START (1), controls both the commitment to enter into DNA synthesis from G1, as well as the commitment to other developmental pathways such as conjugation, quiescence, and meiosis/sporulation. It is now evident that the molecular basis for START is the activation of the G1 form of the Cdc28 protein kinase by G1 cyclins. Although much has been learned about the activation of the Cln/Cdc28 kinase, little is known of the critical substrates. Although the transcriptional activator Swi6 is a good candidate for targeting Cdc28 kinase activity to Cdc28-dependent transcriptional activation, this cannot be a critical target, since deletion of *swi6* is not lethal and does not prevent the execution of START (81). Furthermore, the lack of essentiality of the *MBF1* gene (K. Nasmyth, personal communication), which encodes the MCB binding factor (along with *SWI6*) suggests that neither the SCB nor the MCB transcription systems are the critical targets of START function. Since START controls both morphogenesis (commitment to budding) and commitment to S phase, two disparate cellular processes, it is likely that, rather than a single target, there are multiple targets of phosphorylation that are critical for cell cycle progression. Finding them poses a major challenge for genetic analysis in *Saccharomyces*.

A component of the cdk/cyclin regulatory machinery whose role requires clarification is Cks1/suc1 (see Chapters 5 and 6). This small ubiquitous protein, originally identified in fission yeast (127–130), is highly conserved through the eukaryotic phylogeny (131). For this protein, shown to have a high affinity for most cdks *in vitro* (129, 131, 132), it has been difficult to assign a biochemical role. However, recent genetic data obtained using temperature-sensitive mutant alleles in *S. cerevisiae* indicate that Cks1 is required for both G1 and G2 functions of the Cdc28 kinase but not for activation of histone H1 kinase activity, as assayed *in vitro* (132). This suggests a role for Cks1 in higher-order functions of the kinase, such as complex formation or substrate recognition.

Another issue that requires resolution is the relationship of START control in yeast to G1 control in mammalian cells. A point of commitment to S phase has been identified in fibroblasts and other cell types that has many of the phenomenological trappings of START. In fibroblasts the point of commitment has been called the 'restriction point' or R point; cells past the R point no longer need growth factors or high levels of protein synthesis to enter into S phase (reviewed in refs 133–135). However, unlike in yeast, the molecular basis for this commitment is not yet known. It is tempting to speculate that G1 cyclins and cyclin-dependent kinases are at the heart of this regulation as is the case with yeast. Cyclins that accumulate in G1, in fact, have been identified and shown to activate cdks at the time of the R point (136, 137). Furthermore, these cyclin/cdk activities have been shown to be negatively regulated by signals that confer G1 arrest (138–140). However, causality between activation of these kinases and passage through the R point remains to be demonstrated. Should that be accomplished, the case would be made for a degree of phylogenetic conservation of regulation of the G1/S phase transition parallel to that already demonstrated for control of entry into mitosis (see Chapters 6 and 9).

References

1. Hartwell, L. H., Culotti, J., Pringle, J., and Reid, B. J. (1974) Genetic control of the cell division cycle in yeast. *Science*, **183,** 46.
2. Hartwell, L. H. (1974) *Saccharomyces cerevisiae* cell cycle. *Bacteriol. Rev.*, **38,** 164.
3. Unger, M. W. and Hartwell, L. H. (1976) Control of cell division in *Saccharomyces cerevisiae* by methionyl-tRNA. *Proc. Natl. Acad. Sci. USA*, **73,** 1664.
4. Byers, B. and Goetsch, L. (1975) Behavior of spindles and spindle plaques in the cell cycle and conjugation of *Saccharomyces cerevisiae. J. Bacteriol.*, **124,** 511.
5. Johnston, G. C., Singer, R. A., and McFarlane, E. S. (1977) Growth and cell division during nitrogen starvation of the yeast *Saccharomyces cerevisiae. J. Bacteriol.*, **132,** 723.
6. Pinon, R. (1978) Folded chromosomes in non-cycling yeast cells. Evidence for a characteristic G0 form. *Chromosoma*, **67,** 263.
7. Sumrada, R. and Cooper, T. G. (1978) Control of vacuole permeability and protein degradation by the cell cycle arrest signal in *Saccharomyces cerevisiae J. Bacteriol.*, **136,** 234.
8. Pinon, R. and Pratt, D. (1979) Folded chromosomes of mating-factor arrested cells: Comparison with G0 arrest. *Chromosoma*, **73,** 117.
9. Lillie, S. H. and Pringle, J. R. (1980) Reserve carbohydrate metabolism in *Saccharomyces cerevisiae*: responses to nutrient limitation. *J. Bacteriol.*, **143,** 1384.

10. Pringle, J. R. and Hartwell, L. H. (1981) The *Saccharomyces cerevisiae* cell cycle. In *The Molecular Biology of the Yeast Saccharomyces*. Strathern, J. N., Jones, E. W., and Broach, J. R. (ed.). Cold Spring Harbor Lab., New York, p. 97.
11. Paris, S. and Pringle, J. R. (1983) *Saccharomyces cerevisiae*: heat and gluculase sensitivities of starved cells. *Ann. Microbiol.*, **134**, 379.
12. Iida, H. and Yahara, I. (1984) Specific early-G1 blocks accompanied with stringent response in *Saccharomyces cerevisiae* lead to growth arrest in resting state similar to the G0 of higher eucaryotes. *J. Cell Biol.*, **98**, 1185.
13. Plesset, J., Ludwig, J. R., Cox, B. S., and McLaughlin, C. S. (1987) Effect of cell cycle position on thermotolerance in *Saccharomyces cerevisiae. J. Bacteriol.*, **169**, 779.
14. Throm, E. and Duntze, W. (1970) Mating-type-dependent inhibition of deoxyribonucleic acid synthesis in *Saccharomyces cerevisiae. J. Bacteriol.*, **104**, 1388.
15. Hartwell, L. H. (1973) Synchronization of haploid yeast cells, a prelude to conjugation. *Exp. Cell Res.*, **76**, 111.
16. Bucking-Throm, E., Duntz, W., Hartwell, L. H., and Manney, T. R. (1973) Reversible arrest of haploid yeast cells at the initiation of DNA synthesis by a diffusible sex factor. *Exp. Cell Res.*, **76**, 99.
17. Wilkinson, L. E. and Pringle, J. R. (1974) Transient G1 arrest of *S. cerevisiae* cells of mating type **a** by a factor produced by cells of mating type a. *Exp. Cell Res.*, **89**, 175.
18. Stotzler, D., Kiltz, H.-H., and Duntze, W. (1976) Primary structure of a-factor peptides from *Saccharomyces cerevisiae, Eur. J. Biochem.*, **69**, 397.
19. Betz, R. and Duntze, W. (1979) Purification and partial characterizaton of a-factor, a mating hormone produced by mating-type **a** cells from *Saccharomyces cerevisiae. Eur. J. Biochem.*, **95**, 469.
20. Cross, F. R., Hartwell, L. H., Jackson, C., and Kanopka, J. B. (1988) Conjugation in *Saccharomyces cerevisiae. Annu. Rev. Cell Biol.*, **4**, 429.
21. Marsh, L., Neiman, A. M., and Herskowitz, I. (1991) Signal transduction during pheromone response in yeast. *Annu. Rev. Cell Biol.*, **7**, 699.
22. Reed, S. I. (1991) Pheromone signaling pathways in yeast. *Curr. Opin. Genet. Dev.*, **1**, 391.
23. Johnston, G. C., Pringle, J. R., and Hartwell, L. H. (1977) Coordination of growth and division in the yeast *Saccharomyces cerevisiae. Exp. Cell Res.*, **105**, 79.
24. Hartwell, L. H. and Unger, M. W. (1977) Unequal division in *Saccharomyces cerevisiae* and its implications for the control of cell division. *J. Cell Biol.*, **75**, 422.
25. Adams, J. (1977) The interrelationship of cell growth and division in haploid and diploid cells of *Saccharomyces cerevisiae. Exp. Cell Res.*, **106**, 267.
26. Jagadish, M. N. and Carter, B. L. A. (1977) Genetic control of cell division in yeast cultured at different growth rates. *Nature*, **269**, 145.
27. Johnston, G. C. (1977) Cell size and budding during starvation of the yeast *Saccharomyces cerevisiae. J. Bacteriol.*, **132**, 738.
28. Slater, M. L., Sharrow, S. O., and Gart, J. J. (1977) Cell cycle of *Saccharomyces cerevisiae* in populations growing at different rates. *Proc. Natl. Acad. Sci. USA*, **74**, 3850.
29. Carter, B. L. A. and Jagadish, M. N. (1978) The relationship between cell size and cell division in the yeast *Saccharomyces cerevisiae. Exp. Cell Res.*, **112**, 15.
30. Carter, B. L. A., Lorincz, A. T., and Johnston, G. C. (1978) Protein synthesis, cell division and the cell cycle in *Saccharomyces cerevisiae* following a shift to a richer medium. *J. Gen. Microbiol.*, **106**, 221.
31. Yamada, K. and Ito, M. (1979) Simultaneous production of buds on mother and

daughter cells of *Saccharomyces cerevisiae* in the presence of hydroxyurea. *Plant Cell Physiol.*, **20,** 1471.

32. Tyson, C. B., Lord, P. G., and Wheals, A. E. (1979) Dependency of size of *Saccharomyces cerevisiae* cells on growth rate. *J. Bacteriol.*, **138,** 92.
33. Lord, P. G. and Wheals, A. E. (1980) Asymmetrical division of *Saccharomyces cerevisiae. J. Bacteriol.*, **142,** 808.
34. Lord, P. G. and Wheals, A. E. (1981) Variability of individual cell cycles of *Saccharomyces cerevisiae. J. Cell Sci.*, **50,** 361.
35. Wheals, A. E. (1982) Size control models of *Saccharomyces cerevisiae* cell proliferation. *Mol. Cell Biol.*, **2,** 361.
36. Moore, S. A. (1984) Yeast cells recover from mating pheromone alpha-factor-induced division arrest by desensitization in the absence of alpha-factor destruction. *J. Biol. Chem.*, **259,** 1004.
37. Moore, S. A. (1984) Synchronous cell growth occurs upon synchronizing the two regulatory steps of the *Saccharomyces cerevisiae* cell cycle. *Exp. Cell Res.*, **151,** 542.
38. Brewer, B. J., Chlebowicz-Sledziewska, E., and Fangman, W. L. (1984) Cell cycle phases in the unequal mother/daughter cell cycles of *Saccharomyces cerevisiae. Mol. Cell Biol.*, **4,** 2529.
39. Herefore, L. M. and Hartwell, L. H. (1974) Sequential gene function in the initiation of *Saccharomyces cerevisiae* DNA synthesis. *J. Mol. Biol.*, **84,** 445.
40. Herskowitz, I. (1989) A regulatory hierarchy for cell specialization in yeast. *Nature*, **342,** 749.
41. Hartwell, L. H., Mortimer, R. K., Culotti, J., and Culotti, M. (1973) Genetic control of the cell division cycle in yeast. V. Genetic analysis of cdc mutants. *Genetics*, **74,** 267.
42. Reed, S. I. (1980) The selections of *S. cerevisiae* mutants defective in the start event of cell division. *Genetics*, **95,** 561.
43. Bedard, D. P., Johnston, G. C., and Singer, R. A. (1981) New mutations in the yeast *Saccharomyces cerevisiae* affecting completion of the 'start' event. *Curr. Genet.*, **4,** 205.
44. Jahng, K.-Y., Ferguson, J., and Reed, S. I. (1988) Mutations in a gene encoding the a subunit of a *Saccharomyces cerevisiae* G-protein indicate a role in mating pheromone signaling. *Mol. Cell Biol.*, **8,** 2484.
45. Lorincz, A. T. and Reed, S. I. (1984) Primary structure homology between the product of the yeast division control gene *CDC28* and vertebrate oncogenes. *Nature*, **307,** 183.
46. Reed, S. I., Hadwiger, J. A., and Lorincz, A. T. (1985) Protein kinase activity associated with the product of the yeast cell division cycle gene *CDC28. Proc. Natl. Acad. Sci. USA*, **77,** 2119.
47. Hindley, J. and Phear, G. A. (1984) Sequence of the cell division gene *cdc2* from *Shizosaccharomyces pombe*: patterns of splicing and homology to protein kinases. *Gene*, **31,** 129.
48. Beach, D., Durkacz, B., and Nurse, P. (1982) Functionally homologous cell cycle control genes in budding and fission yeast. *Nature*, **300,** 706.
49. Nurse, P., Thuriaux, P., and Nasmyth, K. A. (1976) Genetic control of the division cycle of the fission yeast *Shizosaccharomyces pombe. Mol. Gen. Genet.*, **146,** 167.
50. Nurse, P. and Bissett, Y. (1981) Gene required in G1 for commitment to cell cycle and in G2 for control of mitosis in fission yeast. *Nature*, **292,** 558.
51. Simon, M. A., Bowtell, D. D. L., Dodson, G. S., Laverty, T. R., and Rubin, G. M. (1991) Ras1 and a putative guanine nucleotide exchange factor perform crucial steps in signaling by the sevenless protein tyrosine kinase. *Cell*, **67,** 701.

52. Wittenberg, C. and Reed, S. I. (1988) Control of the yeast cell cycle is associated with assembly/disassembly of the Cdc28 protein kinase complex. *Cell*, **564,** 1061.
53. Wittenberg, C. Personal communication.
54. Hadwiger, J. A., Wittenberg, C., Richardson, H. E., de Barros Lopes, M., and Reed, S. I. (1989) A family of cyclin homologs that control the G1 phase in yeast. *Proc. Natl. Acad. Sci. USA*, **86,** 6255.
55. Sudbery, P. E., Goodey, A. R., and Carter, B. L. A. (1980) Genes which control cell proliferation in the yeast *Saccharomyces cerevisiae. Nature*, **288,** 401.
56. Cross, F. R. (1988) *DAFI*, a mutant gene affecting size control, pheromone arrest and cell cycle kinetics of *Saccharomyces cerevisiae. Mol. Cell Biol.*, **8,** 4675.
57. Nash, R., Tokiwa, G., Anand, S., Erickson, K., and Futcher, A. B. (1988) The *WHI1*$^{+}$ gene of *Saccharomyces cerevisiae* tethers cell division to cell size and is a cyclin homolog. *EMBO J.*, **7,** 4335.
58. Hunt, T. (1989) Maturation promoting factor, cyclin and the control of M phase. *Curr. Opin. Cell Biol.*, **1,** 268.
59. Lewin, B. (1990) Driving the cell cycle: M phase kinase, its partners, and substrates. *Cell*, **61,** 743.
60. Nurse, P. (1990) Universal control mechanism regulating onset of M-phase. *Nature*, **344,** 503.
61. Doree, M. (1990) Control of M-phase by maturation-promoting factor. *Curr. Opin. Cell Biol.*, **2,** 269.
62 Pines, J. and Hunter, T. (1991) Cyclin-dependent kinases: a new cell cycle motif? *Trends Cell Biol.*, **1,** 117.
63. Maller, J. L. (1991) Mitotic control. *Curr. Opin. Cell Biol.*, **3,** 269.
64. Forsburg, S. and Nurse, P. (1991) Cell cycle regulation in the yeasts *Saccharomyces cerevisiae* and *Shizosaccharomyces pombe. Annu. Rev. Cell Biol.*, **7,** 227.
65. Richardson, H. E., Wittenberg, C., Cross, F. R., and Reed, S. I. (1989) An essential G1 function for cyclin-like proteins in yeast. *Cell*, **59,** 1127.
66. Wittenberg, C., Sugimoto, K., and Reed, S. I. (1990) G1-specific cyclins of *S. cerevisiae*: cell cycle periodicity, regulation by mating pheromone, and association with the p34^{CDC28} protein kinase. *Cell*, **62,** 225.
67. Tyers, M., Tokiwa, G., Nash, R., and Futcher, B. (1992) The Cln3–Cdc28 kinase complex of *S. cerevisiae* is regulated by proteolysis and phosphorylation. *EMBO J.*, **11,** 1773.
68. de Barros Lopes, M. and Reed, S. I. Unpublished work.
69. Rogers, S., Wells, R, and Rechsteiner, M. (1986) Amino acid sequences common to rapidly degraded proteins: the PEST hypothesis. *Science*, **234,** 364.
70. Reed, S. I. and Wittenberg, C. (1990) A mitotic role for the Cdc28 protein kinase of *S. cerevisiae. Proc. Natl. Acad. Sci. USA*, **87,** 5697.
71. Piggott, J. R., Rai, R., and Carter, B. L. A. (1982) A bifunctional gene product involved in two phases of the yeast cell cycle. *Nature*, **298,** 391.
72. Surana, U., Robitsch, H., Price, C., Schuster, T., Fitch, I., Futcher, A. B., and Nasmyth, K. (1991) The role of *CDC28* and cyclins during mitosis in the budding yeast *S. cerevisiae. Cell*, **65,** 145.
73. Ghiara, J. B., Richardson, H. E., Sugimoto, K., Henze, M., Lew, D. J., Wittenberg, C., and Reed, S. I. (1991) A cyclin B homolog in *S. cerevisiae*: chronic activation of the Cdc28 protein kinase by cyclin prevents exit from mitosis. *Cell*, **65,** 163.
74. Fitch, I., Dahman, C., Surana, U., Amon, A., Nasmyth, K., Goetsch, L, Byers, B.,

and Futcher, B. (1992) Characterization of four B-type cyclin genes of the budding yeast *Saccharomyces cerevisiae*. *Mol. Biol. Cell*, **3,** 805.

75. Richardson, H., Lew, D. J., Henze, M., Sugimoto, K., and Reed, S. I. (1992) Cyclin-B homologs in *Saccharomyces cerevisiae* function in S phase and in G2. *Genes Dev.*, **6,** 2021.
76. Cross, F. R. (1990) Cell cycles arrest caused by *CLN* gene deficiency in *Saccharomyces cerevisiae* resembles START-1 arrest and is independent of the mating-pheromone signalling pathway. *Mol. Cell. Biol.*, **10,** 6482.
77. Deshaies, R. and Kirschner, M. Personal communication.
78. Winge, O. and Roberts, C. (1949) A gene for diploidization in yeast. *C. R. Trav. Lab. Carlsberg, Ser. Physiol.*, **24,** 341.
79. Nasmyth, K. (1985) At least 1400 base pairs of 5′ flanking DNA is required for the correct expression of the *HO* gene in yeast. *Cell*, **42,** 213.
80. Nasmyth, K. (1985) A repetitive DNA sequence that confers cell cycle START (*CDC28*)-dependent transcription of the *HO* gene in yeast. *Cell*, **42,** 225.
81. Breeden, L. and Nasmyth, K. (1987) Cell cycle control of the yeast *HO* gene: *cis*- and *trans*-acting regulators. *Cell*, **48,** 389.
82. Ogas, J., Andrews, B. J., and Herskowitz, I. (1991) Transcriptional activation of *CLN1, CLN2,* and a putative new G1 cyclin (*HCS26*) by SWI4, a positive regulator of G1-specific transcription. *Cell*, **66,** 1015.
83. Stern, M., Jensen, R., and Herskowitz, I. (1984) Five *SWI* genes are required for expression of the *HO* gene in yeast. *J. Mol. Biol.*, **178,** 853.
84. Breeden, L. and Nasmyth, K. (1987) Similarity between cell-cycle genes of budding yeast and fission yeast and the *Notch* gene of *Drosophila*. *Nature*, **329,** 651.
85. Andrews, B. J. and Herskowitz, I. (1989) The yeast SWI4 protein contains a motif present in developmental regulators and is part of a complex involved in cell-cycle-dependent transcription. *Nature*, **342,** 830.
86. Primig, M., Sockanathan, S., Auer, H., and Nasmyth, K. (1992) Anatomy of a transcription factor important for the Start of the cell cycle in *Saccharomyces cerevisiae*. *Nature*, **358,** 593.
87. Sidorova, J. and Breeden, L. (1993) Analysis of the SWI4/SWI6 protein complex, which directs G1/S-specific transcription in *Saccharomyces cerevisiae*. *Mol. Cell. Biol.*, **13,** 1069.
88. Andrews, B. J. and Herskowitz, I. (1989) Identification of a DNA binding factor involved in cell-cycle control of the yeast *HO* gene. *Cell*, **57,** 21.
89. Nasmyth, K. and Dirick, L. (1991) The role of SWI4 and SWI6 in the activity of G1 cyclins in yeast. *Cell*, **66,** 995.
90. Andrews, B. J. and Herskowitz, I. (1990) Regulation of cell cycle-dependent gene expression in yeast. *J. Biol. Chem.*, **265,** 14057.
91. Breeden, L. and Mikesell, G. (1991) Cell cycle-specific expression of the SWI4 transcription factor is required for the cell cycle regulation of HO transcription. *Genes Dev.*, **5,** 1183.
92. McIntosh, E. M., Atkinson, T., Storms, R. K., and Smith, M. (1991) Characterization of a short, *cis*-acting DNA sequence which conveys cell cycle stage-dependent transcription in *Saccharomyces cerevisiae*. *Mol. Cell. Biol.*, **11,** 329.
93. Lowndes, N. F., Johnson, A. L., and Johnston, L. H. (1991) Coordination of expression of DNA synthesis genes in budding yeast by a cell-cycle regulated *trans* factor. *Nature*, **350,** 247.
94. Gordon, C. B. and Campbell, J. L. (1991) A cell cycle-responsive transcriptional control

element and a negative control element in the gene encoding DNA polymerase a in *Saccharomyces cerevisiae*. *Proc. Natl. Acad. Sci. USA*, **88,** 6058.

95. Johnston, L. H. and Lowndes, N. F. (1992) Cell cycle control of DNA synthesis in budding yeast. *Nucleic Acids Res.*, **20,** 2403.
96. Marini, N. J. and Reed, S. I. (1992) Direct induction of G1-specific transcripts following reactivation of the Cdc28 kinase in the absence of de novo protein synthesis. *Genes Dev.*, **6,** 557.
97. Dirick, L., Moll, T., Auer, H., and Nasmyth, K. (1992) A central role for *SWI6* in modulating cell cycle Start-specific transcription in yeast. *Nature*, **357,** 508.
98. Lowndes, N. F., Johnson, A. L., Breeden, L., and Johnston, L. H. (1992) *SWI6* protein is required for transcription of the periodically expressed DNA synthesis genes in budding yeast. *Nature*, **357,** 505.
99. White, J. H. M., Green, S. R., Barker, D. G., Dumas, L. B., and Johnston, L. H. (1987) The *CDC8* transcript is cell cycle regulated in yeast and is expressed coordinately with *CDC9* and *CDC21* at a point preceding histone transcription. *Exp. Cell Res.*, **171,** 223.
100. Dirick, L. and Nasmyth, K. (1991) Positive feedback in the activation of G1 cyclins in yeast. *Nature*, **351,** 754.
101. Cross, F. R. and Tinkelenberg, A. H. (1991) A potential positive feedback loop controlling *CLN1* and *CLN2* gene expression at the start of the yeast cell cycle. *Cell*, **65,** 875.
102. Lew, D. J., Marini, N. J., and Reed, S. I. (1992) Different G1 cyclins control the timing of cell cycle commitment in mother and daughter cells of the budding yeast *Saccharomyces cerevisiae*. *Cell*, **69,** 317.
103. Reed, S. I., Wittenberg, C., Lew, D. J., Dulic, V., and Henze, M. (1991) G1 control in yeast and animal cells. *Cold Spring Harbor Symp. Quant. Biol.*, **56,** 61.
104. Elion, E. A., Brill, J. A., and Fink, G. R. (1991) *FUS3* inactivates G1 cyclins and, in concert with *KSS1*, promotes signal transduction. *Proc. Natl. Acad. Sci. USA*, **88,** 9392.
105. Valdivieso, H. M., Sugimoto, K., Jahng, K.-Y., Fernandes, P. M. B., and Wittenberg, C. (1992) *FAR1* is required for post-transcriptional regulation of *CLN2* gene expression in response to mating pheromone. *Mol. Cell. Biol.*, **13,** 1013.
106. Chang, F. and Herskowitz, I. (1990) Identification of a gene necessary for cell cycle arrest by a negative growth factor of yeast: FAR1 is an inhibitor of a G1 cyclin, CLN2. *Cell*, **63,** 999.
107. Chang, F. and Herskowitz, I. (1992) Phosphorylation of *FAR1* in response to a-factor: a possible requirement for cell cycle arrest. *Mol. Biol. Cell*, **3,** 445.
108. Peter, M., Gartner, A., Horecka, J., Ammerer, G., and Herskowitz, I. (1993) FAR1 links the signal transduction pathway to the cell cycle machinery in yeast. *Cell*, **73,** 747.
109. Elion, E. A., Grisafi, P., and Fink, G. R. (1990) *FUS3* encodes a $cdc2^{+}$/CDC28-related kinase required for the transition from mitosis into conjugation. *Cell*, **60,** 649.
110. Baroni, M. D., Martegani, E., Monti, P., and Alberghina, L. (1989) Cell size modulation by *CDC25* and *RAS2* genes in *Saccharomyces cerevisiae*. *Mol. Cell. Biol.*, **9,** 2715.
111. Baroni, M. D., Monti, P., Marconi, G., and Alberghina, L. (1992) cAMP-mediated increase in the critical cell size required for the G1 to S transition in *Saccharomyces cerevisiae*. *Exp. Cell Res.*, **201,** 299.
112. Epstein, C. B. and Cross, F. R. (1992) CLB5: a novel B cyclin from budding yeast with a role in S phase. *Genes Dev.*, **6,** 1695.
113. Basco, R., Segal, M., and Reed, S. I. Negative regulation of G1 and G2 by S phase cyclins of *Saccharomyces cerevisiae*. Submitted.

114. Hartwell, L. H. and Weinert, T. A. (1989) Checkpoints: controls that ensure the order of cell cycle events. *Science*, **246,** 629.
115. Byers, B. (1981) Multiple roles of the spindle pole bodies in the life cycle of *Saccharomyces cerevisiae*. In *Molecular genetics in yeast*: Von Wettstein, D., Friis, J., Kielland-Brandt, M., and Stenderup, A. (ed.). Munksgaard, Copenhagen, p. 119.
116. Storms, R. K., Ord, R. W., Greenwood, M. T., Mirdamadi, B., Chu, F. K., and Belfort, M. (1984) Cell-cycle-dependent expression of thymidylate synthase in *Saccharomyces cerevisiae*. *Mol. Cell. Biol.*, **4,** 2858.
117. Elledge, S. J. and Davis, R. W. (1990) Two genes differentially regulated in the cell cycle and by DNA-damaging agents encode alternative regulatory subunits of ribonucleotide reductase. *Genes Dev.*, **4,** 740.
118. Peterson, T. A., Prakash, L., Prakash, S., Osley, M. A., and Reed, S. I. (1985) Regulation of *CDC9*, the *Saccharomyces cerevisiae* gene that encodes DNA ligase. *Mol. Cell. Biol.*, **5,** 226.
119. Johnston, L. H., White, J., Johnson, A., Lucchini, G., and Plevani, P. (1987) The yeast DNA polymerase I transcript is regulated in both the mitotic cell cycle and in meiosis and is also induced after DNA damage. *Nucleic Acids Res.*, **15,** 5017.
120. Zhou, C. and Jong, A. (1990) *CDC6* mRNA fluctuates periodically in the yeast cell cycle. *J. Biol. Chem.*, **265,** 19904.
121. Bueno, A. and Russell, P. (1992) Dual functions of *CDC6*: a yeast protein required for DNA replication also inhibits nuclear division. *EMBO J.*, **11,** 2167.
122. Kelly, T. J., Martin, G. S., Forsburg, S. L., Stephen, R. J., Russo, A., and Nurse, P. (1993) The fission yeast *cdd8*$^+$ gene product couples S phase to START and mitosis. *Cell*, **74,** 371.
123. Grandin, N. and Reed, S. I. (1993) Differential function and expression of *S. cerevisiae* G-type cyclins in mitosis and meiosis. *Mol. Cell. Biol.*, **13,** 2113.
124. Girard, F., Strausfeld, U., Fernandez, A., and Lamb, N. J. C. (1991) Cyclin A is required for the onset of DNA replication in mammalian fibroblasts. *Cell*, **67,** 1169.
125. Pagano, M., Pepperkok, R., Verde, F., Ansorje, W., and Draetta, G. (1992) Cyclin A is required at two points in the human cell cycle. *EMBO J.*, **11,** 961.
126. Lew, D. J. and Reed, S. I. (1993) Morphogenesis in the yeast cell cycle: regulation by Cdc28 and cyclins. *J. Cell Biol.*, **120,** 1305.
127. Hayles, J., Aves, S., and Nurse, P. (1986) *Suc1* is an essential gene involved in both the cell cycle and growth in fission yeast. *EMBO J.*, **5,** 3373.
128. Hayles, J., Beach, D., Durkacz, B., and Nurse, P. (1986) The fission yeast cell cycle control gene *cdc2*: isolation of a sequence *suc1* that suppresses *cdc2* mutant function. *Mol. Gen. Genet.*, **202,** 291.
129. Brizuela, L., Draetta, G., and Beach, D. (1987) p13 suc1 acts in the fission yeast cell division cycle as a component of the p34^{cdc2} kinase. *EMBO J.*, **6,** 3507.
130. Hadwiger, J. A., Wittenberg, C., Mendenhall, M., and Reed, S. I. (1989) The *S. cerevisiae CKS1* gene, a homolog of the *S. pombe suc1*$^+$ gene, encodes a subunit of the Cdc28 protein kinase complex. *Mol. Cell. Biol.*, **9,** 2034.
131. Richardson, H. E., Stueland, C. S., Thomas, J., Russell, P., and Reed, S. I. (1990) Human cDNAs encoding homologs of the small p34$^{Cdc28/Cdc2}$-associated protein of *Saccharomyces cerevisiae* and *Shizosaccharomyces pombe*. *Genes Dev.*, **4,** 1332.
132. Tang, Y. and Reed, S. I. (1993). The Cdk-associated protein Cks1 functions both in G1 and G2 in *Saccharomyces cerevisiae*. *Genes Dev.*, **7,** 822.

133. Pardee, A. B., Dubrow, R., Hamlin, J. L., and Kletzien, R. (1978) Animal cell cycle. *Annu. Rev. Biochem.*, **47,** 715.
134. Pardee, A. B., Coppock, D. L., and Yang, H. C. (1986) Regulation of cell proliferation at the onset of DNA synthesis. *J. Cell Sci., Suppl.*, **4,** 171.
135. Pardee, A. B. (1989) G1 events and regulation of cell proliferation. *Science*, **246,** 603.
136. Koff, A., Giordano, A., Desai, D., Yamashita, K., Harper, J. W., Elledge, S., Nishimoto, T., Morgan, D. O., Franza, B. R., and Roberts, J. M. (1992) Formation and activation of a cyclin E–cdk2 complex during the G1 phase of the human cell cycle. *Science*, **257,** 1689.
137. Dulic, V., Lees, E., and Reed, S. I. (1992) Association of human cyclin E with a periodic G1–S phase protein kinase. *Science*, **257,** 1958.
138. Koff, A., Ohtsuki, M., Polyak, K., Roberts, J. M., and Massague, J. (1993) Negative regulation of G1 progression in mammalian cells: inhibition of cyclin E-dependent kinase by TGF-β. *Science*, **260,** 536.
139. Slingerland, J. M., Hengst, L., Pan, C.-H., Alexander, D., Stampfer, M., and Reed, S. I. (1994) A novel inhibitor of cyclin-cdk activity detected in transforming growth factor β-arrested epithelial cells. *Mol. Cell. Biol.*, **14,** 3683.
140. Dulic, V., Kaufmann, W., Lees, E., and Reed, S. I. (1993) Negative regulation of G1 progression in human fibroblasts by gamma irradiation: inhibition of cyclin E- and cyclin A-associated Cdk2 kinase activities. *Cell*, **76,** 1013.
141. Schwob, E., and Nasmyth, K. (1993) *CLB5* and *CLB6*, a new pair of B cyclins involved in DNA replication in *Saccharomyces cerevisiae*. *Genes and Dev.*, **7,** 1160.
142. Kühne, C., and Linden, P. (1993) A new pair of B-type cyclins from *Saccharonyces cerevisiae* that function early in the cell cycle. *EMBO Journal*, **12,** 3437.

4 | Controlling entry into mitosis in fission yeast

STUART A. MACNEILL and PETER A. FANTES

1. Introduction

1.1 Universal control over entry into mitosis

As described in Chapter 1 the timing of entry into mitosis in eukaryotic cells is determined by the timing of activation of a protein complex (called MPF) containing one molecule of a cyclin-dependent kinase (cdc2) and one of cyclin B. The genes encoding the cdc2 and cyclin B (cdc13) proteins in the fission yeast *Schizosaccharomyces pombe* were identified genetically as being required for entry into mitosis long before their biochemical function was known or even guessed at. Similarly, two key regulators of cdc2 function, the cdc25 tyrosine phosphatase and the wee1 protein kinase, were first identified and characterized genetically in fission yeast. At present, many other elements implicated in the mitotic control network are known only in this organism. Here we review recent advances in understanding the biochemical basis of the regulatory network that controls entry into mitosis in *S. pombe*, in identifying new elements involved in this regulatory network, and in understanding how entry into mitosis is integrated with other cell cycle events. We begin with a brief introduction to the biology of the fission yeast cell cycle.

1.2 Why study fission yeast?

S. pombe is a unicellular ascomycete fungus with a genome size of 14 Mb that is distantly related to the budding yeast *Saccharomyces cerevisiae* (1). *S. pombe* cells are rod-shaped, between 7 and 14 μm in length depending on cell cycle position, and 3.5 μm in diameter. The cells grow by apical extension and divide by medial fission. The suitability of using *S. pombe* as a model for cell cycle analysis was first noted by Mitchison (2, 3) who observed that since the cells grew only by length extension it was possible to determine at which stage a cell was in the cell cycle simply by measuring its length (see Chapter 2).

S. pombe is particularly amenable to both classical and molecular genetic analyses (4–6). Genetic mapping can be carried out by tetrad dissection or random spore analysis of meiotic products (cells of opposite mating type will undergo mating and meiosis when deprived of nutrients); and complementation analysis performed using zygotically-derived diploids (4, 5). Molecular biological techniques available include high efficiency transformation (allowing genes to be cloned by complementation) and gene replacement (see Chapter 2, and refs 5, 6). (See ref. 7 for a compendium of methods used specifically for analysis of the *S. pombe* cell cycle.)

The organization of the *S. pombe* cell cycle is typically eukaryotic, with discrete G1, S, G2, and M phases occupying around 0.1, 0.1, 0.7, and 0.1 of a cycle respectively (7). Following G2, entry into mitosis is marked by chromosome condensation and by rapid microtubule rearrangement, as the network of cytoplasmic microtubules characteristic of interphase disappears to be replaced by an intranuclear mitotic spindle (8). The condensed chromosomes then become aligned at the metaphase plate before the sister chromatids are pulled apart and septation occurs. The organization of the *S. pombe* cell cycle is therefore very different from that of *S. cerevisiae*, as in this organism a short spindle is formed early in the cell cycle that only elongates at or around the time of cell division (see Chapter 1). This has led to the suggestion that S phase and mitosis overlap in budding yeast (9).

1.3 *S. pombe* cell cycle mutants

Many fission yeast gene functions have been identified by mutational analysis as playing an important part in the processes of mitosis, including a number that are essential for entry into mitosis or that have a role in determining the precise timing of mitotic initiation, as well as functions that are required for the physical processes of mitosis itself.

Cell cycle arrest in *S. pombe* typically leads to cell elongation (the cdc phenotype) (see Chapter 2). This is because there is little or no reduction in cellular growth rate under conditions of cell cycle arrest so that the cells continue to grow by apical extension and become highly elongated (10). For this reason, isolation of cell cycle mutants is a relatively straightforward process, as mutants can be identified by microscopic examination (Fig. 1). Twenty-five genes required for cell cycle progress have been identified in this way. These cdc genes (cdc is an acronym for cell division cycle) fall into a number of classes depending on their arrest point (listed in Table 1), and include genes required for entry in the cell cycle, for the initiation of DNA replication, and for entry into mitosis, as well as genes required for septation (10–12). Other types of cell cycle mutants, such as those blocked in mitosis itself, do not necessarily elongate when arrested. Amongst these gene functions, consideration of which for the most part falls outside the scope of this review, are *nda2*$^{+}$ and *nda3*$^{+}$, which encode α- and β-tubulin respectively (13, 14), *cut1*$^{+}$–*cut19*$^{+}$ (15, 16), *nuc2*$^{+}$ (17), and *top2*$^{+}$, which encodes topoisomerase II (18, 19). In some cases, although nuclear division is blocked, septation occurs on schedule with the result that the mutant cell is cut into two (15, 16). The following section deals specifically with those genes involved at the G2–M transition.

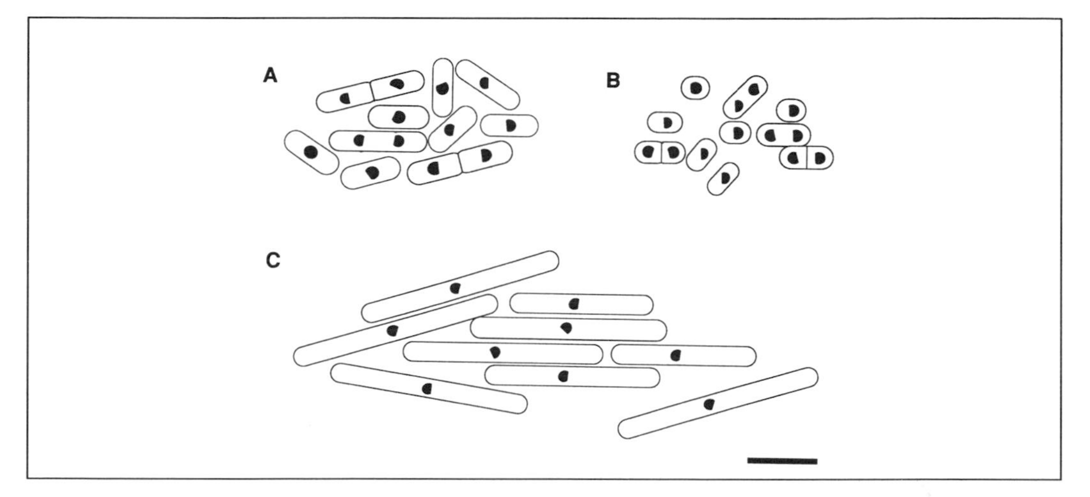

Fig. 1 A schematic representation of wild-type, wee, and interphase arrested fission yeast cells (parts A, B, and C, respectively). Scale bar 10 μm.

Table 1 Cell division cycle (cdc) genes in fission yeast

Function	Genes	Comments
G1–S	*cdc2*$^+$, *cdc10*$^+$	Required in late G1 for entry into S phase (10, 11); cdc10 is part of a transcription complex that controls periodic gene expression at G1–S (145)
S	*cdc17*$^+$, *cdc18*$^+$, *cdc19*$^+$, *cdc20*$^+$ *cdc21*$^+$, *cdc22*$^+$, *cdc23*$^+$, *cdc24*$^+$	Required for DNA replication (10, 11); include DNA ligase (cdc17) and ribonucleotide reductase (cdc22) enzymes (123, 127)
G2	*cdc6*$^+$	Required in mid-G2; function unknown (10)
G2–M	*cdc1*$^+$, *cdc2*$^+$, *cdc13*$^+$, *cdc25*$^+$, *cdc27*$^+$, *cdc28*$^+$	Required in late G2 entry into mitosis (10, 11); see text for functions
Septation	*cdc3*$^+$, *cdc4*$^+$, *cdc7*$^+$, *cdc8*$^+$, *cdc11*$^+$, *cdc12*$^+$, *cdc14*$^+$, *cdc15*$^+$, *cdc16*$^+$	Required for septum formation and cleavage (10–12); products include tropomyosin (cdc8) and a Bub2 homologue (cdc16) (151, 156)

1.4 G2–M mutants

Six genes have been identified whose products are required under normal circumstances for entry into mitosis. Three of these encode the cdc2, cdc13, and cdc25 proteins, which lie at the heart of the regulatory network that controls entry into mitosis in fission yeast and in all likelihood in all eukaryotic cells (20). The remaining three gene functions *cdc1*$^+$, *cdc27*$^+$, and *cdc28*$^+$ (Table 1) are less well understood, although molecular analysis of all three proteins is now underway (detailed in Section 7.2).

In addition to those genes whose functions have been identified as being absolutely required for cell cycle progress, a number of others whose function is to determine

Table 2 Advancing mitosis: the wee phenotype

Recessive loss-of-function mutations	Dominant gain-of-function mutations	By overproduction only
wee1	cdc2 [a]	cdc25
ppa2 [b]		nim1
pyp1 [b]		pyp3
		wis1 [b]

[a] Gross overproduction of cdc2 can also lead to a lethal form of mitotic advancement.
[b] Advancement to a semi-wee value only (cell length at cell division of 10–12 μm, compared to 7–8 μm fully wee and 14 μm wild-type).

the timing of entry into mitosis have been uncovered. This is possible in fission yeast because advancing the timing of entry into mitosis results in G2 being shortened and cells undergoing cell division at a reduced (wee) cell size compared to wild-type cells (21). This is shown schematically in Fig. 1. The first gene to be identified that falls into this category was *wee1*$^+$, which encodes a mitotic inhibitor as loss-of-function mutations in the *wee1* gene bring about mitotic advancement (21, 22). Certain mutations in the *cdc2* gene have a similar effect, although in these cases the mutations are genetically dominant (22, 23). At present at least eight gene functions that can be altered in some way to advance mitosis (either by mutation or overexpression) have been characterized. These functions are listed in Table 2.

While the two classes of genes described above include the majority of those with key functions at the G2–M transition, there are also a number of other gene functions that play important roles at this point in the cycle, in which mutations do not bring about either cell cycle arrest or mitotic advancement. These genes have typically been isolated on the basis of their interaction with known cell cycle genes (24, 25) or by physical methods (26, 27).

2. The mitotic kinase: cdc2/cdc13/suc1

2.1 cdc2

The cdc2 protein kinase is the central catalytic component of the mitotic kinase complex (summarized in Table 3). The *cdc2*$^+$ gene function is required in *S. pombe* both for entry into mitosis from G2 and for entry into S phase from G1 (10, 28). Cells carrying temperature-sensitive mutations in the *cdc2* gene become arrested at one or other of these points in the cycle when shifted to the restrictive temperature, depending on their position prior to the temperature shift—cells in mitosis or G1 become arrested in late G1 (at START) while cells in S or G2 become arrested at the G2–M boundary (10, 28). In addition, *cdc2*$^+$ function is required for the

Table 3 The mitotic kinase in fission yeast: a summary

	Comments
cdc2	Required for both G1–S and G2–M transitions, the activity of the cdc2 protein kinase is regulated by phosphorylation: forms complex with cdc13 and suc1 (10, 28, 35–38)
suc1	Overproduction of suc1 protein delays entry into mitosis: in the absence of suc1, a proportion of the cells arrest mid-mitosis (35, 59, 60)
cdc13	B-type cyclin—required for G2–M transition and later in mitosis: forms complex with cdc2 and is degraded at anaphase (10, 11, 35, 36, 54, 55)

second meiotic nuclear division—in its absence two-spored rather than four-spored asci are formed (29–31)—and determines the timing of entry into mitosis. Two types of dominant wee allele of the *cdc2*$^+$ gene that advance mitosis have been isolated (22, 23). These differ in their interactions with the mitotic regulators cdc25 and wee1 (see below). It was genetic analysis of these mutants, *cdc2-1w* and *cdc2-3w*, and their interactions with *wee1*$^+$ and *cdc25*$^+$, that led to the development of the standard model for the regulation of cdc2 function (32).

2.1.1 cdc2 phosphorylation in fission yeast

The *cdc2*$^+$ gene product, the catalytic component of the mitotic kinase complex, is the 297 amino acid (M_r 34 kDa) protein serine–threonine kinase $p34^{cdc2}$ (33, 34). Functional homologues of $p34^{cdc2}$ have now been identified in a broad range of eukaryotic cell types. Each of these functional homologues is approximately 60 per cent identical at the amino acid sequence level to the *S. pombe* protein. In *S. cerevisiae* the cdc2 homologue is the Cdc28 protein whose function is considered at length in Chapter 3. The functions of cdc2 and the related cdk proteins (cdk is an acronym for cyclin-dependent kinase) in vertebrate cells are considered in Chapter 5 (cdc2) and Chapter 6 (cdk proteins).

The activity of the cdc2 protein kinase measured *in vitro* using histone H1 as an exogenous substrate reaches a maximum at the G2–M transition (35, 36) and is regulated in two ways—by interaction with the other components in the complex, and by phosphorylation of the cdc2 protein itself. In *S. pombe* cdc2 is phosphorylated on both threonine and tyrosine (Fig. 2). The sites of phosphorylation, which are conserved across evolution, are threonine 167 within the central catalytic domain of the enzyme (37), and tyrosine 15 which forms part of the ATP-binding cleft (38). Phosphorylation of neither threonine 14 (found in many higher eukaryotes) nor serine 283 (phosphorylated in chicken cdc2 protein) has been detected in the fission yeast cdc2, earlier reports of serine phosphorylation of the protein having proved erroneous (39, 40).

Understanding how the cdc2 protein was activated at the G2–M transition marked a significant step forward in elucidating the means by which entry into mitosis was effected in *S. pombe* (38). Activation is dependent upon tyrosine 15 dephosphorylation, a reaction catalysed by the cdc25 protein tyrosine phosphatase

(41) and the functionally-related pyp3 enzyme (42). Replacement of tyrosine 15 with non-phosphorylatable phenylalanine has a marked effect on protein function as the mutant protein causes an advancement into mitosis at a small cell size (the wee phenotype, shown in Fig. 1). The cdc25 protein is dispensable under these circumstances (38), a fact that first prompted the suggestion that cdc25 had a role in mediating tyrosine 15 dephosphorylation of cdc2 (see Section 3).

Phosphorylation of threonine 167 is essential for the interaction between cdc2 and cdc13/cyclin B (see below) and is absolutely required for cdc2 protein function (37). Replacing the threonine with a non-phosphorylatable alanine residue abolishes protein function *in vivo*, while substituting a serine for the threonine has no effect on function. Other mutations at threonine 167 (discussed below) give rise to a dominant cell cycle arrest phenotype (7). Recently, the protein kinase responsible for phosphorylation of this threonine residue, called CAK (for cdk-activating kinase), has been purified from both starfish and *Xenopus* (43–45). The *Xenopus* enzyme has been shown to be the previously identified $p40^{MO15}$ protein (46).

2.1.2 cdc2 structure/function

Much has been learned about the function of the cdc2 protein in *S. pombe* from the study of mutant *cdc2* alleles isolated either in genetic screens (10, 11, 22, 23, 47, 48) or by *in vitro* directed mutagenesis of the wild-type gene (37, 38, 40, 49–51). The first of these approaches has led to the isolation of a large number of chromosomal mutant *cdc2* alleles, falling into four categories (Table 4): wee (22, 23), temperature-sensitive lethal (10, 11, 47), cold-sensitive lethal (48), and meiotically-defective (30, 31, 47). Many of these alleles have been cloned and the lesions responsible for their respective mutant phenotypes identified. Each of the cloned alleles encodes a protein that differs from the wild-type cdc2 protein by one amino acid only (47–49, 52), and almost all the mutations affect residues that are conserved across all the cdc2 functional homologues, underlining the importance of these conserved residues for cdc2 function.

Although the mutations are distributed throughout the length of the cdc2 protein (shown in Fig. 2), there are regions in which some clustering of mutations has

Table 4 Chromosomal *cdc2* mutant alleles

Phenotype	Alleles	References
Temperature-sensitive arrest	M35[a], 56/130[b], 33, L7, M26/M55, 17/22[c], 18[c], 45, 48, M63	(10, 11, 47)
Cold-sensitive arrest	A20/B14/D21[c], E8[c], E9[c], 59, r4	(48, 58, 79)
Meiotically-defective	N22	(30, 31, 47)
Wee	1w/2w, 3w, 4w	(22, 23, 47)

[a] M35 is severely meiotically-defective at its permissive temperature as well.
[b] 56/130 is partially wee at its permissive temperature.
[c] G2–M defect only.

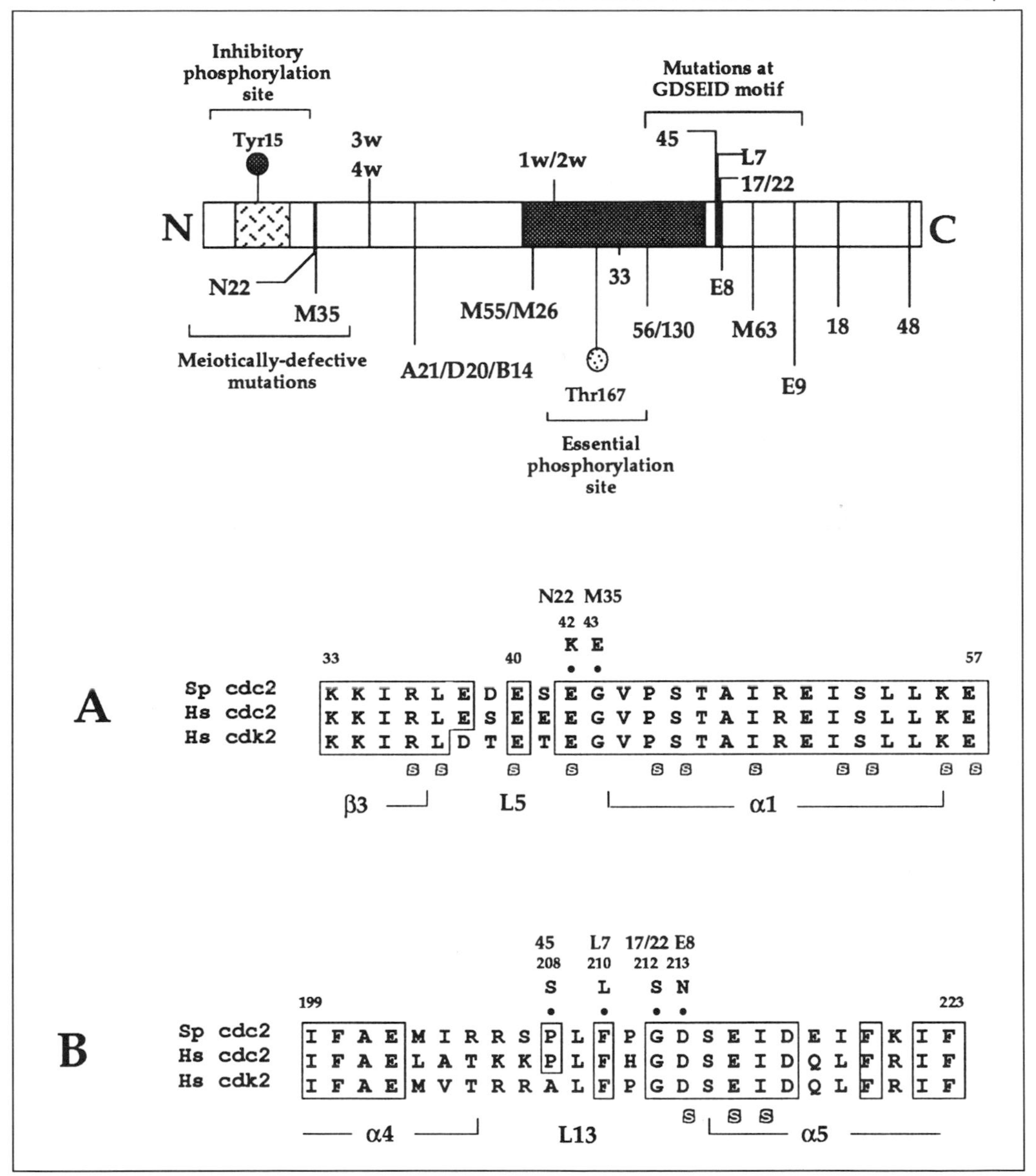

Fig. 2 The cdc2 protein kinase: location of phosphorylation sites and mutations. Top: a schematic representation of the cdc2 protein showing the location of the mutations identified in 23 cloned mutant alleles as well as the two phosphorylation sites, tyrosine 15 in the ATP-binding site (lightly shaded box) and threonine 167 in the central catalytic core of the enzyme (darkly shaded box). Two regions of the protein are shown in detail below—an acidic patch region towards the amino-terminus of the protein (A) and the region encompassing the GDSEID motif (B). In each case the sequence of the fission yeast (Sp) and human (Hs) cdc2 proteins are shown aligned with the human cdk2 protein sequence. Boxed residues are conserved in all three proteins, or in the two cdc2 proteins. Above the aligned sequences the amino acid changes in various mutant alleles are shown (indicated by the •). Predicted cdc2 secondary structure elements, based on the three-dimensional structure of cdk2, are indicated—solvent-exposed residues in cdk2 are indicated by an s beneath the aligned sequences. See text for references and further discussion.

been found, suggesting that these regions may have particular functional importance (47, 48). One such region encompasses the highly-conserved GDSEID motif (amino acids 212–217 in the fission yeast protein) where three of the temperature-sensitive mutations (L7, 17/22, and 45) together with a cold-sensitive mutation (E8) are found within seven residues of one another (see Fig. 2B) (47, 48). The GDSEID motif is found perfectly conserved in all the cdc2 functional homologues but not in many of the cdc2-related (cdk) enzymes (Chapter 6). The mutation in one of two dominant negative cdc2 proteins generated by chemical mutagenesis of the *cdc2*$^+$ gene also maps within this region (40, 50). Similarly two mutations with profound meiotic defects have been mapped to adjacent residues towards the amino-terminus of the cdc2 protein (shown in Fig. 2A): cdc2-N22 is defective for meiosis only and shows no defect in the mitotic cell cycle, while cdc2-M35 is mitotically-defective as well (Table 4). The mutations in these proteins lie at amino acids 42 and 43 respectively, in an acidic patch region. It is tempting to speculate that this part of the protein may be a binding site for a meiosis-specific cdc2 activator (47).

The recently completed X-ray crystal structure of the cdk2 protein (cdk2 is closely related to cdc2 at the primary sequence level and may share overlapping functions *in vivo*—see Chapter 6) provides an excellent framework for modelling the three-dimensional structure of the fission yeast cdc2 protein (53). By superimposing the amino acid sequence of cdc2 onto the tertiary structure of cdk2 it is possible to locate those residues that are mutated in the cdc2 proteins described above with respect to the active site and regulatory phosphorylation sites on the cdc2 enzyme. By this method both the GDSEID region (amino acids 212–217) and the acidic patch around amino acids 42 and 43 are found to be positioned on the surface of the cdc2 protein (see Fig. 2) as residues D213, E215, and I216 in GDSEID, and E40 and E42 in the acidic patch are solvent-exposed, indicating that they may be available for intermolecular interactions with cdc13 or suc1 (see below). Further modelling of the structure of cdc2 will no doubt be informative about the function of this enzyme.

2.2 cdc13

The second component of the mitotic kinase complex is cyclin B, which in *S. pombe* is encoded by the *cdc13*$^+$ gene (54–57). In contrast to the other G2–M *cdc* genes, cells carrying the *cdc13–117* mutation (the best characterized temperature-sensitive *cdc13* mutant allele) arrest at their restrictive temperature with condensed chromosomes but no mitotic spindle (55), and in these cells the cdc2 kinase is active (35). This originally suggested that *cdc13*$^+$ was required for the process of spindle formation rather than mitotic initiation *per se*, but is now known to be a feature peculiar to the *cdc13-117* allele which retains partial function at its restrictive temperature: cells in which the *cdc13*$^+$ gene has been deleted from the chromosome arrest with decondensed chromosomes and inactive cdc2 kinase (54, 55). The *cdc13*$^+$ function is therefore required to activate cdc2.

Biochemical evidence for an interaction between the cdc2 and cdc13 proteins was preceded by several pieces of genetic evidence suggesting such an interaction (58). Firstly, several temperature-sensitive *cdc13* alleles were recovered in a screen for extragenic suppressors of a cold-sensitive *cdc2* allele. Secondly, *cdc13-117* was demonstrated to be lethal in combination with the *cdc2* allele *cdc2-3w*. Thirdly, overexpression of *cdc2*$^{+}$ was shown to suppress the *cdc13-117* defect, but not *cdc13*Δ cells (the symbol Δ indicates that the single haploid chromosomal copy of the gene is either deleted or disrupted). The *cdc13*$^{+}$ gene was subsequently cloned by complementation and shown to have the potential to encode a 482 residue protein (M_r 56 kDa) with significant sequence similarity to the B-type cyclin proteins (54–57). The cdc13/cyclin B protein is approximately 50 per cent identical to cyclin B proteins over the 150–200 amino acids of the cyclin box region.

Like higher eukaryotic B-type cyclins, the cdc13 protein complexes with cdc2 (36) and is degraded during mitotic anaphase (35, 36). Significantly, overproduction of the cdc13 protein has no effect on cell cycle timing of mitosis (54, 55), indicating that cdc13 is not rate-limiting for these events (discussed further in Section 8).

2.3 suc1

The third component of the cdc2–cdc13 complex is the 13 kDa suc1 protein (also known as p13^{suc1}). The *suc1*$^{+}$ gene was identified in two separate screens aimed at identifying *cdc2*$^{+}$-interacting gene functions. Mutations in the chromosomal *suc1*$^{+}$ gene were obtained that suppressed certain temperature-sensitive *cdc2* alleles (29), while the gene was also isolated as a multi-copy suppressor of such mutants (59). Moderate overproduction of suc1 rescues some, but not all, of the temperature-sensitive mutant cdc2 proteins (47, 59), while strong overproduction causes a delay on entry into mitosis in wild-type cells (29, 60). Deletion of the *suc1*$^{+}$ gene results in a form of mitotic arrest associated with a high level of cdc2 kinase activity: a proportion of the deleted cells arrest with mitotic spindles and condensed chromosomes, suggesting that the suc1 protein may have a role in exit from mitosis (35). Recently, it has been suggested (61) that suc1 may also be capable of binding with low affinity to the cdc25 tyrosine phosphatase to inhibit its activity (see Section 3). This may explain why gross overproduction of suc1 causes a delay in G2 in wild-type cells.

2.4 Subcellular localization of the cdc2–cdc13 complex

By using indirect immunofluorescence with antibodies directed against either the cdc2 protein or the cdc13 protein, each protein has been shown to be at least partly nuclear in location (36), with subpopulations of both proteins being associated with the spindle pole bodies, the equivalent of mammalian centrosomes (62). That cdc13 might have a role in spindle formation was previously suggested by the observation that the *cdc13-117* mutant was hypersensitive to the microtubule inhibitor TBZ (58). A substantial subpopulation of cdc13 has also been shown to be present in the nucleolus (see ref. 63 for discussion).

Interestingly, localization of cdc2 to the nucleus is dependent upon cdc13 function: in *cdc13Δ* cells, no nuclear cdc2 is detectable while in *cdc2Δ* cells, cdc13 protein is still nuclear (36). Thus cdc13 may have a role in either actively targeting cdc2 to the nucleus or in ensuring the retention of cdc2 that has entered the nucleus passively. The sequence of the *S. pombe* cdc13 protein does contain a putative nuclear localization signal (54, 55), although to our knowledge its function is yet to be tested.

2.5 Other fission yeast cyclins

In higher eukaryotes a large number of cyclin-like proteins have now been identified, falling into at least five classes (A–E). Some of these proteins interact with cdc2 (specifically cyclins A and B) while others function in combination with the cdc2-related cdk enzymes (see Chapter 6). Ten genes encoding cyclin-like proteins have been found in *S. cerevisiae,* most of which are thought to interact with the cdc2 functional homologue protein Cdc28. These cyclin-like proteins can be grouped into a number of distinct classes on the basis of their sequence similarity and temporal expression patterns, yet there appears to be a degree of functional overlap between each class (see Chapter 3). In fission yeast, in addition to *cdc13*$^+$, four other genes have been identified that encode cyclin-like proteins, *puc1*$^+$, *cig1*$^+$, *cig2*$^+$, and *mcs2*$^+$, although the product of the *mcs2*$^+$ gene (which is most similar to the higher eukaryotic cyclin C proteins) is not thought likely to interact with cdc2 (64–71). The properties of the fission yeast cyclin-like proteins are summarized in Table 5.

The *puc1*$^+$ gene was isolated by virtue of its ability to confer α-factor resistance on Cln3-deficient *S. cerevisiae* cells (see Chapter 2) (64, 65). This screen was intended specifically to isolate fission yeast G1 cyclins and indeed the predicted puc1 protein is most similar to the budding yeast Cln proteins (Table 5). However, to date it has not been possible to demonstrate that puc1 has a positive function at

Table 5 Cyclin-like proteins in fission yeast

	Comments
cdc13	B-type cyclin: essential for G2–M transition, forms complex with cdc2 protein kinase, accumulates in interphase and is destroyed in mitosis (10, 11, 35, 36, 54, 55)
cig1	Non-essential B-type cyclin, function unknown (66, 67)
cig2	Non-essential B-type cyclin, functions early in the cell cycle; mRNA periodically expressed at G1–S (68–70)
puc1	Non-essential cyclin protein, most similar to the budding yeast Cln proteins, possibly involved in exit from the cell cycle (64, 65)
mcs2	Mutants isolated in a screen for mitotic catastrophe suppressors; most similar to cyclin C but possibly first of a new class of cyclins (71)

the G1–S transition in the mitotic cycle in *S. pombe*. Instead a number of lines of evidence point to a role for the puc1 protein in exit from the mitotic cycle. Specifically, expression of the *puc1⁺* gene is increased during nitrogen starvation, *puc1⁺* overexpression blocks sexual development, and *puc1Δ* cells (*puc1⁺* is a non-essential gene) are accelerated into meiosis (65). Overexpression of *puc1⁺* can also affect the mitotic cycle, by delaying entry into mitosis in wild-type cells, and is lethal in combination with *cdc13–117* (at its permissive temperature). This has been suggested to be the result of high levels of the puc1 protein titrating out a component essential for cdc13 function from the cell (64) and in this regard it is noteworthy that a truncated puc1 protein has been shown to co-precipitate with cdc2 when the former is overproduced (65). However, it remains to be seen whether puc1 and cdc2 interact under truly physiological conditions, and if this is the case, what function is performed by the cdc2–puc1 complex.

In contrast to puc1, the *cig1⁺* and *cig2⁺*-encoded proteins are most similar to the B-type cyclins (Table 5). Although *cig1⁺* (which is non-essential) was originally thought to have a role early in the cell cycle, this now appears not to be the case, leaving the cellular role of this protein unclear (65–67). The *cig2⁺* gene was isolated by virtue of its ability to rescue a budding yeast strain deficient for the three redundant G1 cyclins Cln1, Cln2, and Cln3 (68–70). Several classes of cyclin able to rescue this deficiency have been identified, including B-type cyclins that function at G2–M as well as cyclins that are thought to act earlier in the cell cycle. The cig2 protein is most similar to fission yeast cdc13, the two proteins being approximately 65 per cent identical in the cyclin box region, and around 40 per cent overall.

Despite its close sequence similarity to cdc13, cig2 may function early in the cell cycle: *cig2⁺* mRNA levels oscillate through the cell cycle, reaching a peak at G1–S, and *cig2Δ* cells (like *puc1⁺* and *cig1⁺*, the *cig2⁺* gene is non-essential, although *cig2Δ* cells are marginally longer than wild-type at division) display enhanced conjugation frequencies, reminiscent of the phenotype observed with puc1Δ cells (65, 68–70). In addition *cig1Δ cig2Δ* double mutants have been reported to grow much more slowly than either single mutant does (though this point has been disputed: compare refs 68, 69, and 70) and, curiously, *cig1Δ cig2Δ* cultures contain an abnormally high proportion (70 per cent) of binucleate cells, a phenotype that suggests a defect in nuclear division and/or septum formation (discussed in ref. 70). Flow cytometric analysis of these cells indicates that they may also be delayed in the initiation of S phase. Clearly, further work is required to unambiguously define the functions of these proteins.

3. Activating cdc2 by tyrosine 15 dephosphorylation

The timing of activation of the cdc2–cdc13 mitotic kinase complex is determined by the relative activities of two antagonistic pathways—an activatory pathway (described in this section) and an inhibitory one (Section 4). The components of these pathways are summarized in Tables 6 and 7 respectively.

Table 6 Regulating cdc2 activity: elements in the activation pathway

		Phenotype when overproduced	**Phenotype when deleted**
cdc25	protein tyr phosphatase	wee	cdc arrest
pyp3	protein tyr phosphatase	wee	elongated

3.1 Overproduction of the cdc25 protein leads to premature entry into mitosis

The *cdc25*$^+$ gene function is required for, and is a key regulator of, entry into mitosis in fission yeast (72). Cells carrying a temperature-sensitive *cdc25* allele become arrested in late G2 when shifted to the restrictive temperature while overproduction of the cdc25 protein leads to mitotic advancement (the wee phenotype described in Section 1.3) indicating that cdc25 activity is rate-limiting for these events. This advancement is seen when the gene is overexpressed from the strong adh promoter, or when the *cdc25*$^+$ gene is carried on a multi-copy plasmid (72).

When *cdc25-22* cells (*cdc25–22* is the most widely used temperature-sensitive *cdc25* mutant allele) are held at their restrictive temperature, the cdc2 kinase is phosphorylated on tyrosine 15 and is inactive (see below). Activation of the kinase occurs within 10 min after the cells are returned to the permissive temperature (35, 36) and is followed by a highly synchronous mitosis. For this reason the *cdc25-22* mutation is often used as a means of synchronizing cells in G2.

3.2 Regulation of cdc2 function by tyrosine dephosphorylation

Activation of the cdc2–cdc13–suc1 protein complex at the G2–M transition is brought about by dephosphorylation of cdc2 on tyrosine 15. Replacing this tyrosine with a non-phosphorylatable phenylalanine residue leads to premature entry into mitosis and a mutant protein that is no longer dependent upon cdc25 function (38). The failure of initial sequence comparisons to reveal significant sequence similarity between the cdc25 protein and proteins in then-current databases (72) led to the suggestion that cdc25 itself was unlikely to be a tyrosine phosphatase but rather might exert its effects indirectly, through an unidentified enzyme. However, the identification of a new class of tyrosine phosphatase prompted a reassessment of this conclusion and the discovery that the cdc25 protein shares limited sequence similarity to these enzymes (41, 73). Subsequently, the cdc25 protein has been shown to have tyrosine phosphatase activity *in vitro* and to be able to activate the cdc2 protein by tyrosine 15 dephosphorylation (41).

The homology between cdc25 and other protein tyrosine phosphatase enzymes is limited and is confined to a region towards the carboxy-terminus of the fission

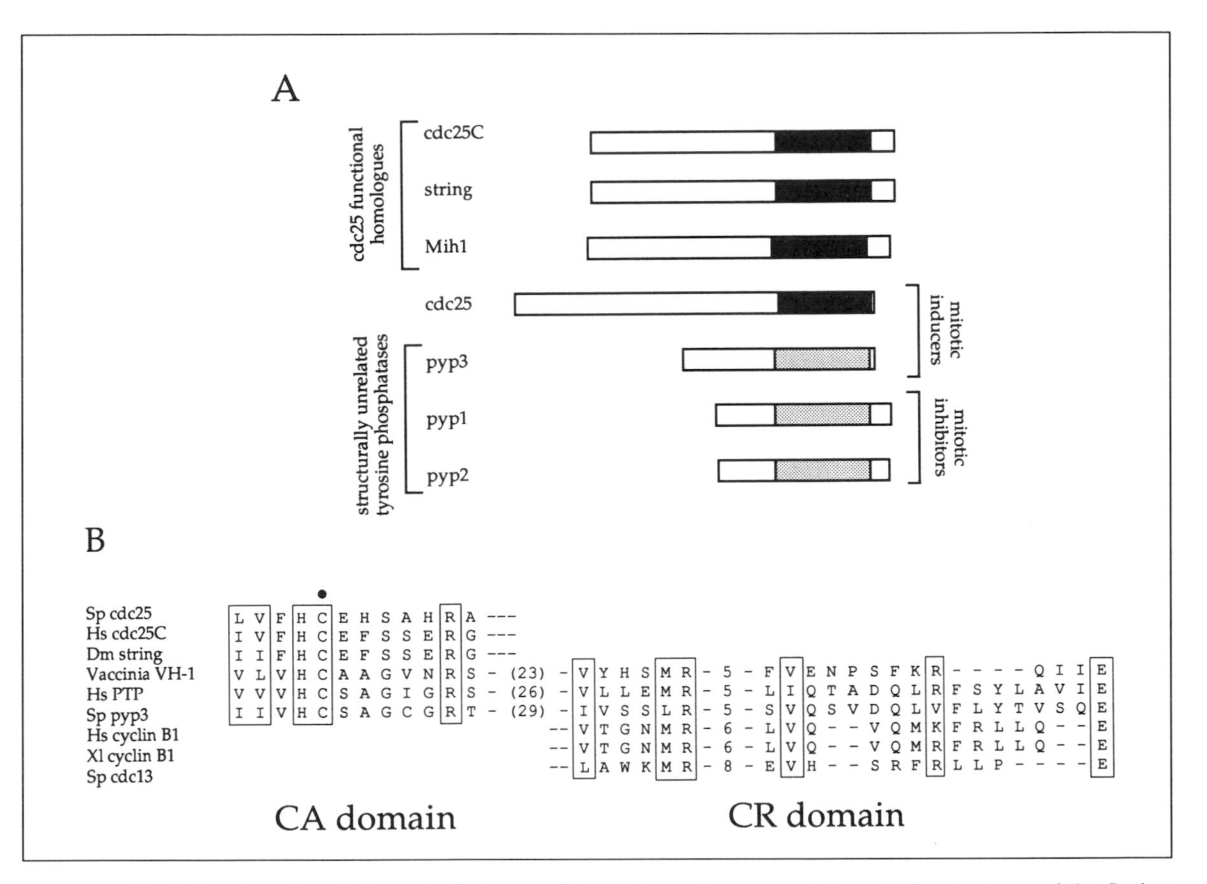

Fig. 3 Tyrosine phosphatases in fission yeast. (A) Schematic representation of the structures of the fission yeast protein tyrosine phosphatases cdc25 and pyp3 (both of which are mitotic inducers) and pyp1 and pyp2 (mitotic inhibitors). The catalytic domain of cdc25 (shaded box) is located towards its carboxy-terminus and is conserved in the cdc25 functional homologues in mammals (only the human cdc25C is shown), *Drosophila* (string), and budding yeast (Mih1 protein). The amino-terminal region of the cdc25 protein is non-essential for function, but may contain regulatory phosphorylation sites (see text). The pyp proteins are only distantly related to the cdc25 family of enzymes in the region of their catalytic domain (lightly shaded box): instead these proteins are more closely related to the higher eukaryotic cytoplasmic tyrosine phosphatases, such as human PTP1 (shown). (B) Conserved residues in the catalytic sites of tyrosine phosphatases, indicating how the catalytic site of cdc25 may be split, with key residues being provided in *trans* by cdc13/cyclin B. Sequences shown are those of *S. pombe* cdc25, human cdc25C, *Drosophila* string, vaccinia virus VH-1 protein, human PTP1, *S. pombe* pyp3, human cyclin B1, *Xenopus* cyclin B1 and *S. pombe* cdc13. Conserved active site residues are boxed. Mutation or chemical modification of cysteine 430 in fission yeast cdc25 (indicated by the • above the sequence) eliminates cdc25 protein function. See text for discussion and references.

yeast protein (illustrated in Fig. 3) (41, 73). This 150 amino acid domain is conserved between the *S. pombe* protein and cdc25 functional homologues identified in a variety of eukaryotic cell types, including humans, *Drosophila*, and *S. cerevisiae* (74–76, reviewed in ref. 77). Expression of this domain alone is sufficient to rescue temperature-sensitive *cdc25-22* cells (41, 72), although the regulation of the protein has not been studied in detail in these cells (see Section 3.5). The amino-terminal domains of the various cdc25 proteins are unrelated at the primary sequence level.

Not all the sequence elements that characterize other higher eukaryotic tyrosine phosphatase enzymes are found in cdc25 and it has been suggested that one of the absent motifs (called the CR motif) is located instead in cdc13 (and in other B-type cyclins), so that part of the active site of the cdc25 enzyme may be provided in *trans* by cdc13 (61). This is shown schematically in Fig. 3. Elements of the CR motif found in the B-type cyclins are typically not seen in G1 cyclins, such as the *S. cerevisiae* Cln proteins, consistent with the idea that cdc25 and/or tyrosine dephosphorylation have no part to play in regulating cdc2 at the G1–S transition. In support of the *trans*-action model, addition of exogenous cyclin B has been reported to stimulate cdc25 tyrosine phosphatase activity three- to five-fold *in vitro* (61, but see ref. 78). Furthermore, residues located within the CR region (also known as the P-box) are essential for cyclin B function (78). It should be noted, however, that an alternative model has been presented to account for the function of the CR motif (P-box) (78) so that further work will be required to unambiguously define its function.

3.3 pyp3

Cells lacking functional cdc25 activity can be kept viable by inactivating the wee1 protein kinase despite the presence of functional mik1 in the cells (see Section 4) (79). The enzyme responsible for tyrosine 15 dephosphorylation of cdc2 in these circumstances is pyp3 (42), a non-essential protein tyrosine phosphatase that was isolated as a multi-copy suppressor of a temperature-sensitive *cdc25* mutation. The 303 amino acid (M_r 34 kDa) pyp3 protein is approximately 30 per cent identical to human PTP1B or fission yeast pyp1 and pyp2 tyrosine phosphatase enzymes at the primary sequence level (see Section 4.3 for a discussion of the role of pyp1 and pyp2) but is largely unrelated to cdc25 (see Fig. 3).

In keeping with its role as a mitotic inducer, overproduction of pyp3 leads to mitotic advancement and cell division at a reduced cell size (7.0–8.5 μm versus 14 μm), while *pyp3*Δ cells are slightly longer than wild-type. This elongation is greatly exacerbated in cells with impaired cdc25 function (42). The pyp3 protein is also capable of efficiently dephosphorylating and activating tyrosine-phosphorylated cdc2 *in vitro* (42).

Are cdc25 and pyp3 the only enzymes capable of dephosphorylating and activating cdc2? One piece of evidence cited in favour of the existence of a third cdc2 tyrosine phosphatase (a putative pyp4 enzyme) is that a *cdc2-3w cdc25*Δ *pyp3*Δ strain is viable despite the absence of both cdc25 and pyp3 activities to counteract the inhibitory phosphorylation of tyrosine 15 (42). However, this assumes that tyrosine 15 dephosphorylation is essential for activation of the mutant cdc2–3w protein, which need not be the case. A further question to be addressed, which may have a crucial bearing on the *in vivo* role of pyp3, is whether this enzyme is able, or is required, to interact with cdc13 for catalytic activity. Since the pyp3 protein contains sequences corresponding to both the CA and CR motifs (shown in Fig. 3) it may be that interaction with cdc13 is not essential for pyp3 protein function. Biochemical analysis of the role of the pyp3 phosphatase will prove informative about the function of this protein in wild-type fission yeast cells.

3.4 Regulation of cdc25 activity by phosphorylation

A key question that remains to be answered is how the activity of the cdc25 protein itself is regulated during the cell cycle in fission yeast. Levels of both the *cdc25*$^+$ mRNA and cdc25 protein fluctuate modestly during the mitotic cell cycle, in each case reaching a maximum at G2–M (80, 81). In *Xenopus*, phosphorylation of the amino-terminal domain of the cdc25 protein is associated with its activation at G2–M (82, 83). The cdc25 protein in fission yeast (which is known to be a phosphoprotein) (80, 81) and its higher eukaryotic homologues (see Chapter 5) all contain numerous potential cdc2 phosphorylation sites, raising the possibility that cdc2 may have a role in activating or inhibiting cdc25 activity via a positive feedback loop (Fig. 4). This model is supported by recent reports showing that the human cdc2 protein is able to phosphorylate cdc25C (one of three cdc25 functional homologues identified in human cells; see Chapter 5) and that the phosphorylation leads to an increase in the ability of cdc25C to activate inactive tyrosine-phosphorylated cdc2 (83, 84).

A curious feature of the activation of cdc2 by the cdc25 tyrosine phosphatase in *S. pombe* is that it is prevented *in vivo* by the microtubule inhibitor thiabendazole (TBZ). When cells carrying a temperature-sensitive *cdc25* mutation are arrested at their restrictive temperature and then released from the arrest, the cells normally undergo mitosis in a highly synchronous manner. The cdc2 kinase is activated within 10 min of release from the block and its activity remains high for approximately 30 min more, by which time almost all the cells in the population have completed mitosis (62). Under these circumstances cdc13 degradation is seen to occur 30 min after release, coincident with mitotic anaphase. In contrast, when the arrested cells are returned to the permissive temperature in the presence of TBZ, mitosis does not occur and, significantly, the cdc2 kinase remains inactive and phosphorylated on tyrosine 15 (59), suggesting that some aspect of microtubule function is required for cdc2 activation by cdc25. How the signal that prevents cdc2 activation is mediated remains an open question. As noted above (Section 2.5), subpopulations of both the cdc2 and cdc13 proteins are known to be localized at the spindle pole bodies, which are presumably major sites of TBZ action (62). *cdc13-117* cells are also more sensitive to TBZ than are wild-type (58). It is conceivable, therefore, that TBZ exerts its effects on cdc25 function via the cdc2 and cdc13 proteins located at the spindle pole bodies, rather than by inhibiting cdc25 directly, so that failure to enter mitosis might result from the inability of the cdc2–cdc13 complex to catalyse amplification of cdc25 activity via the positive feedback loop.

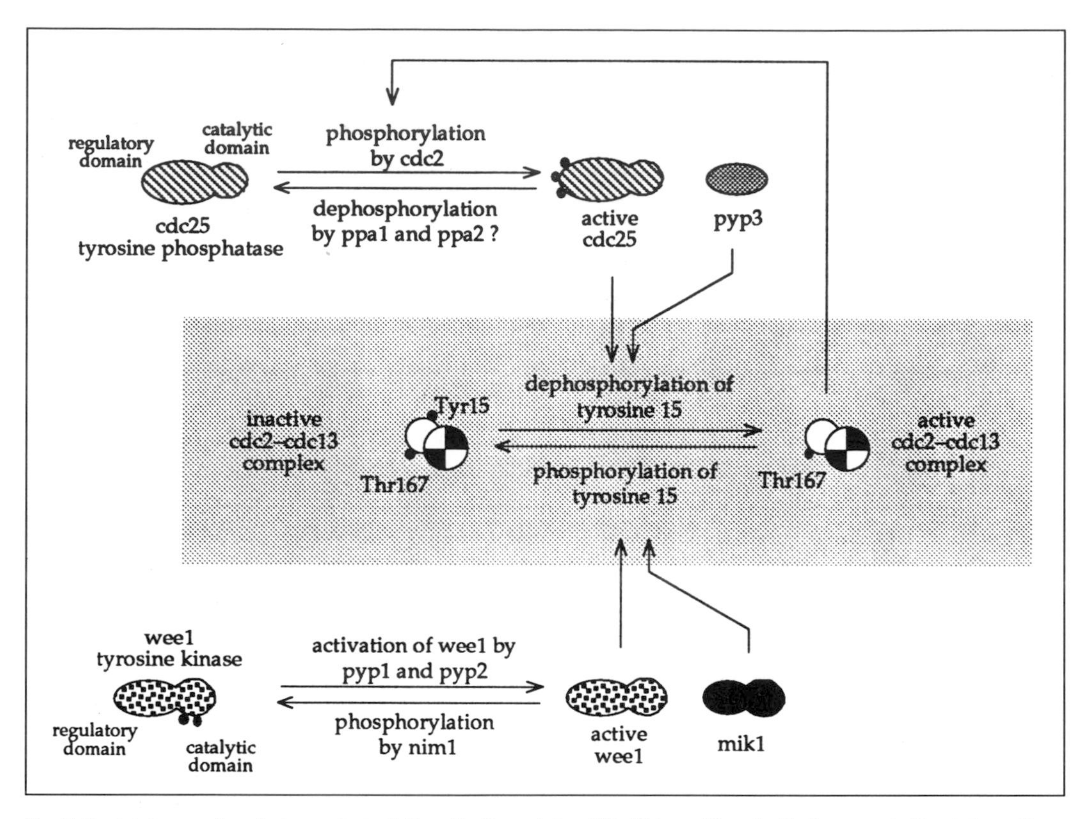

Fig. 4 Regulatory network governing cdc2 activation at the G2–M transition in fission yeast. The interactions shown are based on results obtained in *S. pombe* but also incorporate relationships suggested by observations made in higher eukaryotes. Central (shaded) area: the timing of entry into mitosis is dependent on the balance of activities promoting and inhibiting cdc2–cdc13 complex activation. Activation is achieved by dephosphorylation of tyrosine 15 in cdc2 by the cdc25 and pyp3 tyrosine phosphatase enzymes. Counteracting this is tyrosine 15 phosphorylation catalysed by the wee1 and mik1 enzymes. Upper part: once activated, the cdc2-cdc13 complex may initiate a positive feedback loop by phosphorylating, and activating, more cdc25. The PP-2A phosphatase ppa2 is a negative regulator of mitosis in fission yeast and may act (perhaps along with ppa1) by dephosphorylating and inhibiting cdc25. Lower part: wee1 is phosphorylated and inhibited by nim1. The pyp1 and pyp2 tyrosine phosphatases are also thought to act in this pathway as negative regulators of mitosis, but their targets are unknown. It is possible that wee1 may also phosphorylate nim1 but the effects of this phosphorylation, should it occur, are unknown. Phosphorylation of cdc2 on threonine 167 is achieved by the CAK/CDKK which is yet to be identified in *S. pombe* (not shown). See text for details and references.

4. Regulation of cdc2 activity by tyrosine phosphorylation: the inhibitory pathway

Balancing the activatory activity of the cdc25 and pyp3 enzymes is an inhibitory pathway, five components of which have been identified in fission yeast. Two of these, the wee1 and mik1 protein kinases, are responsible for tyrosine 15 phosphorylation of cdc2 (79, 85).

Table 7 Regulating cdc2 activity: elements in the inhibitory pathway

		Phenotype when overproduced	Phenotype when deleted
wee1	protein kinase	elongated	wee
mik1	protein kinase	elongated	none
nim1	protein kinase	wee	elongated
pyp1	protein tyr phosphatase	elongated	wee
pyp2	protein tyr phosphatase	elongated	none

4.1 wee1

The wee1 protein is a dose-dependent inhibitor of mitosis (Table 7). Loss of wee1 protein function leads to a mitotic advancement with the result that cells divide at approximately half the size of wild-type *S. pombe* cells (7 μm versus 14 μm, see Fig. 1) (21). Wee1 activity is required to prevent lethal premature entry into mitosis (mitotic catastrophe) in cells overproducing the cdc25 protein, or expressing the cdc25-independent *cdc2-3w* allele. This lethality was exploited to clone the *wee1*$^{+}$ gene by complementation (32).

Overproduction of the wee1 protein leads to cell elongation (32)—cells containing one extra copy of the gene divide around 21 μm (50 per cent larger than wild-type), those with four extra copies at 28 μm (100 per cent larger), and those with six extra copies at 35 μm (150 per cent). Cell size at cell division is increased yet further when wee1 is expressed from the adh promoter, with division occurring infrequently at approximately 60 μm (32).

Sequence analysis of the *wee1* gene identified a single, long, open reading frame with the potential to encode an 877 amino acid protein (M_r 107 kDa) with significant sequence similarity in its carboxyl-terminus to protein kinase enzymes. The amino-terminal domains bears no obvious sequence similarity to other proteins but is rich in proline, serine, and threonine residues, often flanked by positively charged amino acids (32), in a manner reminiscent of the PEST sequences found in proteins with short *in vivo* half-lives (86). Consistent with this observation the wee1 protein is present at a low level in *S. pombe* cells, so that analysis of wee1 protein function has depended upon using wee1 overproducer strains (85). Overproduced wee1 protein is phosphorylated only on serine residues *in vivo*, although the enzyme is capable of phosphorylating itself or an exogenously-added synthetic peptide on tyrosine under *in vitro* assay conditions (84). These properties are shared by wee1 protein expressed in and purified from budding yeast or insect cells (87–89), indicating that the wee1 enzyme is a dual-specificity protein kinase. The importance of phosphorylation of the wee1 protein for wee1 function is discussed in Section 4.3.

Much of the biochemical evidence that wee1 is able to phosphorylate cdc2 (Fig. 4) on tyrosine 15 has come from studies on cdc2 and wee1 homologue proteins in higher eukaryotes and is discussed further in Chapter 5 (80–82). Indeed, early work in *S. pombe* failed to provide any support for this notion, as the overproduced wee1 protein was unable to phosphorylate either cdc2 protein expressed in bacteria, or purified active histone H1 kinase from HeLa cells (85). Recently, however, it has been shown that fission yeast cdc2 can be phosphorylated by wee1 on tyrosine 15 (87).

4.2 mik1

The *mik1*$^+$ gene was first isolated as a multi-copy suppressor of the mitotic catastrophe phenotype of a *cdc2-3w wee1-50* strain (where *wee1-50* is a commonly used temperature-sensitive *wee1* mutant allele), and shares many properties with *wee1*$^+$, suggesting that the two genes share an overlapping function (79). Overexpression of the *mik1*$^+$ gene from a multi-copy plasmid results in cell elongation to a level equivalent to that seen when *wee1*$^+$ is overproduced. Overproduction of mik1 protein also delays mitosis in *wee1-50* cells at their restrictive temperature, indicating that *mik1*$^+$ does not act through *wee1*$^+$ (79).

Sequence analysis of the cloned *mik1*$^+$ gene identified an open reading frame encoding a 581 amino acid protein kinase (M_r 66 kDa) with significant sequence similarity to the wee1 protein (around 48 per cent identity over the protein kinase domain). The high degree of sequence similarity between the wee1 and mik1 proteins does not, however, extend beyond the protein kinase catalytic domain although, as is the case with the wee1 enzyme, the amino-terminal domain of mik1 is relatively rich in serine residues (79). The function of this region remains unknown.

Where the *wee1*$^+$ and *mik1*$^+$ functions differ significantly is in the behaviour of cells carrying loss-of-function mutations in each gene. Cells deleted for *mik1*$^+$ (*mik1Δ*) are indistinguishable from wild-type and are not wee, and mik1Δ does not rescue the temperature-sensitive *cdc25* strain. A *mik1Δ wee1-50* strain is wee at the normal permissive temperature for *wee1-50* but lethal at the restrictive temperature, with cells undergoing a severe form of the mitotic catastrophe phenotype previously observed with other mutant combinations. Furthermore, when *mik1Δ wee1-50* cells are shifted to the restrictive temperature, the cdc2 protein becomes rapidly dephosphorylated—no tyrosine-phosphorylated cdc2 is detectable 30 min after the temperature shift. As might be expected, this dephosphorylation is delayed in *cdc25* mutant cells (where only the pyp3 phosphatase is capable of catalysing dephosphorylation) but goes to completion none the less.

In a screen for suppressors of the *mik1Δ wee1-50* lethality, five suppressing mutations that conferred a cold-sensitive cdc phenotype were isolated and all five mapped to the *cdc2* gene (79). The *cdc2-r4* allele was characterized further—this mutation is recessive and grows well at its permissive temperature (Table 3). The cdc2-r4 protein is phosphorylated to wild-type levels on tyrosine 15. When the triple mutant *mik1Δ wee1-50 cdc2-r4* is shifted to the high temperature, cdc2-r4 is rapidly tyrosine-dephosphorylated (79), implying that in the absence of mik1 the wee1 protein kinase alone is responsible for tyrosine 15 phosphorylation, but the cells remain viable owing to the *cdc2-r4* mutation delaying entry into mitosis. Suppression analysis of *cdc2-r4* has led to the identification of the gene encoding the chk1 protein kinase (described in detail in Section 9), an enzyme suggested to have a role in linking the activation of cdc2 to the pre-mitotic checkpoint pathway (90).

4.3 Regulating wee1 activity: nim1/cdr1, pyp1, and pyp2

Three enzymes have been identified that are known or thought to play important roles upstream of the wee1 kinase in the inhibitory pathway—nim1, pyp1, and pyp2 (Table 7). Possible functions of these proteins are illustrated in Fig. 4.

4.3.1 nim1/cdr1

The nim1 protein, a 593 amino acid (M_r 67 kDA) protein kinase, is a mitotic inducer: overexpression of the *nim1*$^+$ gene, which was first identified as a multi-copy suppressor of the temperature-sensitive *cdc25-22* mutation, leads to mitotic advancement (24). At septation, cells deleted for *nim1*$^+$ are slightly longer than wild-type (18 μm versus 14 μm).

Overexpression of *nim1*$^+$, either from a multi-copy plasmid or from the adh promoter, is also capable of suppressing a *cdc25Δ* strain, indicating that the nim1 protein cannot act by boosting the residual cdc25 activity of a temperature-sensitive *cdc25* strain at its restrictive temperature (24). Nor does nim1 act directly on cdc2: none of four *cdc2* alleles tested for suppression by nim1 overproduction was able to grow. Three pieces of evidence point to *nim1*$^+$ acting through *wee1*$^+$. Firstly, *wee1*$^+$ is epistatic to *nim1*$^+$: deleting the *nim1*$^+$ gene in cells already lacking functional wee1 protein has no further effect on cell size at cell division (24). Secondly, overproduction of nim1 suppresses the cell cycle arrest caused by overproduction of wee1. Thirdly, overproduction of nim1 and loss of wee1 activity is not additive, unlike overproduction of cdc25 with loss of wee1 (32).

A step forward in understanding the biological role of nim1 came with the demonstration that *nim1*$^+$ was allelic with *cdr1*$^+$, one of a number of genes identified in an earlier screen for mutants defective in their response to altered nutritional conditions (91, 92). When wild-type *S. pombe* cells are shifted from rich medium to poor, mitosis is advanced with the result that the cell size at division is reduced (see ref. 93 for a discussion). The reverse is true for shifts from poor medium to rich, that is, the cells are delayed in G2 for a period before mitosis is initiated. This control is disrupted in *wee1* and *cdr1* mutants. In *wee1* mutants, nutritional shifts neither accelerate nor delay mitosis, suggesting that the wee1 kinase is responsible for communication signals regarding nutritional status to the mitotic control (93). The *cdr* mutants (cdr is an acronym for changed division response) are slightly longer than wild-type cells under normal growth conditions and undergo a reduced degree of mitotic advancement following starvation with the result that *cdr* mutants arrest as elongated cells (91, 92). The *cdr* mutants are phenotypically similar to cells deleted for *nim1*$^+$ (24).

The 260 amino acid catalytic domain of the nim1 kinase is located at its amino-terminus, the function of the remainder of the protein being unknown (24, 94). However, the carboxy-terminal domain may have a role in monitoring nutritional signals. Altered nim1 proteins lacking this domain can still bring about mitotic advancement and rescue *cdc25* mutants but do not restore the starvation response to *cdr* cells (94). Interestingly, this region of the protein shares some sequence similarity with the *S. cerevisiae* Snf1 kinase which has a role in reversing glucose repression, so that the sequence similarity may reflect a common function in nutritional sensing. The carboxy-terminal domain also shares more limited sequence similarity with the budding yeast Kin1 and Kin2 enzymes of unknown function (94).

The function of the nim1 protein kinase has been investigated biochemically, both in fission yeast itself (95) and *in vitro* using nim1 protein expressed in bacterial or insect cell systems (95–97). In both cases it has been shown that nim1 phosphorylates wee1 directly, so reducing its ability to act on cdc2. Like wee1, the nim1 protein is a dual-specificity protein kinase, capable of autophosphorylation on both serine and tyrosine. Co-expression of wee1 and nim1 in insect cells results in wee1 becoming phosphorylated (on one or more serine residues within its carboxy-terminal catalytic domain) by nim1, with the result that its ability to phosphorylate and inhibit cdc2 on tyrosine 15 in *in vitro* assays is reduced more than ten-fold (95–97). In fission yeast, wee1 phosphorylation is reduced by around 50 per cent in *nim1Δ* cells (which indicates that nim1 is not solely responsible for wee1 phosphorylation *in vivo*), and increased two-fold in a nim1 overproducer (95). Interestingly, wee1 may also phosphorylate nim1, suggesting the possibility of a feedback loop (either positive or negative) involving these two enzymes (96).

4.3.2 pyp1 and pyp2

The pyp1 and pyp2 protein tyrosine phosphatases are also thought to have a role in the inhibitory pathway (Fig. 4), as activators of *wee1*$^{+}$ function (98–100). Overproduction of either protein leads to cell elongation in a wild-type genetic background but has no effect in cells carrying either the temperature-sensitive *wee1-50* mutation (at its restrictive temperature) or the wee1-insensitive *cdc2* mutation *cdc2-1w*. In contrast, cells carrying the *cdc2-3w* mutation become elongated (99, 100). Cells carrying a deletion of *pyp1* are semi-wee, indicating partial advancement into mitosis, while *pyp2Δ* cells are normal. Deleting *pyp1*$^{+}$ suppresses *cdc25-22* as well, while deleting *pyp2* has a similar though much less marked effect (99, 100).

While it is possible that the pyp1 and pyp2 enzymes act directly upon wee1, there is at present no evidence in support of such a model. No tyrosine phosphorylation of the wee1 protein has been detected, although is should be noted that phosphoaminoacid content has been analysed only in protein prepared from a wee1 overproducer strain, so that if only a small fraction of the protein were tyrosine phosphorylated (recall that nim1 has dual-specificity *in vitro*) this might well have escaped detection (discussed further in ref. 85). However, the role of the pyp1 and pyp2 proteins cannot be solely to reverse the effects of the nim1 kinase as a *pyp1Δ nim1Δ* strain enters mitosis at a smaller size than a *nim1Δ* alone (100).

5. wis1

The wis1 protein, like the cdc25, pyp3, and nim1 proteins discussed in the preceding sections, is also a dose-dependent inducer of mitosis, although in this case it has not been possible to locate the action of this enzyme in either the cdc25 (activatory) or wee1 (inhibitory) pathways (25). The *wis1*$^{+}$ gene was isolated as a multi-copy suppressor of the cell cycle arrest defect of a *cdc25-22 wee1-50 win1-1* strain at its restrictive temperature (see ref. 101 for a description of *win1*$^{+}$ function). When overexpressed, the *wis1*$^{+}$ gene causes an advancement of mitosis and a reduction in cell size to a semi-wee value of 10 μm in otherwise wild-type cells, while cells deleted for *wis1*$^{+}$ are elongated, dividing at around 24 μm (25).

Since the mitotic induction brought about by wis1 overproduction cannot be described solely in terms of either the inhibitory or activatory pathways of mitotic regulation (25), the wis1 gene function may either exert effects on both pathways, or directly or indirectly on their common target, the cdc2–cdc13 complex (Fig. 4). Threonine phosphorylation of cdc2 is essential for cell cycle progress (37) while wis1 function is not (25), ruling out the possibility that the wis1 kinase alone could be responsible for cdc2 threonine 167 phosphorylation. Instead, the wis1 protein is likely to be part of a signal transduction pathway, the other components of which are at present unknown. Sequence analysis of the *wis1*$^{+}$ gene shows that it encodes a 605 amino acid protein kinase (predicted M_r 65 kDa) most closely related to a number of proteins that act in architecturally similar but functionally unrelated pathways. These include MAP kinase kinase (MAPKK) from *Xenopus*, Ste7 and Pbs2 in *S. cerevisiae*, and byr1/ste1 in *S. pombe* (discussed in ref. 102). The pathways involving the MAPKK and Ste7 proteins are the best-defined biochemically (see Fig. 5), and both MAPKK and Ste7 have been shown to be dual-specificity enzymes, capable of phosphorylating their substrates on tyrosine and threonine (103, 104).

If the organization of these pathways is conserved across evolution, then wis1 may act by phosphorylating an MAPK homologue (Fig. 5). One such protein has been identified in *S. pombe*, encoded by the *spk1*$^{+}$ gene (105), but there are likely to be more. It is tempting to speculate that wis1 could function as a positive activator of the enzyme that phosphorylates and activates cdc2 on threonine 167, although it should be noted in this respect that the *Xenopus* CAK protein is not an obvious MAPK homologue (43–46). Alternatively, the ultimate target of the wis1 pathway could be a mitotic cyclin such as cdc13. In either case the end result of wis1 action might be to promote interaction between cdc2 and cdc13.

6. Protein serine–threonine phosphatases in fission yeast

Biochemical analysis of phosphoprotein phosphatases in higher eukaryotes has led to the identification of several types of enzyme that can be distinguished on the

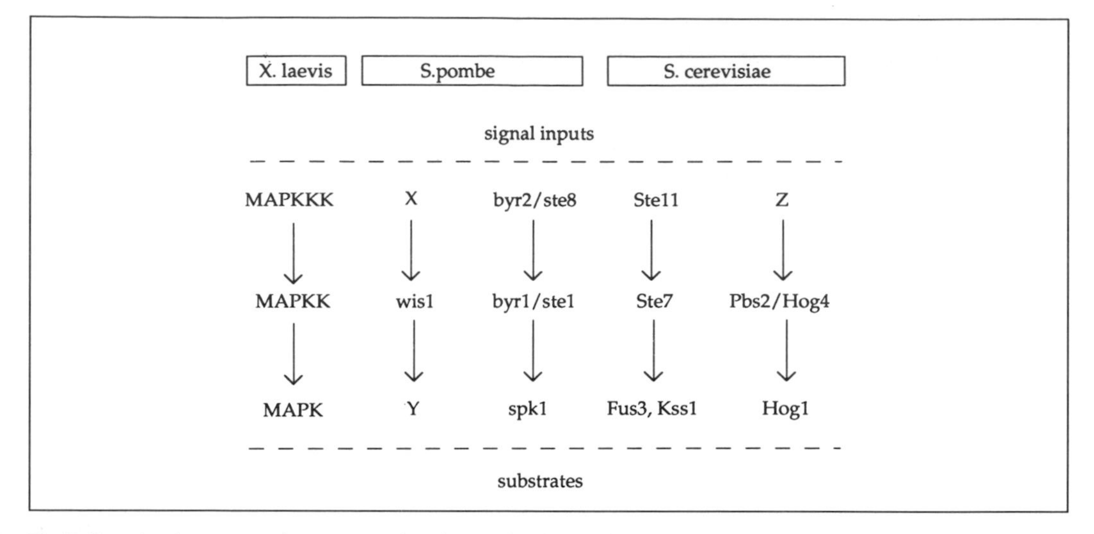

Fig. 5 Structural conservation across signal transduction pathways across evolution. Five such pathways are shown (vertical columns, from left to right): the prototype *X. laevis* MAP kinase activation pathway, the putative wis1 pathway (involving hypothetical kinases X and Y), the mating response pathways in *S. pombe* and *S. cerevisiae* (elements in which are functionally interchangeable with one another, so that byr2/ste8 can substitute for Ste11, byr1/ste1 for Ste7, and spk1 for Kss1 and Fus3), and the osmosensing pathway in *S. cerevisiae* (containing hypothetical kinase Z). Proteins depicted at the same (horizontal) level show closest sequence similarity: thus MAP kinase kinase (MAPKK) is most closely related to fission yeast byr1/ste1 and wis1 proteins, and to budding yeast Ste7 and Pbs2/Hog4. Elements involved in activating each pathway and the eventual substrates of the MAPK homologues are less well defined. See text for discussion.

basis of their substrate specificity and sensitivity to specific phosphatase inhibitors (106). In addition to the four protein tyrosine phosphatases described above, at least seven serine–threonine phosphatases have been identified in *S. pombe*, of which the type 1 and type 2A enzymes (termed PP-1 and PP-2A respectively) are the best characterized (Table 8). While a number of reports have implicated these enzymes in the control of the cell cycle, their precise functions have not as yet been clearly defined. Indeed, as each of the phosphatase enzymes is likely to be involved in other aspects of cell metabolism, the identification of specific cell cycle functions is likely to prove difficult. A solution to this problem may lie in the identification of cell cycle-specific regulatory subunits. In budding yeast, for example, the Cdc55 protein has been identified as a putative PP-2A regulatory subunit (107) and in *S. pombe* the essential sds22 protein may perform a similar function with PP-1 enzymes (108, 109). Of equal importance will be the identification of each enzyme's target substrates.

6.1 Type 1 phosphatases

The two type 1 phosphatase (PP-1) enzymes identified in *S. pombe* are encoded by the *dis2*$^+$ and *sds21*$^+$ genes (see Table 8) (26, 110). Cold-sensitive semi-dominant

Table 8 Protein serine–threonine phosphatases in fission yeast

Protein	Function
dis2	Major PP-1 activity, dis2 mutants cause defects in chromosome disjunction; shares essential overlapping function with sds21 (110)
sds21	Minor PP-1 activity; highly similar to dis2 (110)
ppa1	Minor PP-2A activity; ppa1 is highly similar to ppa2 (26, 27)
ppa2	Major PP-2A activity, cells lacking ppa2 protein are prematurely advanced into mitosis; shares an essential overlapping function with ppa1 (26, 27)
ppe1	Third PP-2A-like enzyme, only 50–60 per cent identical to ppa1 and ppa2; mutants are cold-sensitive and display cell shape defects (114, 116)
ppb1	Recently identified putative PP-2B-like enzyme (157)
ppc1	Recently identified putative PP-2C-like enzyme (157)

mutations in the *dis2* gene result in chromosome non-disjunction, indicative of a defect in the physical processes of mitosis rather than at the G2–M transition (110). Similar behaviour is seen with mutations of PP-1 enzymes in *Aspergillus* (111) and in *Drosophila* (112), suggesting that PP-1 function is conserved across evolution. The dis2 and sds21 proteins (which are nuclear in location) are around 80 per cent identical to one another and between 70–80 per cent to mammalian PP-1 enzymes. While neither gene is essential in itself, simultaneous deletion of both genes is lethal, indicating that the two share at least one essential overlapping function (26).

The *sds22*$^+$ gene encodes a regulatory subunit of the type 1 phosphatase enzymes (109). Like *sds21*$^+$, the *sds22*$^+$ gene was first identified as a multi-copy suppressor of cold-sensitive *dis2* mutants (108). *sds22*$^+$ encodes a 322 amino acid (M_r 40 kDa) that contains 11 tandem copies of a 22-residue leucine-rich repeat sequence. Similar repeats are seen in a number of other proteins (108). Deletion of the *sds22*$^+$ gene is lethal: *sds22*Δ cells are mostly arrested in mitosis, with condensed chromosomes, a short mitotic spindle, and elevated cdc2 kinase activity (109). The sds22 protein interacts physically with both the dis2 and sds21 phosphatases, and in doing so modulates their substrate specificity (109). Deletion of a single leucine-rich repeat has been shown to be sufficient to abolish sds22 protein function as has mutation of individual conserved leucine residues. In both cases, however, the non-functional proteins retain their ability to interact with dis2 and sds21.

Despite apparently functioning in mitosis, *dis2*$^+$ was isolated independently (as *bws1*$^+$) by virtue of the fact that its overexpression was able to cause a *cdc25 wee1* double mutant strain to revert to a cdc phenotype (113). While not fully understood at present, this result may well imply a role for the dis2 protein at the G2–M transition as well as during mitosis. Perhaps under these circumstances dis2 activity is controlled by a different, as yet unidentified, regulatory subunit. Alternatively, overproduction of the dis2 protein may result in the dephosphorylation of target proteins at inappropriate times.

6.2 Type 2A phosphatases

Genes encoding three type 2A (PP-2A) phosphatases have been isolated and characterized in *S. pombe* (Table 8): *ppa1*$^+$, *ppa2*$^+$, and *ppe1*$^+$ (26, 27, 114, 115). The ppa1 and ppa2 proteins are 80 per cent identical to one another at the primary sequence level, and each is between 50–60 per cent identical to ppe1.

As with the type 1 phosphatase genes, neither *ppa1*$^+$ nor *ppa2*$^+$ is essential in itself, though simultaneous deletion of both genes is lethal, indicating that they share an essential overlapping function (26). The ppa2 protein is responsible for the majority of PP-2A activity in cells (26) and, in contrast to dis2 and sds21 PP-1 enzymes, is predominantly cytoplasmic in location (27). Deletion of *ppa2*$^+$ (but not *ppa1*$^+$) causes a reduction in growth rate in complex medium and, significantly, a reduction in cell size at cell division to the semi-wee value of 11 μm (compared to wild-type cells which divide at 14 μm) (26). The ppa2 protein is therefore a negative regulator of entry into mitosis. Deletion of *ppa2*$^+$ is lethal in combination with *wee1–50* and is also able to partially suppress the cell cycle arrest phenotype of the temperature-sensitive *cdc25–22* mutation, while overexpression of *ppa2*$^+$ (or *ppa1*$^+$) protein causes delay in G2 (27). These genetic results are compatible with the ppa2 enzyme functioning as an inhibitor of cdc25 activity in wild-type cells (as shown in Fig. 4), as is thought to be the case for PP-2A enzymes in higher eukaryotes (see Chapter 5). Indeed, there are a number of possible substrates for PP-2A enzymes amongst the mitotic control proteins, since cdc2, cdc25, wee1, and cdc13 are all known to be phosphorylated *in vivo*. Cells deleted for *ppa2*$^+$ are also more sensitive to the specific phosphatase inhibitor okadaic acid (OA) than are wild-type cells or cells carrying the *ppa1Δ* allele. OA treatment of *ppa2Δ* cells leads to a further reduction in cell size at cell division (27).

The *ppe1*$^+$ gene (Table 8) was isolated initially by hybridization (114). Mutations in *ppe1* were uncovered in a screen for suppressors of *pim1* mutant cells, which display hypercondensed chromosomes (115–117). Deletion of *ppe1*$^+$ results in cold-sensitive lethality and altered cell morphology. *ppe1Δ ppa2Δ* double mutant cells are inviable, again suggesting the possibility of a shared essential function (114).

7. Additional gene functions implicated at the G2–M transition

7.1 Genes required for entry into mitosis

In addition to *cdc2*$^+$, *cdc25*$^+$, and *cdc13*$^+$, three other genes have been identified whose functions are required in late G2 for entry into mitosis: *cdc1*$^+$, *cdc27*$^+$, and *cdc28*$^+$ (Table 1) (10, 11). Mutations in any of these three genes arrest cells in G2 and each gene has a transition point (a measure of a gene's final time of action in the cell cycle, discussed in Chapter 2) of approximately 0.65, close to the onset of mitosis at 0.7 (10, 11). Curiously, the timing of both the *cdc1*$^+$ and *cdc27*$^+$ gene

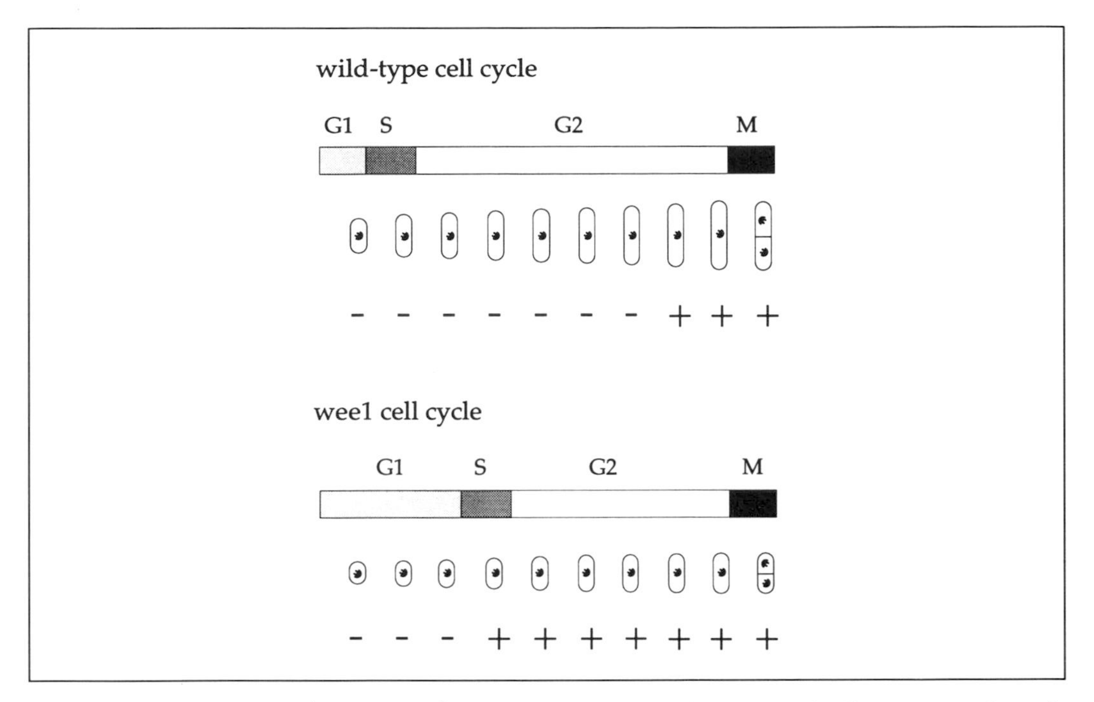

Fig. 6 Advancement of *cdc1*⁺ and *cdc27*⁺ transition point in wee1 mutant cells. Upper part: schematic representation of wild-type cell cycle showing the relative lengths of G1, S, G2, and M phases. When cells carrying a temperature-sensitive *cdc* mutation are shifted to their restrictive temperature, only those cells that have completed the *cdc* gene function at the time of temperature-shift are capable of division. The resulting increase in cell number gives a measure of the final time of action of the gene function in the cycle (called the transition or execution point). For *cdc1*⁺ and *cdc27*⁺ around 25 per cent of the cells go on to divide (indicated as +), giving a transition point of around 0.7 (10, 11), the remainder arresting in late G2 (indicated as −). Lower part: in wee mutant cells the relative lengths of the cell cycle phases are altered, as G2 is shortened by advancement into mitosis, and in these cells the transition points of *cdc1*⁺ and *cdc27*⁺ are advanced also, with around 70 per cent of the cells capable of division (112). See text for further discussion.

functions is regulated in some way by *wee1*⁺, as the transition points of these gene functions are substantially advanced (by 0.4 of a cell cycle) in *wee1* mutant cells, with their new time of action close to the time of completion of DNA replication (118). This is shown schematically in Fig. 6. Similar though less pronounced effects are seen with some but not all temperature-sensitive *cdc2* alleles (93). The reason for this effect is not known. Both *cdc1*⁺ and *cdc27*⁺ have been cloned and sequenced, but as neither of the encoded proteins shares significant sequence similarity to proteins in current databases, their functions are at present unknown (S. A. MacNeill, S. Moreno, P. Nurse, and P. A. Fantes, unpublished work) (119).

7.2 Genes identified by genetic interaction screens

By a variety of genetic methods, a number of other genes have been identified in *S. pombe* that appear to function at the G2–M transition. These include *win1*⁺

(mutants which reverse the suppression of *cdc25* by loss of *wee1*$^+$ function, but only under certain nutritional conditions—see ref. 101), *wis2*$^+$–*wis5*$^+$ (isolated in the same screen as *wis1*$^+$ as suppressors of a *cdc25 wee1 win1-1* triple mutant, although none shares with *wis*$^+$ the ability to bring about mitotic advancement when overexpressed—see ref. 120), *stf1*$^+$ (the *stf1-1* allele interacts with *cdc25* mutants—see ref. 121), and *mcs1*$^+$–*mcs6*$^+$ (suppressors of the mitotic catastrophe phenotype of *cdc2-3w wee1-50* cells—see ref. 122). At present, little is known of the functions of the majority of these genes (almost all have been cloned), so that it is not yet clear how closely they interact with the central elements of the mitotic control network. However, both the *mcs2*$^+$ and *wis2*$^+$ genes encode proteins with sequence similarity to proteins in current databases: *mcs2*$^+$ encodes a cyclin-like protein (discussed in the following section, ref. 71) while the non-essential *wis2*$^+$ gene encodes a novel cyclophilin homologue (R. Weisman and P. A. Fantes, unpublished work).

7.2.1 mcs2

As mentioned above, mutations in *mcs2*$^+$ were isolated as suppressors of the mitotic catastrophe phenotype of a *cdc2–3w wee1–50* strain (122). *mcs2*$^+$ has been cloned, by complementation of an *mcs2* mutant strain, and sequenced to reveal an open reading frame with the potential to encode a 322 amino acid (M_r 38 kDa) cyclin homologue that is most similar to higher eukaryotic cyclin C proteins identified in *Drosophila* and human cells (see Chapter 6) (71). The similarity is weak, however (at only 22 per cent over the cyclin box, while the two higher eukaryotic cyclin C proteins are 72 per cent identical over the same region), suggesting that mcs2 may represent the first of a new class of cyclin. The *mcs2*$^+$ gene is essential—around 60 per cent of *mcs2Δ* cells arrest with a septum and many more show signs of chromosome condensation, suggesting that mcs2 functions in mitosis (71).

Like the B-type cyclin cdc13, the mcs2 protein is nuclear in location but unlike cdc13 does not undergo any apparent changes in localization or abundance through the cell cycle. The mcs2 protein co-precipitates with a protein kinase (not cdc2), and mcs2-associated protein kinase activity is constant through the cycle (71).

One enzyme that may be responsible for the mcs2-associated activity is that encoded by the non-essential *csk1*$^+$ gene. This gene was identified as a multi-copy suppressor of *mcs2* mutants and encodes a 306 amino acid protein kinase (M_r 34 kDa) with greatest similarity to the cdc2 family of enzymes (approximately 30 per cent over the catalytic domain), although many of the structural motifs that are essential for cdc2 function (see Section 2) are absent in the csk1 protein sequence (71). If csk1 is responsible for part of the mcs2-associated protein kinase activity, then it cannot be alone in this, as the activity drops only three-fold in *csk1Δ* cells but is not abolished altogether. The csk1 enzyme could instead be an upstream activator of mcs2 protein function (71).

8. Rate-limiting steps: what determines the timing of entry into mitosis?

It is important to realize that while the mechanism of cdc2 activation at mitosis appears to be conserved across evolution, the specific component of the mitotic control network whose activity determines the precise timing of activation may vary from species to species and from one cell type to the next. In fission yeast the balance of cdc25/pyp3 activity versus wee1/mik1 activity determines the timing of activation of the cdc2–cdc13 complex—overproducing cdc25 or pyp3, or inactivating wee1 or mik1, advances mitosis, while overproducing cdc2 or cdc13 has no effect (Table 2). In other systems the accumulation of cyclin to a critical threshold level may be the key rate-limiting step.

That other components of the system may become rate-limiting for activation under different circumstances is best illustrated by considering the behaviour of *S. pombe* cells in which tyrosine phosphorylation of cdc2 is prevented by inactivation of both wee1 and mik1 activities. Normally, such cells enter mitosis prematurely with fatal results (79). This suggests that there is sufficient cyclin B present in the cell early in G2 to allow premature activation of the cdc2–cdc13 kinase complex when wee1 and mik1 activities are abolished. This lethality can, however, be suppressed in one of two ways: either by mutating the cdc2 kinase to reduce its activity (the cdc2-r4 protein is such a mutant, see Section 4.2) (79), or by inactivating the B-type cyclin cig2 (68). Although the role of cig2 is at present the subject of some discussion (68–70), it has been suggested that in its absence the initiation of mitosis is delayed (recall that *cig2Δ*) cells are marginally longer than wild-type: see Section 2.6) because it takes longer to accumulate sufficient B-type cyclin to proceed. Under these conditions therefore, it may be cdc13 (in the form of the cdc2–cdc13 complex) that is rate-limiting for entry into mitosis. If this is the case then it represents the only demonstration to date that cdc13 levels can have a rate-limiting function for entry into mitosis in *S. pombe*.

9. Checkpoint controls at the G2–M transition

Progression from G2 into mitosis in *S. pombe* is dependent upon the cell reaching a certain minimal size (21), and upon prior completion of DNA replication (10). Understanding how these events generate signals which are then integrated into the mitotic control network, so that cdc2 activation is prevented until the cell has reached the required size and DNA replication and repair is complete, is currently the subject of much investigation. As discussed above, mutations in the *wee1*$^+$ and *nim1*$^+$/*cdr1*$^+$ genes can influence the cell size control over entry into mitosis and its modulation by nutrients (see Section 4.3). The following sections deal specifically with the controls that ensure that cells with incompletely replicated or damaged DNA do not attempt to enter mitosis until replication or repair is complete.

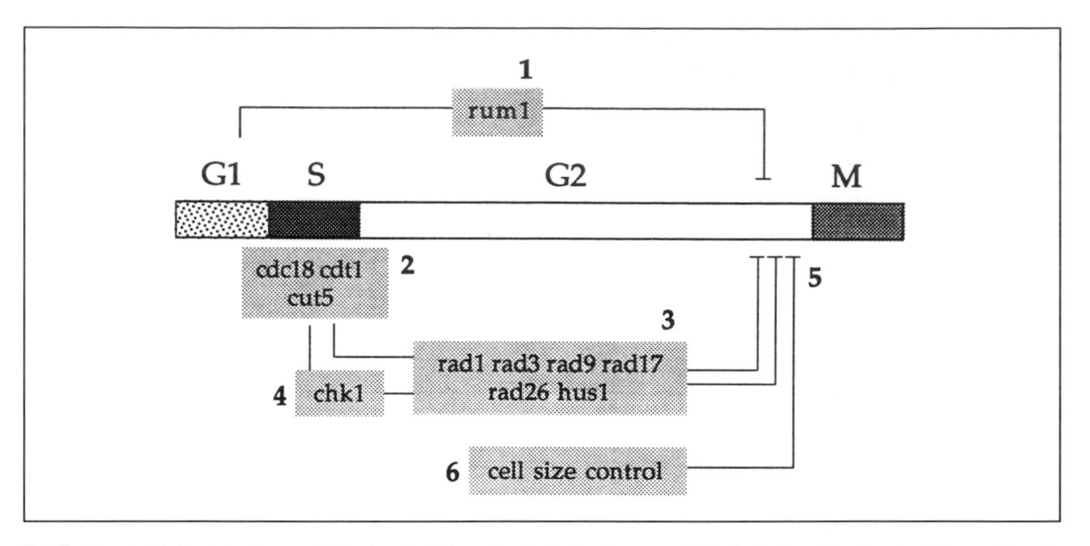

Fig. 7 Checkpoint controls regulating entry into mitosis in fission yeast. Cells in the G1 phase of the cell cycle are prevented from entering mitosis by the action of the rum1 protein (top, labelled **1**). Early in G2, the cdc18, cut5, and cdt1 proteins (**2**) are required to establish a state that can be monitored by the rad/hus checkpoint (**4,5**). In the absence of any of these three proteins, cells are unable to replicate their DNA but nevertheless enter mitosis, with lethal consequences, indicating that the rad/hus checkpoint is inoperative under these circumstances. Mutations in any of the *rad1*$^+$, *rad3*$^+$, *rad9*$^+$, *rad17*$^+$, *rad26*$^+$, and *hus1*$^+$ genes (**3**) are unable to arrest cell cycle progress in response to either DNA damage or incomplete DNA replication brought about by inactivating DNA ligase or by treating cells with hydroxyurea. Cells deleted for *chk1*$^+$/*rad27*$^+$ (**4**) are sensitive to DNA damage and unligated DNA, but are still able to arrest cell cycle progress in response to hydroxyurea. How these signals are integrated into the mitotic control (**5**) is not known, but the DNA replication and DNA damage signals appear to differ. Entry into mitosis is also dependent upon the cell reaching a certain minimal size (**6**). See text for details and references.

9.1 DNA replication defects and DNA damage trigger checkpoint activation and block entry into mitosis

S. pombe cells arrested during DNA replication, such as those treated with the DNA synthesis inhibitor hydroxyurea or carrying mutations in any of a number genes required for completion of DNA replication (see Table 1 and below), display none of the features indicative of entry into mitosis, such as chromosome condensation or spindle formation (10). Entry into mitosis is prevented by a checkpoint (see Chapter 1 for a discussion of the concept of checkpoints) which senses that replication is incomplete and delays activation of the cdc2–cdc13 complex accordingly. A second checkpoint (but one that shares many common gene functions with the DNA replication checkpoint, see below) senses DNA damage, causing X-ray or ultraviolet-irradiated *S. pombe* cells to undergo cell cycle arrest in G2 to allow time for DNA repair to occur.

The molecular nature of the signal that triggers activation of the DNA replication checkpoint is not known, but inactivation of any of a wide range of replication gene functions causes activation of the checkpoint and cell cycle arrest. These genes include several that are known to encode components of the replication machinery, such as *cdc17*$^{+}$, *pol3*$^{+}$, *pcn1*$^{+}$, and *cdc21*$^{+}$ (which encode DNA ligase, the catalytic subunit of DNA polymerase δ, the polymerase δ processivity factor PCNA, and a protein of the mcm family respectively—see Chapter 7), as well as genes encoding enzymes involved in nucleotide precursor synthesis, such as *cdc22*$^{+}$ (which encodes the large subunit of ribonucleotide reductase) (123–127). As noted above, treatment of cells with hydroxyurea, a potent inhibitor of ribonucleotide reductase activity, also brings about checkpoint activation and cell cycle arrest. However, this does not necessarily mean that the checkpoint can sense defects of nucleotide precursor synthesis *per se*. Analysis of DNA structures in hydroxyurea-treated cells indicates that replication has been initiated in these circumstances, and it may well be the stalled (precursor-starved) replication complexes and/or partially replicated DNA structures that are sensed by the checkpoint (125). Further investigation will be required to resolve these possibilities.

9.2 Checkpoint gene functions

9.2.1 *rad/hus* genes

Six genes have been identified that play essential roles in the operation of both the DNA replication and DNA damage checkpoints: *rad1*$^{+}$, *rad3*$^{+}$, *rad9*$^{+}$, *rad17*$^{+}$, *rad26*$^{+}$, and *hus1*$^{+}$ (90, 128–131) (Fig. 7). The *rad* genes were isolated amongst a larger collection of radiation-sensitive (rad) fission yeast mutants (132). In X-ray- or UV-treated wild-type cells, mitosis is delayed in order to allow time for DNA damage repair to occur and four of the initial collection of *rad* genes, *rad1*$^{+}$ (128, 129), *rad3*$^{+}$, *rad9*$^{+}$, and *rad17*$^{+}$ (128), along with the independently-isolated *rad26*$^{+}$ gene (131), define functions that are specifically required for this delay. These five *rad* mutants are also unable to detect incomplete DNA replication caused by hydroxyurea treatment or by a temperature-sensitive *cdc17* mutation (128, 129). Similarly, mutations in the *hus1* gene, isolated primarily on the basis of their sensitivity to hydroxyurea, are also sensitive to UV (130). All six checkpoint *rad/hus* genes are non-essential for mitotic growth under normal conditions.

Each of these six genes has been cloned and sequenced (ref. 131, and references cited therein, 133–135). The predicated rad1, rad3, rad9, rad26, and hus1 proteins show no homology to proteins in current databases, while the rad17 protein has been reported to display limited sequence similarity to proteins that are components of replication factor C (RF-C), a DNA polymerase δ accessory factor (see ref. 131, and references cited therein). This exciting observation provides the first link between the checkpoint function and the machinery of DNA replication, and raises the speculative possibility that the rad17 protein is able to sense blocked DNA replication and/or DNA damage by virtue of being a component of the DNA polymerase complex.

9.2.2 *chk1*

The *chk1*$^{+}$ gene also has an important part to play in the operation of the premitotic checkpoints. Like mutants of the six checkpoint *rad/hus* genes described above, cells deleted for the *chk1* gene are deficient for checkpoint arrest when replication is blocked by a *cdc17* (DNA ligase) mutation, and are sensitive to UV irradiation (though the degree of sensitivity is less than that displayed by *rad1* cells) (90). However, *chk1*Δ cells are markedly different from the *rad/hus* mutants in their response to hydroxyurea treatment, which leads to cell cycle arrest rather than the lethal entry into mitosis seen with the other checkpoint *rad/hus* mutants. This result indicates that UV-induced DNA damage and unligated DNA are recognized by a different means than the DNA replication block imposed by hydroxyurea, and suggests that the chk1 protein kinase may be involved only in the former (90). On this basis, it is reasonable to place the $chk1^+$ gene function upstream of the *rad/hus* genes in the putative checkpoint pathway.

Paradoxically, however, available evidence can be taken to indicate that chk1 functions downstream of the *rad/hus* genes. The $chk1^+$ gene was identified as a multi-copy suppressor of the cold-sensitive *cdc2-r4* mutation (see Section 4.2), and overexpression of $chk1^+$ can also partially rescue the UV sensitivity of *rad1* mutants, suggesting that $chk1^+$ is one of several elements downstream of $rad1^+$, linking the checkpoint pathway to the mitotic control. This paradox is yet to be resolved. As the $chk1^+$ gene encodes a 496 amino acid protein serine–threonine kinase homologue, identifying the substrates and regulators of this protein will doubtless shed light on its function (90).

9.3 Integration into the mitotic control network

9.3.1 DNA replication checkpoint

An important objective now is to link the checkpoint gene functions with the mitotic control and cdc2. The cell cycle arrest caused by incomplete DNA replication can be circumvented by mutations in the mitotic control network (136, 137). In contrast to wild-type cells treated with hydroxyurea, cells overproducing cdc25 enter mitosis on schedule following hydroxyurea treatment but rapidly lose viability as the cell attempts to divide the incompletely replicated DNA (136). Cells carrying the cdc25-independent *cdc2-3w* allele show similar behaviour to this, whilst *wee1* and *cdc2-1w* mutant cells are still capable of responding to incomplete DNA replication by arresting cell cycle progress. This indicates that the sensitivity of cdc25-overproducing or *cdc2-3w* cells to hydroxyurea is not merely a reflection of the fact that the cells are advanced into mitosis at a small cell size, since the *wee1* and *cdc2-1w* mutations have equal or greater effects on cell size.

These observations initially led to the suggestion that the DNA replication checkpoint signal fed into the mitotic control via the cdc25 protein, and that wee1 had no role in the checkpoint mechanism (136). Since then, however, this model has been revised to take account of the observations that *cdc25*Δ *wee1* cells are not arrest-deficient, as would be expected if the signal fed solely through cdc25, and that overproduction of wee1 can reverse the hydroxyurea-sensitivity of *cdc2-3w*, implying that wee1 levels can influence checkpoint arrest (137). There are two alternative explanations for these data: either the checkpoint signal feeds into the mitotic control by two or more redundant pathways (simultaneously inhibiting cdc25 activity and stimulating wee1/mik1 activity, for example), or the signal feeds into cdc2 independently of tyrosine 15 phosphorylation but that its effectiveness in inhibiting cdc2 activity is greatly affected by the tyrosine phosphorylation state of the cdc2 protein (137). Further investigation will be required to resolve these possibilities.

Interestingly, the *cdc2-3w* mutant is considerably less sensitive to hydroxyurea than are the *rad/hus* mutants (130). One reason for this might be that complete loss of *rad/hus* function has a greater effect on the checkpoint system than does *cdc2-3w*, which encodes a protein with perhaps only modestly altered function. However, the difference appears to be more fundamental, as *cdc2-3w* cells treated with hydroxyurea lose viability when they enter the mitosis following the blocked S phase while *rad* or *hus* mutants lose viability in S phase itself (130), suggesting that the *rad* and *hus* gene functions described above may be required for the process of recovery from hydroxyurea arrest as well as functioning in the checkpoint pathway.

9.3.2 DNA damage checkpoint

Although cells carrying the *cdc2-3w* mutation are sensitive to hydroxyurea and to the inhibition of replication caused by inactivating *cdc17*$^{+}$, these cells still retain the capacity to undergo cell cycle arrest when UV-irradiated. This result suggests strongly that the DNA replication and DNA damage checkpoints feed into the mitotic control by distinct mechanisms to prevent entry into mitosis, and calls into question the role of tyrosine phosphorylation in influencing DNA damage-mediated checkpoint arrest (see previous section).

Whether wee1 is required for DNA damage-induced arrest has been the subject of some debate (compare refs 138 and 139): *wee1Δ* cells are UV-sensitive, but still seem capable of delaying mitosis in response to irradiation, while *cdc2-1w* cells (the cdc2-1w protein is insensitive to inhibition by wee1) are not (128). This has been taken to indicate that wee1 performs a second function, aside from its role in regulating entry into mitosis, that is required for the normal cellular response to DNA damage. The nature of this function is not known.

9.4 *cdc18*$^{+}$, *cut5*$^{+}$/*rad4*$^{+}$, and *cdt1*$^{+}$

Not all mutations that arrest cell cycle progress during S-phase bring about activation of the pre-mitotic rad/hus checkpoint. Cells deleted for the *cdc18*$^{+}$ or *cdt1*$^{+}$ genes (140, 141), or carrying a temperature-sensitive mutant allele of the *cut5*$^{+}$/*rad4*$^{+}$ gene (142), are blocked in early S phase with a 1C DNA content, yet these cells attempt to undergo mitosis, indicating that the rad/hus checkpoint is inoperative under these conditions (Fig. 7). This has led to the suggestion that the cdc18, cdt1, and cut5/rad4 proteins may be required to establish the molecular structure, perhaps a replication complex, that the rad/hus checkpoint can recognize (140–142). Interestingly, the phenotype observed with *cdc18Δ* cells is different from that seen when cells carrying the temperature-sensitive *cdc18-K46* mutant are shifted to the restrictive temperature. These latter cells arrest with a 2C DNA content (though DNA replication is not completed normally, as judged by the behaviour of the *cdc18-K46* chromosomes in a pulsed field gel electrophoresis assay) and do not attempt to enter mitosis, indicating that the checkpoint control is intact in these cells and has been activated. One attractive hypothesis to explain these results is that the temperature-sensitive cdc18-K46 protein is competent for replication complex formation (so providing a structure that can be monitored by the rad/hus checkpoint) but is defective for its specific function in the complex (with the result that replication cannot be completed). The temperature-sensitive cut5-T401 protein, on the other hand, may cause dissociation of such a complex when transferred to the restrictive temperature, so that the rad/hus checkpoint cannot be activated and entry into mitosis ensues. Hydroxyurea treatment of *cut5* mutant cells fails to prevent their entry into mitosis at the restrictive temperature, again indicative of the absence of checkpoint control under these conditions. As cut5 mutant cells are also sensitive to UV irradiation at their permissive temperature, it is perhaps unsurprising that mutants in the gene had previously been identified, defining the *rad4*$^+$ gene (142). The cut5/rad4 protein displays limited homology to the human XRCC1 DNA repair protein (143, 144), suggesting the possibility of a shared function.

The *cdc18*$^+$ and *cdt1*$^+$ genes are transcriptionally regulated, with maximal transcript levels being reached at G1–S. In both cases transcription is dependent upon *cdc10*$^+$ function, indicating a role for the cdc10-containing DSC1Sp transcription complex (145). Indeed, ectopic expression of either *cdc18*$^+$ or *cdt1*$^+$ is able to rescue the temperature-sensitive *cdc10* mutant allele *cdc10-129* under restrictive or semi-restrictive conditions respectively, indicating that these genes are important targets of the DSC1Sp complex. The cdc18 protein displays sequence similarity to the budding yeast Cdc6 protein, which has been shown to be required for replication initiation in *S. cerevisiae* and to delay entry into mitosis when expressed in *S. pombe* (146). Additional evidence from the budding yeast suggests that the Cdc6 protein may interact, directly or indirectly, with replication origins (147).

9.5 Preventing entry into mitosis from G1

In addition to preventing the initiation of mitosis under conditions of incomplete DNA replication or DNA damage, it is also important for the cell to prevent mitotic initiation while in the pre-START G1 phase of the cell cycle. A single gene, called *rum1*$^+$, has been identified that is thought to have a key role in this process (Fig. 7): in contrast to single *cdc10* mutants, double mutant *cdc10 rum1*Δ cells do not arrest in G1 but rather proceed into mitosis with fatal consequences (148). Thus *rum1*$^+$ function is required to prevent entry into mitosis when *cdc10* mutant cells are arrested in G1. Deletion of the *rum1*$^+$ gene also shortens or eliminates G1 altogether under conditions where cells are deprived of nutrients or advanced into mitosis (148). Note, however, that the replication checkpoint is still able to operate in the absence of *rum1*$^{+}$ function: rum1Δ cells treated with hydroxyurea undergo normal cell cycle arrest.

The *rum1*$^{+}$ gene was isolated by virtue of the phenotype conferred on wild-type cells by its overexpression (148). Overproduction of the rum1 protein prevents entry into mitosis and also acts to return cells repeatedly to the pre-START G1 phase of the cycle with the result that they over-replicate their DNA. Under these circumstances, therefore, the usual dependency of S-phase initiation on prior completion of mitosis is broken. This dependency is also disrupted when cells expressing certain mutant cdc2 proteins are heat-shocked, indicating that cdc2 has an important role to play in this process (see ref. 149), and *cdc2*$^{+}$ is indeed required for *rum1*$^{+}$-induced re-replication, as is *cdc10*$^{+}$ (148).

The biochemical function of the *rum1*$^{+}$ gene product remains to be established (see ref. 150 for discussion): the rum1 protein is 238 residues long and has no significant sequence similarity to proteins in current databases. The protein does contain a putative cdc2 phosphorylation site, suggesting that cdc2 may have a part to play in regulating rum1 function, as well as five putative MAP kinase sites (see Section 5) (148).

10. Maintaining high cdc2 activity during mitosis

In *S. pombe* cells arrested in mitosis by mutation of the *nda3* gene (which encodes β-tubulin), activity of the cdc2 kinase remains high. This is taken to indicate the presence of a checkpoint that only allows cdc2 inactivation once certain aspects of mitosis (relating specifically to spindle assembly and disassembly) are complete. Recent data point to the product of the *S. pombe cdc16*$^{+}$ gene, which was originally identified as affecting the regulation of septum formation, having a role in this checkpoint mechanism (151). This was demonstrated by examining the level of cdc2 activity in *cdc16* mutant cells that were firstly pre-synchronized in mitosis by arrest using a cold-sensitive *nda3* β-tubulin mutation and then released from the β-tubulin block in the presence of TBZ. Normally under these circumstances cdc2 activity, high at the *nda3* arrest point, remains high as the cells are unable to form a spindle in the presence of TBZ. In the absence of *cdc16*$^{+}$ function, however, cdc2 kinase activity falls rapidly, indicating that cdc16 protein function is required for maintaining high levels of cdc2 activity in mitotically arrested cells. In *S. cerevisiae* the non-essential *BUB2* gene function has been implicated as part of a similar checkpoint system (152, reviewed in ref. 153)—like *cdc16*$^{+}$, *BUB2* is required to maintain high levels of cdc2 activity in cells treated with microtubule inhibitors. The predicted cdc16 protein displays significant sequence similarity to Bub2, and expression of the *BUB2* gene in *cdc16* mutant cells rescues their cell cycle defect (151).

11. Proteolysis and exit from mitosis

Temperature-sensitive *S. pombe mts2* mutant cells are defective in chromosome segregation, undergoing a transient cell cycle arrest when shifted to the restrictive temperature, with condensed chromosomes and a short metaphase spindle present in a significant fraction of the cells (154). This transient arrest is also associated with high cdc2 kinase activity.

The *mts2⁺* gene has been cloned and sequenced, and encodes a 448 amino acid protein that is >70 per cent identical to the S4 subunit of the human 26S proteosome complex that is responsible for ubiquitin-mediated proteolysis in the cell (154). The mts2 and S4 proteins are true functional homologues, since expression of the human gene in fission yeast will rescue *mts2Δ* cells as well as temperature-sensitive *mts2* mutant cells. Evidence that *mts2* is a component of the fission yeast proteosome is provided from the observation that *mts2* mutant cells accumulate ubiquitinated proteins at the restrictive temperature (154).

These results suggest that at least one protein must be degraded to facilitate exit from mitosis in *S. pombe*. In *Xenopus* ubiquitin-mediated degradation of cyclin B has been shown to be required for the metaphase–anaphase transition. Although other possibilities cannot be ruled out at this stage, it is possible that the 26S proteosome may function to promote exit from mitosis in fission yeast via cdc13 degradation. However, although equivalent mutations in budding yeast have similar phenotypes to *mts2* mutant cells and display elevated levels of the B-type cyclins Clb2 and Clb3 (see ref. 155, and Chapter 2), there are no gross changes in cdc13 levels in *mts2* mutant cells. This result does not of course rule out the possibility that only a particular subpopulation of cdc13 (perhaps that which is complexed with cdc2) is being stabilized in *mts2* mutants. Alternatively, either or both of the two B-type cyclins cig1 or cig2 could be targets for proteolysis, as could other unrelated proteins. Further work will distinguish between these possibilities.

12. Summary

The success of the fission yeast *S. pombe* as a genetic model for events at the G2–M boundary has led to the identification of around 20–25 gene functions that are likely to act at this point in the cycle, with the result that the network of interactions governing entry into mitosis has increased in complexity considerably in recent years (Fig. 3). Future progress in understanding how entry into mitosis is regulated in eukaryotic cells will continue to make full use of genetic and biochemical analyses in yeast, coupled with investigation of the corresponding functional homologue proteins in higher eukaryotic cells.

Acknowledgements

We should like to thank our colleagues in Edinburgh, London, and Kyoto for their help in the preparation of this work, and Dr S. Mackie for his comments on the manuscript. S.A.M. is funded by the Medical Research Council and the Wellcome Trust: P.A.F. acknowledges support from the Cancer Research Campaign, the Medical Research Council, and the Wellcome Trust.

References

1. Sipiczki, M. (1989) Taxonomy and phylogenesis. In *Molecular biology of the fission yeast*. Nasim, A., Young, P., and Johnston, B. F. (ed.). Academic Press, London, p. 431.
2. Mitchison, J. M. (1957) The growth of single cells. I. *Schizosaccharomyces pombe. Exp. Cell Res.*, **13,** 244.
3. Mitchison, J. M. (1990) The fission yeast *Schizosaccharomyces pombe. BioEssays*, **12,** 189.
4. Gutz, H., Heslot, H., Leupold, U., and Loprieno, N. (1974) *Schizosaccharomyces pombe*. In *Handbook of genetics*, Vol. 1, King, R. C. (ed.). Plenum, New York, p. 395.
5. Alfa, C., Fantes, P., Hyams, J., McLeod, M., and Warbrick, E. (1993) *Experiments with fission yeast: a laboratory course manual*. Cold Spring Harbor Laboratory Press, New York.
6. Moreno, S., Klar, A., and Nurse, P. (1991) Molecular genetic analysis of fission yeast *Schizosaccharomyces pombe. Methods Enzymol.*, **94,** 795.
7. MacNeill, S. A. and Fantes, P. A. (1993) Methods for analysis of the fission yeast cell cycle. In *The cell cycle—a practical approach*. Brooks, R. and Fantes, P. A. (ed.). IRL Press, Oxford, p. 93.
8. Hagan, I. M. and Hyams, J. S. (1988) The use of cell cycle division mutants to investigate the control of microtubule distribution in the fission yeast *Schizosaccharomyces pombe. J. Cell Sci.*, **89,** 343.
9. Nurse, P. (1985) Cell cycle control genes in yeast. *Trends Genet.*, **1,** 51.
10. Nurse, P., Thuriaux, P., and Nasmyth, K. (1976) Genetic control of the cell division cycle in the fission yeast *Schizosaccharomyces pombe. Mol. Gen. Genet.*, **146,** 167.
11. Nasmyth K. and Nurse, P. (1981) Cell division cycle mutants altered in DNA replication and mitosis in the fission yeast *Schizosaccharomyces pombe. Mol. Gen. Genet.*, **182,** 119.
12. Minet, M., Nurse, P., Thuriaux, P., and Mitchison, J. M. (1979) Uncontrolled septation in a cell division cycle mutant of the fission yeast *Schizosaccharomyces pombe. J. Bacteriol.*, **137,** 440.
13. Umesono, K., Hiraoka, T., Toda, T., and Yanagida, M. (1983) Two cell division cycle genes *nda2* and *nda3* of the fission yeast *Schizosaccharomyces pombe* control microtubular organisation and sensitivity to anti-mitotic benzimidazole compounds. *J. Mol. Biol.*, **168,** 271.
14. Hirano, T., Toda, T., and Yanagida, M. (1984) The *nda3* gene of fission yeast encodes β-tubulin: a cold-sensitive mutation reversibly blocks spindle formation and chromosome movement in mitosis. *Cell*, **39,** 349.
15. Hirano, T., Funahashi, S., Uemura, T., and Yanagida, M. (1986) Isolation and characterisation of *Schizosaccharomyces pombe* cut mutants that block nuclear division but not cytokinesis. *EMBO. J.*, **5,** 2973.
16. Samejima, I., Matsumoto, T., Nakaseko, Y., Beach, D., and Yanagida, M. (1993) Identification of seven new cut genes involved in *Schizosaccharomyces pombe* mitosis. *J. Cell Sci.*, **105,** 135.
17. Hirano, T., Hiraoka, Y., and Yanagida, M. (1988) A temperature-sensitive mutation of the *Schizosaccharomyces pombe* gene *nuc2*$^{+}$ that encodes a nuclear scaffold-like protein blocks spindle elongation in mitotic anaphase. *J. Cell Biol.*, **106,** 1171.

18. Uemura, T. and Yanagida, M. (1986) Mitotic spindle pulls but fails to separate chromosomes in type II topoisomerase mutants: uncoordinated mitosis. *EMBO J.*, **5,** 1003.
19. Uemura, T., Ohkura, H., Adachi, Y., Morino, K., Shiozaki, K., and Yanagida, M. (1987) DNA topoisomerase is required for condensation and separation of mitotic chromosomes in *S. pombe*. *Cell,* **50,** 917.
20. Nurse, P. (1990) Universal control mechanism regulating onset of M-phase. *Nature,* **344,** 503.
21. Nurse, P. (1975) Genetic control of cell size at division in yeast. *Nature,* **256,** 547.
22. Nurse, P. and Thuriaux, P. (1980) Regulatory genes controlling mitosis in the fission yeast *Schizosaccharomyces pombe*. *Genetics,* **96,** 627.
23. Fantes, P. A. (1981) Isolation of cell size mutants of a fission yeast by a new selective method: characterisation of mutants and implications for division control mechanisms. *J. Bacteriol.,* **146,** 746.
24. Russell, P. and Nurse, P. (1987) The mitotic inducer *nim1*$^{+}$ functions in a regulatory network of protein kinase homologs controlling the initiation of mitosis. *Cell,* **49,** 569.
25. Warbrick, E. and Fantes, P. A. (1991) The wis1 protein kinase is a dosage dependent regulator of mitosis in *Schizosaccharomyces pombe*. *EMBO J.,* **302,** 4291.
26. Kinoshita, N., Ohkura, H., and Yanagida, M. (1990) Distinct essential roles of type 1 and 2A protein phosphatases in the control of the fission yeast cell division cycle. *Cell,* **63,** 405.
27. Kinoshita, N., Yamano, H., Niwa, H., Yoshida, T., and Yanagida, M. (1993) Negative regulation of mitosis by the fission yeast protein phosphatase ppa2. *Genes Dev.,* **7,** 1059.
28. Nurse, P. and Bissett, Y. (1981) Gene required in G1 for commitment to cell cycle and in G2 for control of mitosis in fission yeast. *Nature,* **292,** 558.
29. Hayles, J., Aves, S., and Nurse, P. (1986) suc1 is an essential gene involved in both cell cycle and growth in fission yeast. *EMBO J.,* **5,** 3373.
30. Niwa, O. and Yanagida, M. (1988) Essential and universal role of MPF/*cdc2*$^{+}$. *Nature,* **336,** 430.
31. Nakaseko, Y., Niwa, O., and Yanagida, M. (1984) A meiotic mutant of the fission yeast *Schizosaccharomyces pombe* that produces mature asci containing two diploid spores. *J. Bacteriol.,* **157,** 334.
32. Russell, P. and Nurse, P. (1987) Negative regulation of mitosis by *wee1*$^{+}$, a gene encoding a protein kinase homolog. *Cell,* **49,** 559.
33. Beach, D., Durkacz, B., and Nurse, P. (1982) Functionally homologous cell cycle control genes in budding and fission yeast. *Nature,* **300,** 706.
34. Simanis, V. and Nurse, P. (1986) The cell cycle control gene *cdc2*$^{+}$ of fission yeast encodes a protein kinase potentially regulated by phosphorylation. *Cell,* **45,** 261.
35. Moreno, S., Hayles, J., and Nurse, P. (1989) Regulation of the $p34^{cdc2}$ protein kinase during mitosis. *Cell,* **60,** 321.
36. Booher, R. N., Alfa, C. E., Hyams, J. S., and Beach, D. H. (1989) The fission yeast cdc2/cdc13/suc1 protein kinase: regulation of catalytic activity and nuclear location. *Cell,* **58,** 485.
37. Gould, K. L., Moreno, S., Owen, D. J., Sazer, S., and Nurse, P. (1991) Phosphorylation at Thr 167 is required for *Schizosaccharomyces pombe* $p34^{cdc2}$ function. *EMBO J.,* **10,** 3297.
38. Gould, K. L. and Nurse, P. (1989) Tyrosine phosphorylation of the fission yeast *cdc2*$^{+}$ protein kinase regulates entry into mitosis. *Nature,* **342,** 39.

39. Potashkin, J. A. and Beach, D. H. (1988) Multiple phosphorylated forms of the fission yeast cell division cycle gene *cdc2*$^{+}$. *Curr. Genet.*, **14,** 235.
40. Fleig, U. N. and Nurse, P. (1991) Expression of a dominant negative allele of cdc2 prevents activation of the endogenous p34^{cdc2} kinase. *Mol. Gen. Genet.*, **226,** 432.
41. Millar, J. B. A., McGowan, C. H., Lenaers, G., Jones, R., and Russell, P. (1991) p80^{cdc25} mitotic inducer is the tyrosine phosphatase that activates cdc2 kinase in fission yeast. *EMBO J.*, **10,** 4301.
42. Millar, J. B. A., Lenaers, G., and Russell, P. (1992) pyp3 PTPase acts as a mitotic inducer in fission yeast. *EMBO J.*, **11,** 4933.
43. Fesquet, D., Labbé, J.-C., Derancourt, J., Capony, J.-C., Galas, S., Girard, F., Lorca, T., Shuttleworth, J., Doreé, M., and Cavadore, J.-C. (1993) The MO15 gene encodes the catalytic subunit of a protein kinase that activates cdc2 and other cyclin-dependent kinases (CDKs) through phosphorylation of Thr161 and its homologues. *EMBO J.*, **12,** 3111.
44. Poon, R. Y. C., Yamashita, K., Adamczewski, J. P., Hunt, T., and Shuttleworth, J. (1993) The cdc2-related protein p40^{MO15} is the catalytic subunit of a protein kinase that can activate p33^{cdk2} and p34^{cdc2}. *EMBO J.*, **12,** 3123.
45. Solomon, M. J., Harper, J. W., and Shuttleworth, J. (1993) CAK, the p34cdc2 activating kinase, contains a protein identical or closely related to p40MO15. *EMBO J.*, **12,** 3133.
46. Shuttleworth, J., Godfrey, R., and Colman, A. (1990) pMO15, a cdc2-related protein kinase involved in negative regulation of meiotic maturation of *Xenopus* oocytes. *EMBO J.*, **9,** 3233.
47. MacNeill, S. A., Creanor, J., and Nurse, P. (1991) Isolation, characterisation and molecular cloning of the fission yeast p34^{cdc2} protein kinase gene: identification of temperature-sensitive G2-arresting alleles. *Mol. Gen. Genet.*, **229,** 109.
48. Ayscough, K., Hayles, J., MacNeill, S. A., and Nurse, P. (1992) Cold-sensitive mutants of p34^{cdc2} that suppress a mitotic catastrophe phenotype in fission yeast. *Mol. Gen. Genet.*, **232,** 344.
49. Booher, R. and Beach, D. (1986) Site-specific mutagenesis of *cdc2*$^{+}$, a cell cycle control gene of the fission yeast. *Schizosaccharomyces pombe. Mol. Cell. Biol.*, **6,** 3523.
50. Fleig, U. N., Gould, K. L., and Nurse, P. (1992) A dominant negative allele of p34^{cdc2} shows altered phosphoamino acid content and sequesters p56^{cdc13} cyclin. *Mol. Cell. Biol.*, **12,** 2295.
51. MacNeill, S. A. and Nurse, P. (1993) Mutational analysis of the fission yeast p34^{cdc2} protein kinase gene. *Mol. Gen. Genet.*, **236,** 415.
52. Carr, A. M., MacNeill, S. A., Hayles, J., and Nurse, P. (1989) Molecular cloning and sequence analysis of mutant alleles of the fission yeast cdc2 protein kinase gene: implications for cdc2^{+} protein structure and function. *Mol. Gen. Genet.*, **218,** 41.
53. DeBrandt, H. L., Rosenblatt, J., Jancarik, J., Jones, H. D., Morgan, D. O., and Kim, S.-U. (1993) Crystal structure of cyclin-dependent kinase 2. *Nature*, **363,** 595.
54. Booher, R. and Beach, D. (1988) Involvement of *cdc13*$^{+}$ in mitotic control in *Schizosaccharomyces pombe*: possible interaction of the gene product with microtubules. *EMBO J.*, **7,** 2321.
55. Hagan, I., Hayles, J., and Nurse, P. (1988) Cloning and sequencing of the cyclin-related *cdc13*$^{+}$ gene and a cytological study of its role in mitosis. *J. Cell Sci.*, **91,** 587.
56. Solomon, M., Booher, R., Kirschner, M., and Beach, D. (1988) Cyclin in fission yeast. *Cell*, **54,** 738.

57. Goebl, M. and Byers, B. (1988) Cyclin in fission yeast. *Cell*, **54,** 739.
58. Booher, R. and Beach, D. (1987) Interaction between *cdc2*$^+$ and *cdc13*$^+$ in the control of mitosis in fission yeast: dissociation of the G1 and G2 roles of the *cdc2*$^+$ protein kinase. *EMBO J.*, **6,** 3441.
59. Hayles, J., Beach, D., Durkacz, B., and Nurse, P. (1986). The fission yeast cell cycle control gene cdc2: isolation of a sequence that suppresses cdc2 mutants. *Mol. Gen. Genet.*, **202,** 291.
60. Hindley, J., Phear, G., Stein, M., and Beach, D. (1987) *suc1*$^+$ encodes a predicted 13-kilodalton protein that is essential for cell viability and is directly involved in the division cycle of *Schizosaccharomyces pombe*. *Mol. Cell. Biol.*, **7,** 504.
61. Galaktionov, K. and Beach, D. (1991) Specification activation of cdc25 tyrosine phosphatases by B-type cyclins: evidence for multiple roles of mitotic cyclins. *Cell*, **67,** 1181.
62. Alfa, C. E., Ducommun, B., Beach, D., and Hyams, J. S. (1990) Distinct nuclear and spindle pole body populations of cyclin–cdc2 in fission yeast. *Nature*, **347,** 680.
63. Gallagher, I. M., Alfa, C. E., and Hyams, J. S. (1993) p63^{cdc13}, a B-type cyclin, is associated with both the nucleolar and chromatin domains of the fission yeast nucleus. *J. Cell Sci.*, **4,** 1087.
64. Forsburg, S. L. and Nurse, P. (1991) Identification of a G1-type cyclin *puc1*$^+$ in the fission yeast *Schizosaccharomyces pombe*. *Nature*, **351,** 245.
65. Forsburg, S. L. and Nurse, P. (1994) Analysis of the *Schizosaccharomyces pombe* cyclin *puc1*; evidence for a role in cell cycle exit. *J. Cell Sci.*, **107,** 601.
66. Bueno, A., Richardson, H., Reed, S. I., and Russell, P. (1991) A fission yeast B-type cyclin functioning early in the cell cycle. *Cell*, **66,** 149.
67. Bueno, A., Richardson, H., Reed, S. I., and Russell, P. (1991) A fission yeast B-type cyclin functioning early in the cell cycle (erratum). *Cell*, **73,** 1050.
68. Bueno, A. and Russell, P. (1993) Two fission yeast B-type cyclins, cig2 and cdc13, have different functions in mitosis. *Mol. Cell. Biol.*, **13,** 2286.
69. Bueno, A. and Russell, P. (1994) Two fission yeast B-type cyclins, cig2 and cdc13, have different functions in mitosis (correction). *Mol. Cell. Biol.*, **14,** 869.
70. Connolly, T. and Beach, D. (1994) Interaction between the cig1 and cig2 B-type cyclins in the fission yeast cell cycle. *Mol. Cell. Biol.*, **14,** 768.
71. Molz, L. and Beach, D. (1993) Characterisation of the fission yeast mcs2 cyclin and its associated protein kinase activity. *EMBO J.*, **12,** 1723.
72. Russell, P. and Nurse, P. (1986) *cdc25*$^+$ functions as an inducer in the mitotic control of fission yeast. *Cell*, **45,** 145.
73. Moreno, S. and Nurse, P. (1991) Clues to the action of the cdc25 protein. *Nature*, **351,** 194.
74. Sadhu, K., Reed, S. I., Richardson, H., and Russell, P. (1990) Human homolog of cdc25 mitotic inducer is primarily expressed in G2. *Proc. Natl. Acad. Sci. USA*, **87,** 5139.
75. Edgar, B. A. and O'Farrell, P. H. (1989) Genetic control of cell division patterns in the *Drosophila* embryo. *Cell*, **57,** 177.
76. Russell, P., Moreno, S., and Reed, S. I. (1989) Conservation of mitotic controls in fission and budding yeasts. *Cell*, **57,** 295.
77. Millar, J. B. A. and Russell, P. (1992) The cdc25 M-phase inducer: an unconventional protein phosphatase. *Cell*, **68,** 407.
78. Zheng, X.-F. and Ruderman, J. V. (1993) Functional analysis of the P box, a domain required for the activation of cdc25. *Cell*, **75,** 155.

79. Lundgren, K., Walworth, N., Booher, R., Dembski, M., Kirschner, M., and Beach, D. (1991) mik1 and wee1 cooperate in the inhibitory tyrosine phosphorylation of cdc2. *Cell*, **64,** 1115.
80. Moreno, S., Nurse, P., and Russell, P. (1990) Regulation of mitosis by cyclic accumulation of $p80^{cdc25}$ mitotic inducer in fission yeast. *Nature*, **344,** 549.
81. Ducommun, B., Draetta, G., Young, P., and Beach, D. (1990) Fission yeast cdc25 is a cell-cycle regulated protein. *Biochem. Biophys. Res. Commun.*, **167,** 301.
82. Kumagai, A. and Dunphy, W. G. (1992) Regulation of the cdc25 protein during the cell cycle in *Xenopus* extracts. *Cell*, **70,** 139.
83. Hoffmann, I., Clarke, P. R., Marcote, M. J., Karsenti, E., and Draetta, G. (1993) Phosphorylation and activation of human cdc25C by cdc2–cyclin B and its involvement in the self-amplification of MPF at mitosis. *EMBO J.*, **12,** 53.
84. Izumi, T. and Maller, J. L. (1993) Elimination of cdc2 phosphorylation sites in the cdc25 phosphates blocks initiation of M-phase. *Mol. Biol. Cell*, **4,** 1337.
85. Featherstone, C. and Russell, P. (1991) Fission yeast $p107^{wee1}$ mitotic inhibitor is a tyrosine/serine kinase. *Nature*, **349,** 808.
86. Rogers, S., Wells, R., and Rechsteiner, M. (1989) Amino acid sequences common to rapidly degraded proteins: the PEST hypothesis. *Science*, **234,** 364.
87. McGowan, C. H. and Russell, P. R. (1993) Human wee1 kinase inhibits cell division by phosphorylating $p34^{cdc2}$ exclusively on Tyr 15. *EMBO J.*, **12,** 75.
88. Parker, L. L. and Piwnica-Worms, H. (1992) Inactivation of the $p34^{cdc2}$–cyclin B complex by the human wee1 tyrosine kinase. *Science*, **257,** 1955.
89. Parker, L. L., Atherton-Fessler, S., and Piwnica-Worms, H. (1992) $p107^{wee1}$ is a dual specificity protein kinase that phosphorylates $p34^{cdc2}$ on tyrosine 15. *Proc. Natl. Acad. Sci. USA*, **89,** 2917.
90. Walworth, N., Davey, S., and Beach, D. (1993) Fission yeast chk1 protein kinase links the rad checkpoint pathway to cdc2. *Nature*, **363,** 368.
91. Young, P. G. and Fantes, P. A. (1984) Changed division response mutants function as allosuppressors. In *Growth, cancer and the cell cycle*. Skehan, P. and Friedman, S. J. (ed.). Humana Press, New Jersey, USA.
92. Young, P. G. and Fantes, P. A. (1987) *Schizosaccharomyces pombe* mutants affected in their division response to starvation. *J. Cell Sci.*, **88,** 295.
93. Fantes, P. A., Warbrick, E., Hughes, D. A., and MacNeill, S. A. (1991) New elements in the mitotic control of the fission yeast *Schizosaccharomyces pombe. Cold Spring Harbor Symp. Quant. Biol.*, **61,** 605.
94. Feilotter, H., Nurse, P., and Young, P. G. (1991) Genetic and molecular analysis of *cdr1/nim1* in *Schizosaccharomyces pombe. Genetics*, **127,** 309.
95. Wu, L. and Russell, P. (1993) nim1 kinase promotes mitosis by inactivating wee1 tyrosine kinase. *Nature*, **363,** 738.
96. Coleman, T. R., Tang, Z., and Dunphy, W. G. (1993) Negative regulation of the wee1 protein kinase by direct action of the nim1/cdr1 mitotic inducer. *Cell*, **72,** 919.
97. Parker, L. L., Walter, S. A., Young, P. G., and Piwnica-Worms, H. (1993) Phosphorylation and inactivation of the mitotic inhibitor wee1 by the nim1/cdr1 kinase. *Nature*, **363,** 736.
98. Ottilie, S., Chernoff, J., Hannig, G., Hoffman, C. S., and Erikson, R. L. (1991) A fission yeast gene encoding a protein with features of protein-tyrosine-phosphatases. *Proc. Natl. Acad. Sci. USA*, **88,** 3455.
99. Ottilie, S., Chernoff, J., Hannig, G., Hoffman, C. S., and Erikson, R. L. (1992) The

fission yeast genes *pyp1*$^{+}$ and *pyp2*$^{+}$ encode protein tyrosine phosphatases that negatively regulate mitosis. *Mol. Cell. Biol.*, **12,** 5571.

100. Millar, J. B. A., Russell, P., Dixon, J. E., and Guan, K. L. (1992) Negative regulation of mitosis by two functionally overlapping PTPases in fission yeast. *EMBO J.*, **11,** 4943.
101. Ogden, J. E. and Fantes, P. A. (1986) Isolation of a novel type of mutation in the mitotic control of *Schizosaccharomyces pombe* whose phenotypic expression is dependent on the genetic background and nutritional environment. *Curr. Genet.*, **10,** 509.
102. Neiman, A. M., Stevenson, B. J., Xu, H.-P., Sprague, G. F., Herskowitz, I., Wigler, M., and Marcus, S. (1993) Functional homology of protein kinase required for sexual differentiation in *Schizosaccharomyces pombe* and *Saccharomyces cerevisiae* suggests a conserved signal transduction module in eukaryotic organisms. *Mol. Biol. Cell*, **4,** 107.
103. Kosako, H., Nishida, E., and Gotoh, Y. (1993) cDNA cloning of MAP kinase reveals kinase cascade pathways in yeasts to vertebrates. *EMBO J.*, **12,** 787.
104. Gartner, A., Nasmyth, K., and Ammerer, G. (1992) Signal transduction in *Saccharomyces cerevisiae* requires tyrosine and threonine phosphorylation of FUS3 and KSS1. *Genes Dev.*, **6,** 1280.
105. Toda, T., Shimanuki, M., and Yanagida, M. (1991) Fission yeast genes that confer resistance to staurosporine encode an AP-1-like transcription factor and a protein kinase related to mammalian ERK/MAP2 and budding yeast FUS3/KSS1 kinases. *Genes Dev.*, **6,** 60.
106. Cohen, P. (1989) The structure and regulation of protein phosphatases. *Annu. Rev. Biochem.*, **54,** 453.
107. Healy, A. M., Zolnierowicz, S., Stapleton, A. E., Goebl, M., DePaoli-Roach, A. A., and Pringle, J. R. (1991) *CDC55*, a *Saccharomyces cerevisiae* gene involved in cellular morphogenesis; identification, characterisation, and homology to the B subunit of mammalian type 2A protein phosphatase. *Mol. Cell. Biol.*, **11,** 5767.
108. Ohkura, H. and Yanagida, M. (1991) Fission yeast *sds22*$^{+}$ gene is essential for a mid-mitotic transition and encodes a leucine-rich protein which positively modulates type 1 protein phosphatase. *Cell*, **64,** 149.
109. Stone, E. M., Yamano, H., Kinoshita, N., and Yanagida, M. (1993) Mitotic regulation of protein phosphatases by the fission yeast sds22 protein. *Curr. Biol.*, **3,** 13.
110. Ohkura, H., Kinoshita, N., Miyatani, S., Toda, T., and Yanagida, M. (1989) The fission yeast *dis2*$^{+}$ gene required for chromosome disjoining encodes one of two putative type 1 protein phosphatases. *Cell*, **57,** 997.
111. Doonan, J. H. and Morris, N. R. (1989) The *bimG* gene of *Aspergillus nidulans*, which is required for completion of anaphase, encodes a homolog of mammalian phosphoprotein phosphatase 1. *Cell*, **57,** 996.
112. Axton, J. M., Dombradi, V., Cohen, P. T. W., and Glover, D. (1990) One of the protein phosphatase 1 isozymes in *Drosophila* is essential for mitosis. *Cell*, **63,** 33.
113. Booher, R. and Beach, D. (1989) Involvement of the type 1 protein phosphatase encoded by *bws1*$^{+}$ in fission yeast mitotic control. *Cell*, **57,** 1009.
114. Shimanuki, M., Kinoshita, N., Ohkura, H., Yoshida, T., Toda, T., and Yanagida, M. (1993) Isolation and characterisation of the fission yeast protein phosphatase gene *ppe1*$^{+}$ involved in cell shape control and mitosis. *Mol. Biol. Cell*, **4,** 303.
115. Matsumoto, T. and Beach, D. (1993) Interaction of the pim1/spi1 mitotic checkpoint with a protein phosphatase. *Mol. Biol. Cell*, **4,** 337.

116. Matsumoto, T. and Beach, D. (1991) Premature initiation of mitosis in yeast lacking RCC1 or an interacting GTPase. *Cell*, **66,** 347.
117. Sazer, S. and Nurse, P. (1993) A fission yeast RCC1-related protein is required for the mitosis to interphase transition. *EMBO J.*, **13,** 606.
118. Fantes, P. A. (1983) Control of timing of cell cycle events in fission yeast by the *wee1*$^{+}$ gene. *Nature*, **302,** 153.
119. Hughes, D. A., MacNeill, S. A., and Fantes, P. A. (1992) Molecular cloning and sequence analysis of *cdc27*$^{+}$ required for G2–M transition in the fission yeast *Schizosaccharomyces pombe*. *Mol. Gen. Genet.*, **231,** 401.
120. Warbrick, E. and Fantes, P. A. (1992) Five novel elements involved in the regulation of mitosis in fission yeast. *Mol. Gen. Genet.*, **232,** 440.
121. Hudson, J., Feilotter, H., and Young, P. G. (1990) *stf1*: non-wee mutations epistatic to cdc25 in the fission yeast *Schizosaccharomyces pombe*. *Genetics*, **126,** 309.
122. Molz, L., Booher, R., Young, P., and Beach, D. (1989) *cdc2* and the regulation of mitosis: six interacting *mcs* genes. *Genetics*, **122,** 773.
123. Nasmyth, K. A. (1977) Temperature-sensitive lethal mutations in the structural gene for DNA ligase in the yeast *Schizosaccharomyces pombe*. *Cell*, **12,** 1109.
124. Francesconi, S., Park, H., and Wang, T. S. F. (1993) Fission yeast with DNA polymerase δ temperature-sensitive alleles exhibits cell division cycle phenotype. *Nucleic Acids Res.*, **21,** 3821.
125. Waseem, N. H., Labib, K., Nurse, P., and Lane, D. P. (1992) Isolation and analysis of the fission yeast gene encoding polymerase δ accessory protein PCNA. *EMBO J.*, **11,** 5111.
126. Coxon, A., Maundrell, K., and Kearsey, S. E. (1992) Fission yeast *cdc21*$^{+}$ belongs to a family of proteins involved in an early step of chromosome replication. *Nucleic Acids Res.*, **20,** 5571.
127. Fernandez-Sarabia, M.-J., McInerny, C. J., Harris, P., Gordon, C., and Fantes, P. A. (1993) The cell cycle genes *cdc22*$^{+}$ and *suc22*$^{+}$ of the fisson yeast *Schizosaccharomyces pombe* encode the large and small subunits of ribonucleotide reductase. *Mol. Gen. Genet.*, **238,** 241.
128. Al-Khodairy, F. and Carr, A. M. (1992) DNA repair mutants defining G2 checkpoint pathways in *Schizosaccharomyces pombe*. *EMBO J.*, **11,** 1343.
129. Rowley, R., Subramani, S., and Young, P. G. (1992) Checkpoint controls in *Schizosaccharomyces pombe:* rad1. *EMBO J.*, **11,** 1335.
130. Enoch, T., Carr, A. M., and Nurse, P. (1992) Fission yeast genes involved in coupling mitosis to completion of DNA replication. *Genes Dev.*, **6,** 2035.
131. Sheldrick, K. S. and Carr, A. M. (1993) Feedback controls and G2 checkpoints: fission yeast as a model system. *BioEssays*, **15,** 775.
132. Phipps, J. A., Nasim, A., and Miller, D. R. (1985) Recovery, repair and mutagenesis in *Schizosaccharomyces pombe*. *Adv. Genet.*, **23,** 1.
133. Sunnerhagen, P., Seaton, B. L., Nasim, A., and Subramani, S. (1990) Cloning and analysis of a gene involved in DNA repair and recombination, the *rad1* gene of *Schizosaccharomyces pombe*. *Mol. Cell. Biol.*, **10,** 3750.
134. Seaton, B. L., Yucel, J., Sunnerhagen, P., and Subramani, S. (1992) Isolation and characterisation of the *Schizosaccharomyces pombe rad3*$^{+}$ gene which is involved in DNA repair and DNA synthesis checkpoints. *Gene*, **119,** 83.
135. Murray, J. M., Carr, A. M., Lehmann, A. R., and Watts, F. Z. (1991) Cloning and characterisation of the DNA repair gene *rad9* from *Schizosaccharomyces pombe*. *Nucleic Acid Res.*, **19,** 3525.

136. Enoch, T. and Nurse, P. (1990). Mutation of fission yeast cell cycle control genes abolishes dependence of mitosis on DNA replication. *Cell*, **60,** 665.
137. Enoch, T., Gould, K. L., and Nurse, P. (1991) Mitotic checkpoint control in fission yeast. *Cold Spring Harbor Symp. Quant. Biol.*, **61,** 409.
138. Rowley, R., Hudson, J., and Young, P. G. (1992) The wee1 protein kinase is required for radiation-induced mitotic delay. *Nature*, **356,** 353.
139. Barbet, N. C. and Carr, A. M. (1993) Fission yeast wee1 protein kinase is not required for DNA damage-dependent mitotic arrest. *Nature*, **364,** 824.
140. Kelly, T. J., Martin, G. S., Forsburg, S. L., Stephen, R. J., Russo, A., and Nurse, P. (1993) The fission yeast *cdc18*$^{+}$ gene product couples S-phase to START and mitosis. *Cell*, **74,** 371.
141. Hofmann, J. F. X. and Beach, D. (1994) cdt1 is an essential target of the cdc10/sct1 transcription factor: requirement for DNA replication and inhibition of mitosis. *EMBO J.*, **13,** 425.
142. Saka, Y. and Yanagida, M. (1994) Fission yeast *cut5*$^{+}$, required for S-phase onset and M-phase restraint, is identical to the radiation-damage gene *rad4*$^{+}$. *Cell*, **74,** 383.
143. Fenech, M., Carr, A. M., Murray, J., Watts, F. Z., and Lehmann, A. R. (1991) Cloning and characterisation of the *rad4*$^{+}$ gene of *Schizosaccharomyces pombe*: a gene showing short regions of sequence similarity to the human XRCC1 gene. *Nucleic Acids Res.*, **19,** 6737.
144. Lehmann, A. R. (1993) Duplicated region of sequence similarity to the human *XRCC1* DNA repair gene in the *Schizosaccharomyces pombe rad4/cut5* gene. *Nucleic Acids Res.*, **21,** 5274.
145. Lowndes, N., McInerny, C. J., Johnson, A. L., Fantes, P. A., and Johnston, L. H. (1992) Control of DNA synthesis genes in fission yeast by the cell cycle gene *cdc10*$^{+}$. *Nature*, **355,** 449.
146. Bueno, A. and Russell, P. (1992) Dual functions of CDC6: a yeast protein required for DNA replication also inhibits nuclear division. *EMBO J.*, **11,** 2167.
147. Hogan, E. and Koshland, D. (1992) Addition of extra origins of replication to a minichromosome suppresses its mitotic loss in *cdc6* and *cdc14* mutants of *Saccharomyces cerevisiae*. *Proc. Natl. Acad. Sci. USA*, **89,** 3098.
148. Moreno, S. and Nurse, P. (1994) Regulation of progression through the G1 phase of the cell cycle by the *rum1*$^{+}$ gene. *Nature*, **367,** 236.
149. Broek, D., Bartlett, R., Crawford, K., and Nurse, P. (1991) Involvement of p34^{cdc2} in establishing the dependence of S-phase on mitosis. *Nature*, **349,** 388.
150. Murray, A. W. (1994) Rum tale of replication. *Nature*, **367,** 21.
151. Frankhauser, C., Marks, J., Reymond, A., and Simanis, V. (1993) The *S. pombe cdc16* gene is required for both maintenance of p34^{cdc2} kinase activity and regulation of septum formation: a link between mitosis and cytokinesis? *EMBO J.*, **12,** 2697.
152. Hoyt, M. A., Totis, L., and Roberts, B. T. (1991) *Saccharomyces cerevisiae* genes required for cell cycle arrest in response to loss of microtubule function. *Cell*, **66,** 507.
153. Chang, F. and Nurse, P. (1993) Finishing the cell cycle: control of mitosis and cytokinesis on fission yeast. *Trends Genet.*, **9,** 333.
154. Gordon, C., McGurk, G., Dillon, P., Rosen, C., and Hastie, N. (1993) Defective mitosis due to a mutation in the gene for a fission yeast 26S protease subunit. *Nature*, **366,** 355.
155. Ghislain, M., Udvardy, A., and Mann, C. (1993) *S. cerevisiae* 26S protease mutants arrest cell division in G2/metaphase. *Nature*, **366,** 358.

156. Balusubramanian, M. K., Helfman, D. M., and Hemmingsen, S. M. (1992) A new tropomyosin essential for cytokinesis in the fission yeast *S. pombe. Nature,* **360,** 84.
157. Yanagida, M., Kinoshita, N., Stone, E. M., and Yamano, H. (1992) Protein phosphatases and cell division cycle control. In *Regulation of the eukaryotic cell cycle,* Ciba Symposium 170, Hunter, T. (ed.). Wiley, Chichester, UK, p. 130.

5 The cdc2 kinase: structure, activation, and its role at mitosis in vertebrate cells

GABRIELE BASI and GUILIO DRAETTA

1. Introduction

In the current view of eukaryotic cell cycle control, the existence of factors which positively regulate the progression through different phases of the cell cycle is assumed. In the last few years much progress has been made towards the molecular characterization of a factor which promotes entry into mitosis called MPF (M-phase promoting factor). The first evidence for the existence of such a factor dates back to studies in the slime mould *Physarum polycephalum* (1). *P. polycephalum* is a plasmodial organism (myxomycete) in which nuclear division is naturally synchronous, and therefore it is particularly suitable for studying the regulatory mechanisms controlling entry into mitosis. Experiments in which plasmodia at different stages of the cell cycle were fused, revealed the existence of a positive regulator of mitotic initiation, the levels of which fluctuated during the cell cycle, reaching a peak just prior to mitosis. The results also indicated that protein synthesis was required for the appearance of the mitotic factor and suggested that a translocation of this factor from the cytoplasm to the nucleus occurred shortly before the onset of mitosis. More than twenty years after this initial observation, MPF was characterized and it was confirmed that the view proposed for the regulation of mitosis in *P. polycephalum* is essentially correct for all eukaryotes.

In human cells, evidence for the existence of a positive regulator of mitosis came from experiments in which mitotic and interphase cells were fused (2). When cells synchronized in G1, S, or G2 phase were fused to cells blocked in mitosis, premature chromosome condensation was induced in the interphase cells (mitosis was, however, not completed because the mitotic spindle did not form). The same authors also obtained evidence for the existence of a positive regulator for S phase by experiments of fusion between cells synchronized in G1 and S phases (3). Soon after fusion with an S-phase cell there was a rapid induction of DNA synthesis in the nucleus of a G1 cell, indicating the existence of an S-phase promoting factor (SPF).

The molecular characterization of SPF has proven to be difficult. Very recently, however, substantial progress has been made towards the identification of molecules regulating the G1/S transition in higher eukaryotes (4) (see also Chapter 6). While the molecular components of MPF have been conserved throughout evolution in all eukaryotes, the complexity of SPF seems to vary in different organisms. Regulation of SPF requires the integration of external growth signals into the cell cycle engine, which is a fundamental step in proliferation. The signal transduction pathways which regulate growth are likely to have diverged during evolution, since cells from different organisms (and within the same organism, different cell types) have to respond to different external stimuli. As a consequence of this divergence, the cell cycle machinery has evolved specific forms of SPF, which respond to different growth signals. Activation of MPF, in contrast, is less subject to regulation from external growth signals: usually, cells that have started S phase are committed to complete a further round of mitotic division (5). This may explain why MPF has been conserved during evolution while SPF has not. The convergence of studies on different organisms has been crucial for the molecular characterization of MPF. Studies on the regulation of mitosis have also been facilitated by the existence of specific cytological markers which characterize M phase, and are absent in other phases of the cell cycle.

2. Conservation of the regulatory components of MPF in eukaryotes

2.1 A brief history

As experiments in somatic cells demonstrated that mitosis is regulated by an M-phase promoting factor, microinjection experiments using frog oocytes brought to light the existence of a factor which promotes entry into meiosis.

Immature frog oocytes are arrested at the G2/M boundary of meiosis I. Maturation of oocytes, which is triggered *in vivo* by progesterone, consists of the completion of the first meiotic division and a subsequent arrest of the eggs at the metaphase of meiosis II (arrest that is relieved upon fertilization). Maturation can be easily monitored by the appearance of a white spot on the animal pole (the dark pigmented oocyte pole), which is caused by the breakdown of the oocyte nucleus (germinal vesicle) upon entry into meiosis I. This event is often referred as germinal vesicle breakdown (GVBD). When an extract from metaphase-arrested eggs was microinjected into immature oocytes, maturation (GVBD) could be induced in the absence of progesterone (6, 7). These experiments revealed the existence of a factor which positively regulates maturation of the oocytes (called maturation promoting factor or MPF). The distinction between M-phase promoting factor and maturation promoting factor was purely nominal, since it was found that extracts from blastomeres of amphibian embryos undergoing mitosis or from human mitotic cells could induce maturation when microinjected into frog oocytes (8, 9). Extracts from mitotic cells of other organisms, including yeast, could also induce maturation of amphibian or marine oocytes (10–13).

These findings strongly suggested the evolutionary conservation of MPF in all eukaryotes. Separate lines of research, initially focusing on the study of mitotic control in different organisms, have since converged allowing identification of the molecular mechanisms of mitotic regulation in eukaryotes.

While biochemists were attempting the purification of MPF by fractionating extracts and injecting those fractions into oocytes, geneticists had started the identification of yeast mutants and genes involved in cell cycle control (14, 15) (see also Chapters 2 and 5). In particular, studies on mitotic control in the fission yeast *Schizosaccharomyces pombe* had identified a gene, called *cdc2*, whose product was required at two points of the cell cycle, in G1 and in G2 (16). *cdc2* mutants entering mitosis at a smaller cell size than wild-type cells had also been isolated (17). These mutants (called 'wee' mutants) acted in a dominant fashion, suggesting a positive role for cdc2 in mitotic initiation. *Saccharomyces cerevisiae* and human *cdc2* homologous genes were subsequently isolated by screening for DNA sequences that could suppress the temperature-sensitive defect of an *S. pombe cdc2* mutant (18, 19). The *cdc2* homologue in *S. cerevisiae* turned out to be a gene that was already known, called *CDC28*. The existence of functional homologues of *cdc2* in distantly related organisms (phylogenetically *S. pombe* is not closer to *S. cerevisiae* than to human), was again suggestive that the mechanisms regulating the onset of mitosis in eukaryotes had been conserved throughout evolution.

Three other genes involved in regulating the timing of mitosis had been identified in *S. pombe*: *cdc25*, *wee1*, and *nim1* (see Chapter 5). *cdc25* and *wee1* operate upstream of *cdc2*, the first being an activator and the second an inhibitor of cdc2 function (20, 21). *nim1* acts upstream of *wee1*, negatively regulating its function (21). *cdc2* had also been shown to interact with two additional genes, *suc1* and *cdc13* (22, 23). The two gene products interact physically with the cdc2 protein and are regulators of the cdc2 function (24–26).

Meanwhile, biochemists had developed a simple bioassay (far simpler than microinjection of oocytes) for testing MPF activity. This assay made use of cell-free egg extracts in which the effect of MPF, added from an extract or a purified fraction, could be easily monitored (27). By taking advantage of this new cell-free assay, MPF was purified for the first time. The major components of MPF, purified from *Xenopus laevis* eggs, were two proteins of 32 and 45 kDa (28). The smaller polypeptide was shown to crossreact with antibodies raised against a conserved region of cdc2, indicating that a component of MPF is a *Xenopus* homologue of cdc2 (29). A similar conclusion was reached by showing that the product of the *S. pombe suc1* gene could bind a *Xenopus* cdc2 homologue and that MPF activity could be depleted by passing a crude preparation of MPF through a suc1 column (30).

Shortly after, a separate line of investigation converged in the identification of cdc2 as a key regulator of mitosis in eukaryotes. Investigators were looking for enzymatic activities which displayed periodicity in the cell cycle. Among those, a histone H1 kinase activity was present upon entry into mitosis. The M-phase specific histone H1 kinase had been initially identified in *P. polycephalum* (31) and was subsequently shown to be present in a variety of organisms. When this protein kinase was purified from starfish oocytes, one of its components was found to be a protein of 34 kDa that, similarly to the 32 kDa component of MPF, crossreacted with antibodies raised against cdc2 and interacted tightly with suc1 coupled to Sepharose beads (32, 33). The correlation between cdc2, MPF, and M-phase specific histone H1 kinase was even more striking, taking into account that cdc2 was already known to be a protein kinase specifically activated during the *S. pombe* cell division cycle (34) and that purified MPF from *Xenopus* eggs had been shown to display histone H1 kinase activity (28). All these experiments supported the idea that MPF represented the mitotic form of the cdc2 kinase and that cdc2 was the M-phase specific histone H1 kinase.

These findings had a strong impact on cell cycle research. After many years of independent work, biochemists and geneticists realized they had been working on the same molecular entity. From there on, biochemists and yeast geneticists working on the cell cycle showed much greater interest in each other's research.

2.2 The mitotic cdc2 kinase is a complex of multiple subunits

cdc2 is the catalytic subunit of the MPF/histone H1 kinase. A second component of MPF, with an important regulatory function, is cyclin B. Cyclins were initially identified in sea urchin and surf clam fertilized eggs as proteins which are synthesized continuously during interphase and are rapidly degraded at mitosis (35, 36). The cyclic accumulation of these proteins was synchronous to the periodic activation and inactivation of MPF, indicating a possible correlation between MPF activity and cyclin levels (Fig. 1). The rationale of these thoughts originated from the observation that, although the eggs of marine invertebrates contain stockpiles of proteins sufficient for thousands of cells, they require new protein synthesis before each mitosis, suggesting that in these cells requirement for protein synthesis has a regulatory function (37). Genes encoding cyclins were first cloned from clam and sea urchin (38, 39). The original classification of cyclins, based on a conserved

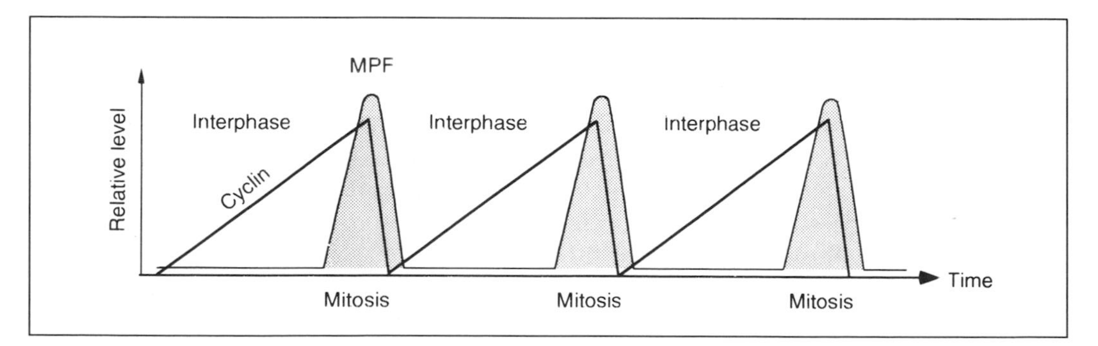

Fig. 1 Fluctuation of cyclin levels and MPF activity during the early embryonic cell cycle. The correlation between cyclin synthesis and MPF activation during the initial cell cycles of fertilized sea urchin eggs was the first indication that cyclins could play an important role in the activation of MPF.

region in the centre of the cyclin sequence (called cyclin box), distinguished two cyclins subfamilies named A and B types. Subsequently, additional members of the cyclin family (cyclins C–H) have been discovered (see Chapter 6).

A direct involvement of cyclins in the control of mitosis was initially suggested by the finding of sequence similarity between the *S. pombe cdc13* gene product and other known cyclins (40, 41). Genetic evidence had shown that interaction between *cdc13* and *cdc2* was important for the cdc2 function at the G2/M transition (23). The sequence homology between cdc13 and cyclins suggested a possible physical interaction between cyclins and cdc2, and consequently the presence of cyclins in MPF.

The first evidence that cyclins physically interact with the cdc2 kinase was provided from experiments on embryos of the clam *Spisula solidissima*. In this organism, cyclins were shown to associate with a cdc2 homologue by forming two different complexes, one between cdc2 and cyclin A, the other between cdc2 and cyclin B (42). Both complexes were found to display histone H1 kinase activity at mitosis but not in interphase.

Subsequently, the existence of a physical complex between the *cdc2* and *cdc13* gene products was demonstrated in *S. pombe* and, as in starfish, the complex was shown to form an active histone H1 kinase at mitosis (25). Evidence was also provided that cdc13 accumulates during G2 and is degraded during the metaphase to anaphase transition. Coincident with degradation of cdc13 there was an abrupt decline of the cdc2 kinase activity, indicating that degradation of cyclins in late mitosis could be responsible for the periodic inactivation of MPF.

A functional role of cyclins in MPF activation was demonstrated by showing that after depletion of all endogenous mRNA from a *Xenopus* egg extract, addition of sea urchin cyclin B mRNA alone was sufficient to promote entry into mitosis (43). Consistent with this result, ablation of endogenous cyclin B mRNA from such extracts with antisense oligonucleotides, was sufficient to block entry into mitosis (44). By using this system it was also possible to establish that cyclin B degradation has an important regulatory function both in mitosis and in meiosis (45). Truncation of the first 90 amino acids of cyclin B generates a proteolysis-resistant mutant (see Section 2.2.2 and Fig. 3). When the mRNA encoding the truncated form of cyclin B was added to a cell-free extract, inactivation of MPF and exit from mitosis were blocked. Injection of this mRNA into mature oocytes could also inhibit the completion of meiosis II in eggs stimulated by calcium, proving that cyclin B degradation is necessary for inactivation of MPF *in vitro* and *in vivo*. B-type cyclins were eventually identified in purified preparations of starfish and *Xenopus* MPF, and were shown to correspond to the 45–50 kDa polypeptides originally found associated with cdc2 (46, 47).

Also associated with MPF is a small protein of 13 kDa, which corresponds to the product of the *suc1* gene (24, 48) (see Section 2.2.3). In non-transformed mammalian cells, it has been reported that PCNA (proliferating cell nuclear antigen) and a 21 kDa protein encoded by the CIP1/WAF1/SDI1 gene are present in a complex with cdc2–cyclin B (49) (see Section 2.2.4).

2.2.1 cdc2

Experiments with mammalian cells first suggested that the cdc2 kinase is specifically activated at mitosis (48). Evidence in support of a requirement for the mammalian cdc2 kinase at the onset of mitosis was subsequently provided by antibody microinjection experiments in mouse cells (50). After serum stimulation of quiescent mouse fibroblasts, polyclonal anti-cdc2 antibodies were microinjected in the cytoplasm of the cells and their effect on DNA synthesis and mitosis was analysed. Although no effect on DNA synthesis was observed, injection of the antibodies effectively inhibited cell division, suggesting that the cdc2 function in mammalian cells is required in G2/M but not in G1/S.

Additional evidence for a specific role of cdc2 at mitosis was provided by the isolation of a temperature-sensitive mouse cell line (FT210) defective for the cdc2 function. The temperature-sensitive G2 cell cycle block in this cell line could be suppressed by expressing the human cdc2 gene. The mouse cdc2 gene isolated from the FT210 cells was shown to contain two point mutations responsible for the temperature-sensitive defect of the endogenous cdc2 protein (51). A role of cdc2 in an earlier phase of the cell cycle than mitosis, however, cannot be ruled out from these experiments, since both cyclin A and cyclin B are known to form complexes with the cdc2 kinase. The cdc2–cyclin A complex is known to become active in G2 before the cdc2–cyclin B complex, which is maximally active at mitosis (52, 53). Therefore, the cell cycle arrest observed in the FT210 cell line might occur in G2 rather than at the G2/M boundary (the same argument is valid for the microinjection experiments with anti-cdc2 antibodies).

Unlike the yeasts *S. pombe* and *S. cerevisiae*, in which the cdc2/Cdc28 kinase is required both in G1/S and in G2/M, in vertebrates, the cdc2 kinase does not seem to play a role during the early phases of the cell cycle. Increasing evidence indicates that in higher eukaryotes, progression through the G1 and S phases is regulated by cdc2-related kinases, the cyclin-dependent kinases (cdks) (4, 54) (see also Chapter 6). For 'historical' reasons, cdc2, which in the new nomenclature corresponds to cdk1, is here referred to by its original name. At least five additional cdks have been identified so far in human cells (cdk2–cdk6). Two members of this family, cdk2 and cdk3, have been shown to be required for the G1/S progression in mammalian cells (55, 56).

B-type cyclins are necessary for the activation of the cdc2 kinase at the onset of mitosis, and cyclin B degradation is required for the inactivation of the histone H1 kinase in late mitosis. But is cyclin B synthesis the rate-limiting factor which regulates MPF activation and entry into mitosis? Studies on meiotic induction in *Xenopus* oocytes had demonstrated the existence of an inactive form of MPF, called preMPF. Injection into oocytes of a small amount of MPF can activate a much larger amount of MPF in the absence of any protein synthesis, indicating that the mechanism of MPF self-amplification is due to a post-translational modification of preMPF (13, 57). The idea that cyclin B first forms an inactive complex with cdc2 and that such a complex (preMPF) is activated by a post-translational modification at the onset of mitosis was also suggested by the kinetics of MPF activation. While the rate of cyclin B synthesis during the cell cycle, from interphase to mitosis, is linear, activation of MPF at the G2/M transition is exponential, suggesting that cyclin B accumulation alone does not trigger the initiation of mitosis (see Fig. 1).

The first indication of the nature of the post-translational modification of preMPF was obtained in human cells, where it was shown that cdc2 is a phosphoprotein and that the state of cdc2 phosphorylation is subjected to cell cycle regulation (48). cdc2 is phosphorylated on tyrosine and threonine residues. The levels of tyrosine phosphorylation vary during the cell cycle, increasing from G1 to G2 and disappearing during mitosis, coincident with activation of the histone H1 kinase (58). Studies in vertebrates and in *S. pombe* strongly suggest a functional role of cdc2 tyrosine phosphorylation in mitotic regulation. Exposure of mouse fibroblasts to vanadate or addition of purified suc1 protein to *Xenopus* extracts blocks entry into mitosis. The block correlates, in both cases, with inhibition of cdc2 tyrosine dephosphorylation and kinase activation, supporting the idea that cdc2 tyrosine dephosphorylation is required for the mitotic activation of the cdc2 kinase (59, 60). In *S. pombe*, an inactive cdc2–cdc13 complex can be detected in a *cdc25* temperature-sensitive mutant arrested at the G2/M transition. When the *cdc25* mutant is released at the permissive temperature, the cdc2 kinase is rapidly activated and cells can enter in mitosis (25). This activation correlates with tyrosine dephosphorylation on a conserved residue of cdc2 (Tyr15). Substitution of this residue with phenylalanine (Tyr15Phe) induces premature entry into mitosis (61). The latter finding is definitive proof that, in *S. pombe*, tyrosine dephosphorylation of cdc2 is a key event in the regulation of the timing of mitosis. Major regulators of the cdc2 tyrosine phosphorylation are the wee1 kinase and the cdc25 phosphatase (see Section 2.3).

In vertebrates, in addition to Tyr15, another conserved threonine residue in cdc2 (Thr14) is phosphorylated (62). Phosphorylation of this residue is cell cycle regulated and, similarly to Tyr15, is maximal in late G2 and absent in mitosis. Dephosphorylation of Tyr15 and Thr14 is required for the full activation of the mitotic cdc2 kinase, indicating that in higher eukaryotes both phosphorylation sites are important for mitotic initiation (63, 64). Consistent with these results, transfection of human cells with the single mutant Tyr15Phe can induce premature entry into mitosis to a lesser extent than the double mutant Thr14Ala/Tyr15Phe. However, the single mutant Thr14Ala does not affect the timing of mitosis, indicating that in higher eukaryotes Tyr15 also represents the major regulatory site of the cdc2 kinase (64).

Another residue in cdc2 (Thr167 in *S. pombe* and Thr161 in higher eukaryotes) is phosphorylated *in vivo* (62, 65, 66). Evidence that phosphorylation on Thr161 positively regulates the cdc2 function has been obtained by mutagenesis analysis. Substitution of Thr161 with alanine results in an inactive kinase which is unable to rescue an *S. pombe* temperature-sensitive cdc2 mutant (65, 67). The kinase which phosphorylates cdc2 on T161 (also known as CAK=Cdc2 activating kinase) was purified from *Xenopus* and starfish, and corresponds to the previously identified

MO15 kinase (68–70). *In vitro*, CAK phosphorylates and activates cdc2 associated with cyclins but not monomeric cdc2 (69). Purified CAK has a bigger molecular size than monomeric MO15 (70, 71), and at least one additional polypeptide has been shown to co-purify with the active MO15 kinase (69). Interestingly, bacterially-expressed *Xenopus* MO15 is inactive and can be activated by incubation with a *Xenopus* extract and ATP (71). These findings suggest that MO15 is the catalytic subunit of CAK, and that its activity is regulated by subunit association and by phosphorylation. One of the MO15-associated subunits has been recently characterized and corresponds to cyclin H, a protein homologous to other known cyclins (184). In *Xenopus* extracts, dephosphorylation of cdc2 on Thr161 occurs in late mitosis after cyclin degradation (72). It has been proposed that Thr161 dephosphorylation is necessary, besides cyclin B degradation, for inactivation of the cdc2 kinase and exit from mitosis (72). The phosphatase which hydrolyses the Thr161 phosphate of cdc2 has not yet been identified.

Taking these results into account, the pattern of mitotic regulation in vertebrates can be summarized as follows (Fig. 2; see also Fig. 4). During interphase, cdc2 and cyclin B associate and are readily phosphorylated on Thr14, Tyr15, and Thr161. This inactive complex (preMPF) accumulates during the S and G2 phases of the cell cycle. Dephosphorylation of cdc2 on Thr14 and Tyr15 is the rate-limiting step controlling entry into mitosis. After dephosphorylation of these residues, the cdc2–cyclin B complex becomes active and mitosis can start. Ubiquitination and the subsequent degradation of cyclin B (see below) brings about the inactivation of MPF, which is necessary for exiting mitosis. After dephosphorylation of Thr161, which probably occurs after cyclin degradation, monomeric unphosphorylated cdc2 is formed and the cycle can restart.

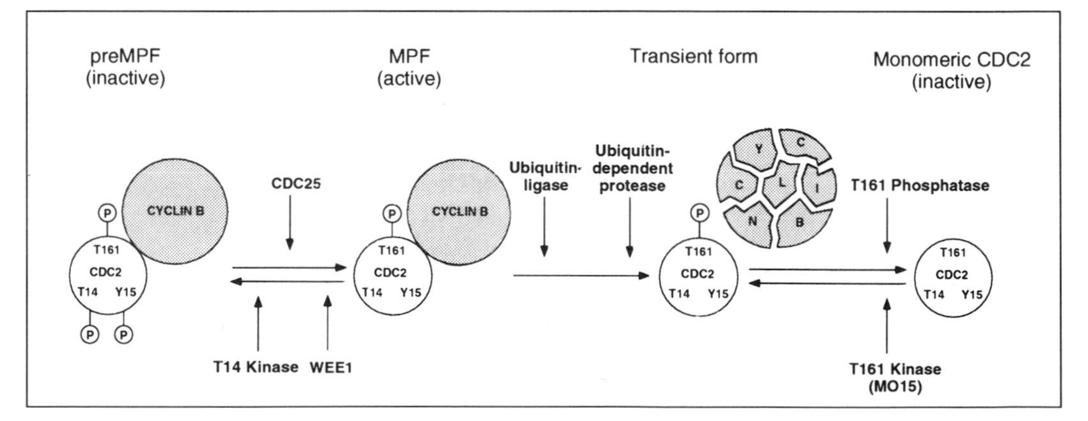

Fig. 2 Activation and inactivation of cdc2 at mitosis. Dephosphorylation of cdc2 on Thr14 and Tyr15 promotes activation of the cdc2–cyclin B complex formed during interphase (preMPF) and is the rate-limiting step for mitotic initiation. The Thr161 phosphate regulates positively the cdc2 kinase and is also present in the active M-phase promoting factor (MPF). At the metaphase–anaphase transition, cyclin B is ubiquitinated and subsequently degraded by an ubiquitin-dependent protease. Following degradation of cyclin B, the cdc2 kinase is dephosphorylated on Thr161 and MPF is inactivated. After formation of inactive monomeric cdc2, cells can exit from mitosis and start a new cycle.

In chicken cells, phosphorylation of cdc2 on a serine residue (Ser277) during the G1 phase of the cycle has also been reported (62). The relevance of this phosphorylation event in the regulation of the cdc2 kinase is not clear. Mutation of this residue to aspartate, and expression of the mutant chicken cdc2 in *S. pombe*, causes a cold-sensitive cell cycle arrest or a wee phenotype, depending on growth conditions (73).

2.2.2 Cyclins

The cdc2 kinase has been shown to associate with regulatory subunits (cyclins) important for the cyclic activation of the cdc2 kinase. The mitotic form of the cdc2 kinase is associated with cyclin B, which in mammals was first identified as a 62 kDa protein associated with the human cdc2 kinase at mitosis (48, 74). Another protein, cyclin A, has been shown to associate with the human cdc2 and to activate the kinase shortly before mitosis. The cdc2–cyclin A complex is probably required for progression through the cell cycle in late G2 (53). Cyclin A also associates with a cdc2-like protein (cdk2), and this complex has been shown to possess kinase activity throughout S phase and to be required for progression through S phase (53, 75, 76). Other members of the cyclin family (cyclins C–G) are thought to play a role in earlier phases of the cell cycle (4) (see also Chapter 6).

Cyclins contain at least two functional domains (Fig. 3). One is the **cyclin box**, a conserved stretch of about 100 amino acids, which also define the subfamily to which each cyclin belongs. The cyclin box mediates the binding and the activation of the associated catalytic subunit, and accounts for the specificity of different cyclin–kinase complexes (77, 78). The second domain in A- and B-type cyclins (but not in cyclins acting in earlier phases of the cell cycle) is the **destruction box**, located at the N-terminus of the molecule. This region contains a small cluster of conserved residues which is followed by a lysine-rich stretch. It has been demonstrated that B-type cyclins are degraded by ubiquitin-dependent proteases (79). The destruction box probably contains a recognition site for a ubiquitin-conjugating enzyme, while the lysine-rich stretch contains putative ubiquitination sites (ubiquitination occurs at lysine residues). Fusion of the region containing the destruction box and the adjacent lysine-rich stretch to a bacterial protein is sufficient to confer cell cycle-dependent proteolysis to the fusion product in a cell-free system (79). In agreement with these findings, expression of a cyclin B mutant deleted of this region, or containing a single amino acid substitution in the first position of the destruction box (R to C) causes a mitotic arrest in *Xenopus* egg extracts and in human cells (45, 79, 80). Interestingly, the destruction box is very similar between B- and A-type cyclins except for the residue in position 6, which in B-type cyclins is charged or polar, but in A-type cyclins is non-polar (79). Since cyclin A is degraded at the end of the G2 phase, while cyclin B is degraded at the metaphase/anaphase transition, this difference in the destruction box might account for the different timing of cyclin degradation. It is likely that other domains at the C-terminus of cyclins acting in early phases of the cell cycle are responsible for the cell cycle-dependent degradation of these proteins. Data in support of this idea

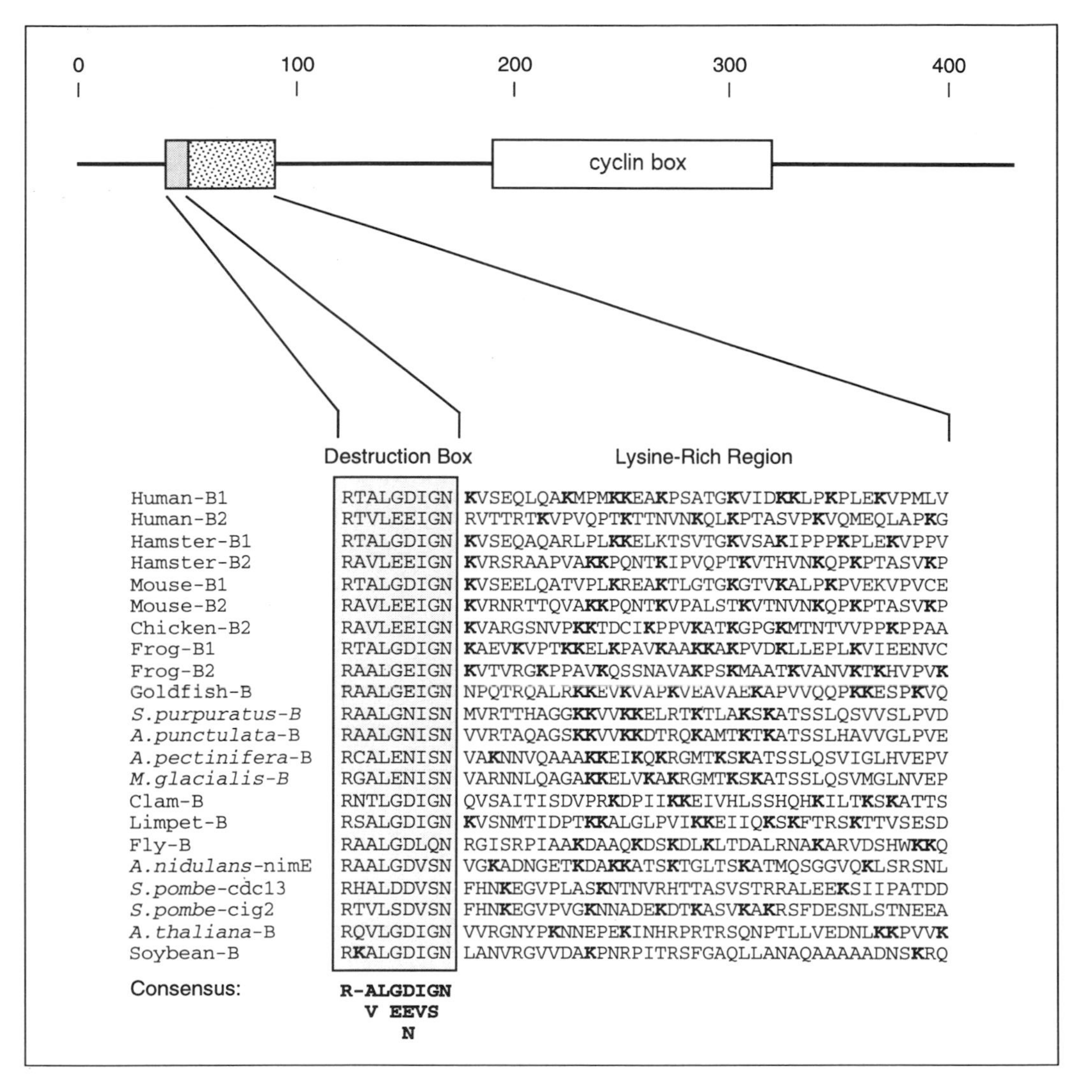

Fig. 3 The N-terminal region of B-type cyclins contains two domains important for ubiquitination and cyclin degradation. A schematic representation of the functional domains of B-type cyclins is shown at the top. Numbers are indicative for the location of the domains in the amino acid sequence of human cyclin B1. At the bottom, the region required for cyclin degradation is shown for B-type cyclins of different species. In this region, two adjacent domains are present: the destruction box and a lysine-rich region. The first domain consists of nine amino acids which are relatively conserved in all B-type cyclins, while the second region is not conserved at the amino acid level and is characterized by the high content of lysine residues (in boldface). The destruction box probably contains a recognition sequence for a ubiquitin-conjugating enzyme, while the lysine-rich stretch contains several putative ubiquitination sites.

come from the yeast *S. cerevisiae,* in which the C-terminal region of CLN3 (a G1 cyclin) has been shown to be required for the cell cycle-dependent degradation of the protein (81).

Cyclins are thought to play multiple roles: they are not only required for the cyclic activation of the associated catalytic subunit, but they are probably also important in determining the specific subcellular localization of the associated kinase during the cell cycle. It is thought that each cyclin can target the associated catalytic subunit to specific subcellular compartments, where the function of the kinase is required. This hypothesis is mainly supported by immunofluorescence studies. In mammals, the localization of cyclin B is cytoplasmic during interphase and becomes nuclear at the onset of mitosis. In contrast, cyclin A is nuclear throughout S and G2 (82, 53) and specifically localizes at DNA replication sites during S phase (83). During interphase cyclin B is mainly located in the perinuclear region and is associated with the centrosomes, while during metaphase it is associated with the chromosomes and the mitotic spindle (82).

Another function of cyclins is probably to cooperate with the catalytic subunit in substrate recognition. This is suggested by two *in vitro* findings:

(a) the cdc2–cyclin B complex can efficiently phosphorylate human cdc25 C protein, while the cdc2–cyclin A complex cannot (84);

(b) the cyclin A–cdc2 (or cdk2) complex can phosphorylate the retinoblastoma-related p107 protein, while the cyclin B–cdc2 (or cdk2) complex cannot (85).

Finally, it is interesting to note that in *S. cerevisiae*, several B-type cyclins have been identified which have a partially redundant function at mitosis (86). In human, two B-type cyclins are known (87). Whether the two human B-type cyclins are functionally distinct has not yet been established.

2.2.3 suc1

Another protein which interacts with cdc2 and is thought to regulate MPF activity is suc1. The *suc1* gene was first identified in *S. pombe* by its ability to rescue the temperature-sensitive defect of yeast strains carrying certain mutant alleles of *cdc2*, when overexpressed on a multicopy plasmid (**suc1** stands for **su**ppressor of **c**ell cycle block) (22). Subsequently, homologous genes have been isolated in *S. cerevisiae* (88) and in higher eukaryotes (89). The degree of evolutionary conservation between the yeast and the human suc1 homologues is quite high (more than 60 per cent identity). The suc1 protein does not share significant sequence similarity to other proteins in database searches.

suc1 has been shown to associate with a small fraction of the cdc2 protein both in *S. pombe* and in human (24, 48). In *S. pombe*, the affinity of suc1 coupled to Sepharose beads for cdc2 is higher in G2/M than in earlier phases of the cell cycle, indicating that suc1 may specifically interact with the mitotic form of the cdc2 kinase (90).

Until now the role of suc1 in the regulation of the cdc2 kinase at mitosis has been poorly understood. *In vitro*, unlike the small inhibitor $p21^{CIP1/WAF1/SDI1}$ (see Section 2.2.4), suc1 and its human homologues have no inhibitory effect on the active cdc2–cyclin B complex (26, 25, 91). Addition of bacterially expressed suc1 to *Xenopus* oocyte extracts blocks entry into M phase and tyrosine dephosphorylation of cdc2 (60). Overexpression of suc1 in *S. pombe* causes a delay in cell cycle progression during the G2 phase (92). These studies have suggested a role for suc1 in inhibiting cdc2 tyrosine dephosphorylation and mitotic initiation. Experiments of loss of suc1 function in fission yeast indicate, however, that at physiological concentrations suc1 does not inhibit the entry into mitosis (93). Microinjection of bacterially produced suc1 or anti-suc1 antibodies in human cells inhibits cell division and causes formation of multiple micronuclei (50). This finding is quite difficult to interpret. In *S. pombe*, high-level *suc1* expression under the control of an inducible promoter causes a block in G2, without mitotic abnormalities (93). Loss of suc1 function, in contrast, causes cell cycle block in mitosis with a G2 DNA content, condensed chromosomes, mitotic spindles, 15-fold increase in the activity of the mitotic cdc2 kinase, and high levels of cdc13 protein (26, 183). These findings indicate that in *S. pombe* the suc1 function is required for progression through mitosis (183). In contrast, in *S. cerevisiae* loss of suc1 function seems to block progression through G1/S and G2/M (94).

In human, not only cdc2, but also cdk2 and cdk3, have conserved the ability to bind suc1 *in vitro* (95). Two human homologues of the *suc1* gene, ckshs1 and ckshs2, have been identified (89). The structure of ckshs2 has recently been solved, and it has been proposed that ckshs2 exists in three different forms: monomeric, dimeric, and hexameric (96). Multimerization of ckshs2 is influenced, *in vitro*, by calcium or pH changes, and it has been suggested that it could play an important role in the assembly of higher order complexes containing multiple cdc2–cyclin heterodimers (96).

2.2.4 PCNA and p21

All the members of the cdk family associate with cyclins. In addition, several cdk–cyclin complexes are found associated with PCNA and with small inhibitor peptides, which range in molecular size between 14 and 27 kDa (49, 97, 185). In mammalian cells, the cdc2–cyclin B complex is found associated with PCNA and with a protein of 21 kDa (p21). This association is absent in transformed cells (49). PCNA is a cofactor of DNA polymerase δ, and it is required during DNA replication and repair. p21 was cloned almost simultaneously by different groups, which named the gene according to the approach they used to clone it. Investigators who were screening for cdk2 interacting proteins named the gene CIP1 (cdk2 interacting protein-1) (98), those who were screening for genes induced by p53 chose the name WAF1 (wild-type p53-activated fragment-1) (99), and others who were looking for genes specifically expressed in senescent cells called the gene SDI1 (senescent cell-derived inhibitor-1) (100). The gene was also cloned by a fourth group, which microsequenced a 21 kDa protein that co-immunoprecipitated with cyclin D (101). *In vitro*, $p21^{CIP1/WAF1/SDI1}$ inhibited the catalytic activity of all the cdk–cyclin complexes tested (98, 101). This inhibition was also observed when p21 was added together with Ckshs1 (a human suc1 homologue), suggesting that the two proteins do not compete for binding to cdc2. It is important to note that, *in vitro*, the cdc2–cyclin B complex is inhibited by p21 less effectively than the cdk2–cyclin A complex. This suggests that p21 might not play a primary role in the regulation of MPF activity *in vivo*. The presence of PCNA in the cdc2–cyclin B complex is not well understood. It has been suggested that PCNA could function as a signalling molecule activated in the presence of DNA damage or other signals. The fact that this complex does not exist in transformed cells, in which the basic cell cycle still functions correctly, indicates that the association between PCNA, p21, and cdc2–cyclin B might be important in the process of cellular transformation, but that is not essential for the regulation of MPF.

2.3 Upstream regulators of the cdc2 function: wee1 and cdc25

Regulation of cdc2 tyrosine phosphorylation is probably the best understood mechanism of mitotic control. The phosphorylation state of Tyr15, regulated by the balance between kinases and phosphatases, is critical in determining the timing of mitosis in *S. pombe* and in higher eukaryotes (see also Section 2.2.1).

The *wee1* gene encodes a protein kinase which phosphorylates the Tyr15 residue of cdc2. The properties of *wee1* have been extensively studied in *S. pombe*, in which the gene was first isolated (see Chapter 5). A human homologue of the *wee1* gene has also been cloned (102). Genetic studies established that in *S. pombe* wee1 is a dose-dependent inhibitor of mitosis operating upstream of cdc2 (17, 21). In agreement with this finding, overexpression of the human wee1 homologue in HeLa cells inhibits cell division (103). Wee1 displays sequence homology with the family of Ser/Thr protein kinases. However, *in vivo* and *in vitro* experiments have proven that the wee1 kinase is capable of auto-phosphorylation and of phosphorylating exogenous substrates on tyrosine residues (104). Using a baculovirus expression system, it was demonstrated that co-expression of wee1 and cdc2 results in the phosphorylation of cdc2 on Tyr15. Although the wee1 kinase has been described as a kinase with dual specificity (for Ser/Thr and for Tyr), apparently it cannot phosphorylate the Thr14 residue of cdc2 (103, 105). This implies that another, as yet unidentified, inhibitory kinase is responsible for the Thr14 phosphorylation. The ability of the wee1 kinase to phosphorylate cdc2 in baculovirus-infected cells is highly increased by co-expression of cyclin A or B and, indeed, the purified wee1 kinase phosphorylates with much greater affinity cdc2 associated with cyclins than monomeric cdc2 (106–108). This is consistent with the finding that in crude extracts (109) or in intact cells (107), cdc2 is tyrosine-phosphorylated only when associated with cyclins.

In *S. pombe* the *wee1* gene is not essential. An additional gene, *mik1*, has been isolated which is partially redundant to *wee1*, and is also involved in regulating the state of tyrosine-phosphorylation of cdc2 (110). *mik1* encodes a putative protein kinase which is about 50 per cent homologous to wee1. Similarly to wee1, mik1 can directly phosphorylate and inhibit the cdc2 kinase. It is likely that homologues of the mik1 kinase will soon be identified in higher eukaryotes.

In *Xenopus* extracts the kinase activity of wee1 is high in interphase and low at mitosis (94). These mitotic extracts contain a kinase which inhibits the activity of wee1 by phosphorylating the protein at its N-terminus. It has been suggested that such kinase could be encoded by *nim1*, a gene that was originally identified as a negative regulator of *wee1* in *S. pombe* (111). *nim1* encodes a protein kinase which can efficiently phosphorylate and inactivate the wee1 kinase *in vitro* (112). However, nim1 phosphorylates wee1 at the C-terminus but not at the N-terminus, indicating that nim1 is not the kinase that phosphorylates wee1 in *Xenopus* extracts (94). Depletion of the cdc2–cyclin B complex from a *Xenopus* mitotic extract can prevent phosphorylation and inactivation of wee1 in the absence but not in the presence of okadaic acid (a phosphatase inhibitor) (94). This suggests the presence of a phosphatase, probably a type-2A phosphatase (see below), which keeps wee1 in a dephosphorylated and active form (Fig. 4). At mitosis, such a phosphatase would be in turn inactivated, directly or indirectly, by the cdc2–cyclin B kinase. Okadaic acid can inhibit both type-2A and type-1 phosphatases. The suggestion that the phosphatase inhibiting wee1 might be a type-2A phosphatase is indirect and comes from experiments which have shown that type-2A but not

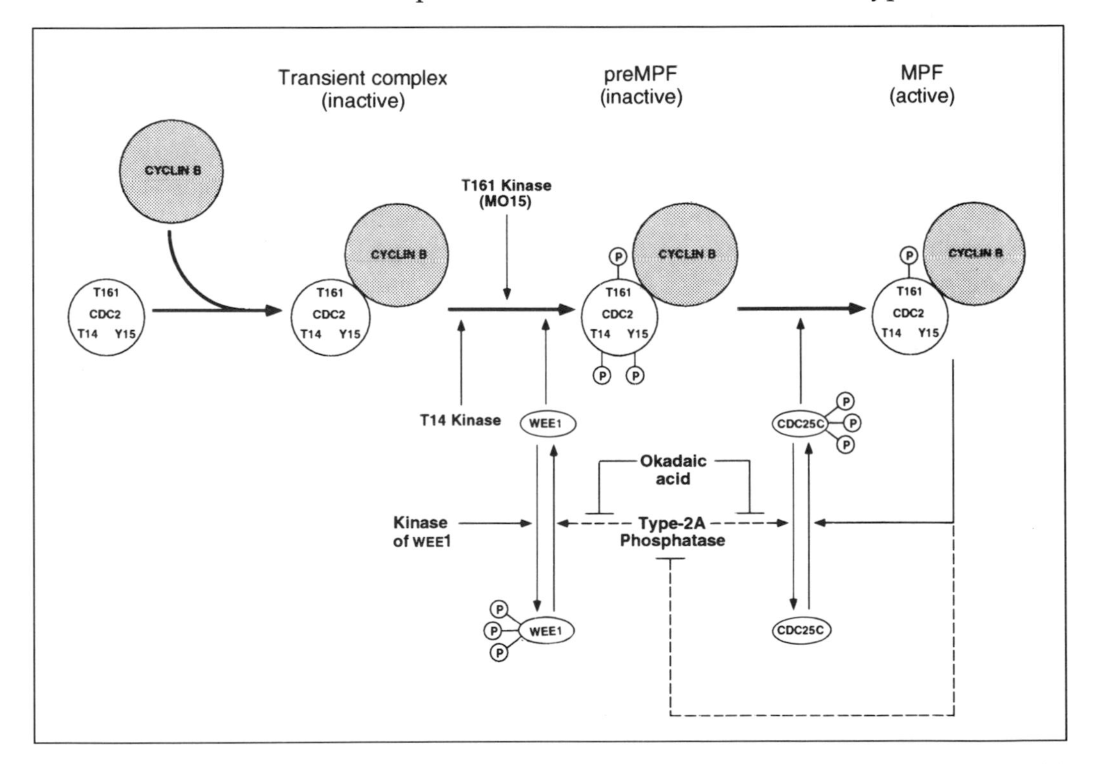

Fig. 4 Regulation of cdc25 and wee1 activity by phosphorylation. Cyclin B accumulates during S and G2 phases and associates with monomeric cdc2. As soon as cyclin B associates with cdc2, the complex is phosphorylated on Thr14, Tyr15, and Thr161. Pre-accumulation of inactive MPF (preMPF) during interphase allows the rapid activation of MPF at the onset of mitosis by the cdc25 phosphatase. Both wee1 and cdc25 are regulated by phosphorylation, the active form of wee1 being unphosphorylated and the active form of cdc25 hyperphosphorylated. The dashed lines indicate hypothetical regulatory pathways which might be direct or indirect. The role of type-2A phosphatase and of the cdc2–cyclin B complex in the regulation of wee1 and cdc25 is discussed in the text. (—|: inhibitory pathway. →: activatory pathway).

type-1 phosphatases inhibit the activation of the cdc2 kinase in *Xenopus* extracts (113–115). Additional evidence comes from *S. pombe*, where a null mutant strain for *ppa2* (a gene encoding a type-2A phosphatase) shows an advancement of mitosis (116), while overexpression of *ppa2* causes a delay in entry into mitosis (117). Taken together, these results indicate that a type-2A phosphatase negatively regulates the activation of cdc2 at mitosis, and suggest that dephosphorylation and activation of wee1 might be controlled by a type-2A phosphatase.

The *cdc25* gene encodes a phosphatase which catalyses the tyrosine dephosphorylation of cdc2 at the onset of M phase (see also Chapter 4). *cdc25* was first isolated in *S. pombe* (20) and subsequently, homologous genes have been cloned in other eukaryotes (118–123). In human, three different *cdc25* homologues, named respectively cdc25 A, B, and C, have been reported (124–126). All these genes encode functional homologues of cdc25, since they can suppress the temperature-sensitive defect of a *cdc25 S. pombe* mutant strain. Biochemical evidence has established that cdc25 is a phosphatase highly specific for cdc2, which can dephosphorylate cdc2 not only on Tyr15 (127, 128) but also on Thr14 (129). A second tyrosine phosphatase has also been described (pyp3) in fission yeast, which has a partially redundant function to cdc25 and is also involved in cdc2 tyrosine dephosphorylation (130) (see Chapter 4).

The different forms of cdc25 (A, B, and C) are activated at specific phases of the cell cycle by phosphorylation. In human, cdc25 A is phosphorylated and activated in S phase (131), while cdc25 C is phosphorylated and activated at mitosis (84). Purified cdc2–cyclin B kinase from human cells can phosphorylate cdc25 C at the same sites that are phosphorylated *in vivo*, and induce a 15-fold increase of cdc25 activity when preMPF is used as a substrate (84) (Fig. 4). Phosphorylation and activation of cdc25 C at mitosis has also been observed in *Xenopus* egg extracts (121, 122). In *Xenopus*, dephosphorylation and inactivation of cdc25 C can be observed after addition of phosphorylated cdc25 to an interphase extract (84). This event can be inhibited with okadaic acid but not with a specific inhibitor of type-1 phosphatases, indicating that dephosphorylation of cdc25 is probably due to a type-2A phosphatase (115) (Fig. 4). Phosphorylation of cdc25 C is an important regulatory event which is part of the auto-amplificatory loop required for the rapid activation of preMPF at mitosis (see Section 4.2). Consistent with a role for cdc25 C at mitosis, microinjection of anti-cdc25 C antibodies blocks entry into mitosis in human cells (132). Immunofluorescence studies have shown that a fraction of cdc25 C is nuclear (132). For this reason it has been proposed that activation of preMPF to MPF occurs in the nucleus and not in the cytoplasm.

3. The cdc2 kinase: structure–function relationship and substrate specificity

The protein kinase subunit encoded by the *cdc2* gene is the catalytic core of MPF. The activity of the cdc2 kinase is regulated by phosphorylation events and by its association with regulatory subunits. Mutagenesis analysis and three-dimensional modelling of cdc2 (based on the crystal structure of the cyclic AMP-dependent protein kinase) have allowed the identification of domains which are important for its interactions with the regulatory subunits (67, 133, 134). The crystal structure of cdk2, a member of the cdk family closely related to cdc2, has also been determined and has revealed some structural features which are specific to the cdk family (135). Information on the molecular anatomy of cdc2, and of the other proteins which regulate and interact with cdc2, will lead to a better understanding of the regulatory mechanisms which control mitosis and the cell cycle.

Additional information about the regulation of M phase is derived from an analysis of the substrates for the mitotic cdc2 kinase. Some of the potential substrates which have been identified suggest that the cdc2 kinase might directly trigger some relevant mitotic events.

3.1 Conserved residues in the cdc2 kinase family, three-dimensional modelling of cdc2, and structure of cdk2

Functional homologues of cdc2 have been isolated in many different eukaryotes. Figure 5 shows a sequence comparison of cdc2 homologues from ten different species. The degree of homology ranges between 58 per cent (between Sccdc28 and Hscdc2) and 97 per cent (between murine cdc2 and Hscdc2).

The structure of the catalytic core of cdc2, and in general of all protein kinases, consists of two lobes (Figs 6 and 7) (134–136): a small one, derived from the N-terminus, and a larger one, derived from the central and the C-terminal part of the molecule. In cdc2 the link between the two lobes is provided by amino acids 80–85. The ATP binding domain is located in the cleft between the two lobes. Most of the residues which bind ATP are situated in the small lobe. The bigger lobe is involved in the binding of the γ-phosphate of ATP, in catalysis (transfer of the γ-phosphate to the P site of the substrate), and also contains the substrate binding domain. The structure of cdk2 (135) is similar to the general structure of cAPK, but shows some important differences. One major difference is in the conformation of the ATP binding pocket. The alignment between the β- and γ-phosphate of ATP in cdk2 is almost perpendicular to the alignment in cAPK. Both kinases have been crystallized in a complex with Mg^{2+}ATP. However, cdk2 has been crystallized in an inactive conformation, while cAPK has been crystallized in its active form. The different conformation in the phosphates groups of ATP would explain the lack of activity in cdk2.

The cyclin binding region, located in the small lobe, includes the loop L4 and the α-helix 1 (containing a conserved stretch of amino acids called PSTAIR motif), and the N-terminal part of the β-strands 1 and 3. Since cyclins bind to a region which is close to the substrate recognition domain, it seems likely that their binding might play a role in the recognition of the substrate (see also Section 2.2.2).

The suc1 binding region is distributed between the two halves of the molecule.

```
Hscdc2     MEDYTKIEKIGEGTYGVVYKGRHKTTGQ---VVAMKKIRLESEEEGVPSTAIREIS   53
Ggcdc2     MEDYTKIEKIGEGTYGVVYKGRHKTTGQ---VVAMKKIRLESEEEGVPSTAIREIS   53
Xlcdc2     MDEYTKIEKIGEGTYGVVYKGRHKATGQ---VVAMKKIRLENEEEGVPSTAIREIS   53
Dmcdc2     MEDFEKIEKIGEGTYGVVYKGRNRLTGQ---IVAMKKIRLESDDEGVPSTAIREIS   53
Spcdc2     MENYQKVEKIGEGTYGVVYKARHKLSGR---IVAMKKIRLEDESEGVPSTAIREIS   53
Sccdc2  MSGLANYKRLEKVGEGTYGVVYKALDLRPGQGQRVVALKKIRLESEDEGVPSTAIREIS   60
Mscdc2     ---GENVEKIGEGTYGVVYKARDRVTNE---TIALKKIRLEQEDEGVPSTAIREIS   50
Atcdc2     MDQYEKVEKIGEGTYGVVYKARDKVTNE---TIALKKIRLEQEDEGVPSTAIREIS   53
Zmcdc2     MEQYEKVEKIGEGTYGVVYKALDKATNE---TIALKKIRLEQEDEGVPSTAIREIS   53
Oscdc2     MEQYEKEEKIGEGTYGVVYRARDKVTNE---TIALKKIRLEQEDEGVPSTAIREIS   53

Hscdc2  LLKEL----RHPNIVSLQDVLMQDS-RLYLIFEFLSMDLKKYLD--SIPPGQYMDSSLVK  106
Ggcdc2  LLKEL----HHPNIVCLQDVLMQDA-RLYLIFEFLSMDLKKYLD--TIPSGQYLDRSRVK  106
Xlcdc2  LLKEL----QHPNIVCLLDVLMQDS-RLYLIFEFLSMDVKKYLD--SIPSGQYIDTMLVK  106
Dmcdc2  LLKEL----KHENIVCLEDVLMEEN-RIYLIFEFLSMDLKKYMD--SLPVDKHMESELVR  106
Spcdc2  LLKEVNDENNRSNCVRLLDILHAES-KLYLVFEFLDMDLKKYMDRISETGATSLDPRLVQ  112
Sccdc2  LLKEL----KDDNIVRLYDIVHSDAHKLYLVFEFLDLDLKRYME--GIPKDQPLGADIVK  114
Mscdc2  LLKEM----QHRNIVRLQDVVHSDK-RLYLVFEYLDLDLKKHMD--SSP-EFIKDPRQVK  102
Atcdc2  LLKEM----QHSNIVKLQDVVHSEK-RLYLVFEYLDLDLKKHMD--STP-DFSKDLHMIK  105
Zmcdc2  LLKEM----NHGNIVRLHDVVHSEK-RIYLVFEYLDLDLKKFMD--SCP-EFAKNPTLIK  105
Oscdc2  LLKEM----HHGNIVRLHDVIHSEK-RIYLVFEYLDLDLKKFMD--SCP-EFAKNPTLIK  105

Hscdc2  SYLYQILQGIVFCHSRRVLHRDLKPQNLLIDDK-GTIKLADFGLARAFGIPIRVYTHEVV  165
Ggcdc2  SYLYQILQGIVFCHSRRVLHRDLKPQNLLIDDK-GVIKLADFGLARAFGIPVRVYTHEVV  165
Xlcdc2  SYLYQILQGIVFCHSRGVLHRDLKPQNLLIDNK-GVIKLADFGLARAFGIPVRVYTHEVV  165
Dmcdc2  SYLYQITSAILFCHRRRVLHRDLKPQNLLID-KSGLIKVADFGLGRSFGIPVRIYTHEIV  165
Spcdc2  KFTYQLVNGVNFCHSRRIIHRDLKPQNLLID-KEGNLKLADFGLARSFGVPLRNYTHEIV  171
Sccdc2  KFMMQLCKGIAYCHSHRILHRDLKPQNLLIN-KDGNLKLGDFGLARAFGVPLRAYTHEIV  173
Mscdc2  MFLYQMLCGIAYCHSHRVLHRDLKPQNLLIDRRTNSLKLADFGLARAFGIPVRTFTHEVV  162
Atcdc2  TYLYQILRGIAYCHSHRVLHRDLKPQNLLIDRRTNSLKLADFGLARAFGIPVRTFTHEVV  165
Zmcdc2  SYLYQILHGVAYCHSHRVLHRDLKPQNLLIDRRTNALKLADFGLARAFGIPVRTFTHEVV  165
Oscdc2  SYLYQILRGVAYCHSHRVLHRDLKPQNLLIDRRTNALKLADFGLARAFGIPVRTFTHEVV  165

Hscdc2  TLWYRSPEVLLGSARYSTPVDIWSIGTIFAELATKKPLFHGDSEIDQLFRIFRALGTPNN  225
Ggcdc2  TLWYRSPEVLLGSALYSTPVDIWSIGTIFAELATKKPLFHGDSEIDQLFRIFRALGTPNN  225
Xlcdc2  TLWYRAPEVLLGSVRYSTPVDVWSVGTIFAEIATKKPLFHGDSEIDQLFRIFRSLGTPNN  225
Dmcdc2  TLWYRAPEVLLGSPRYSCPVDIWSIGCIFAEMATRKPLFQGDSEIDQLFRMFRILKTPTE  225
Spcdc2  TLWYRAPEVLLGSRHYSTGVDIWSVGCIFAEMIRRSPLFPGDSEIDEIFKIFQVLGTPNE  231
Sccdc2  TLWYRAPEVLLGGKQYSTGVDTWSIGCIFAEMCNRKPIFSGDSEIDQIFKIFRVLGTPNE  233
Mscdc2  TLWYRAPEILLGSRHYSTPVDVWSVGCIFAEMANRRPLSPGDSEIDELFKIFRILGTPNE  222
Atcdc2  TLWYRAPEILLGSHHYSTPVDIWSVGCIFAEMISQKPLFPGDSEIDQLFKIFRIMGTPYE  225
Zmcdc2  TLWYRAPEILLGARQYSTPVDVWSVGCIFAEMVNQKPLFPGDSEIDELFKIFRILGTPNE  225
Oscdc2  TLWYRAPEILLGSRQYSTPVDMWSVGCIFAEMVNQKPLFPGDSEIDELFKIFRVLGTPNE  225

Hscdc2  EVWPEVESLQDYKNTFPKWKPGSLASHVKNLDENGLDLLSKMLIYDPAKRISGKMALNHP  285
Ggcdc2  DVWPDVESLQDYKNTFPKWKPGSLGTHVQNLDEDGLDLLSKMLIYDPAKRISGKMALNHP  285
Xlcdc2  EVWPEVESLQDYKNTFPKWKGGSLSSNVKNIDEDGLDLLSKMLVYDPAKRISARKAMLHP  285
Dmcdc2  DIWPGVTSLPDYKNTFPCWSTNQLTNQLKNLDANGIDLIQKMLIYDPVHRISAKDILEHP  285
Spcdc2  EVWPGVTLLQDYKSTFPRWKRMDLHKVVPNGEEDAIELLSAMLVYDPAHRISAKRALQQN  291
Sccdc2  AIWPDIVYLPDFKPSFPQWRRKDLSQVVPSLDPRGIDLLDKLLAYDPINRISARRAAIHP  293
Mscdc2  DTWPGVTSLPDFKSTFPRWPSKDLATVVPNLEPAGLDLLNSMLCLDPTKRITARSAVEHE  282
Atcdc2  DTWRGVTSLPDYKSAFPKWKPTDLETFVPNLDPDGVDLLSKMLLMDPTKRINARAALEHE  285
Zmcdc2  QSWPGVSCLPDFKTAFPRWQAQDLATVVPNLDPAGLDLLSKMLRYEPSKRITARQALEHE  285
Oscdc2  QSWPGVSSLPDYKSAFPKWQAQDLATIVPTLDPAGLDLLSKMLRYEPNKRITARQALEHE  285

Hscdc2  YFNDLDNQIKKM                                                 297
Ggcdc2  YFDDLDKSTLPANLIKKF                                           303
Xlcdc2  YFDDLDKSSLPANQIRN                                            302
Dmcdc2  YFNGFQSGLVRN                                                 297
Spcdc2  YLRDFH                                                       297
Sccdc2  YFQES                                                        298
Mscdc2  YFKDIKFVP                                                    291
Atcdc2  YFKDLGGMP                                                    294
Zmcdc2  YFKDLEVVQ                                                    294
Oscdc2  YFKDLEMVQ                                                    294
```

Fig. 5 Conserved residues in the cdc2 kinase family. Sequence comparison between members of the cdc2 kinase family. Conserved residues, are boxed. Hs = *Homo sapiens*, Xl = *Xenopus laevis*, Gg = *Gallus gallus*, Dm = *Drosophila melanogaster*, Sp = *Schizosaccharomyces pombe*, Sc = *Saccharomyces cerevisiae*, Ms = *Medicago sativa* (partial clone), At = *Arabidopsis thaliana*, Zm = *Zea mays*, Os = *Oryza sativa*.

One of the two suc1 binding domains is formed by the α-helix 5, in the big lobe. The second binding domain is represented by the loop L5 and the β-strand 2, located in the small lobe. This model predicts that suc1 is a small elongated molecule which bridges the two lobes of the cdc2 kinase, and is consistent with the reported structure of ckshs2 (96).

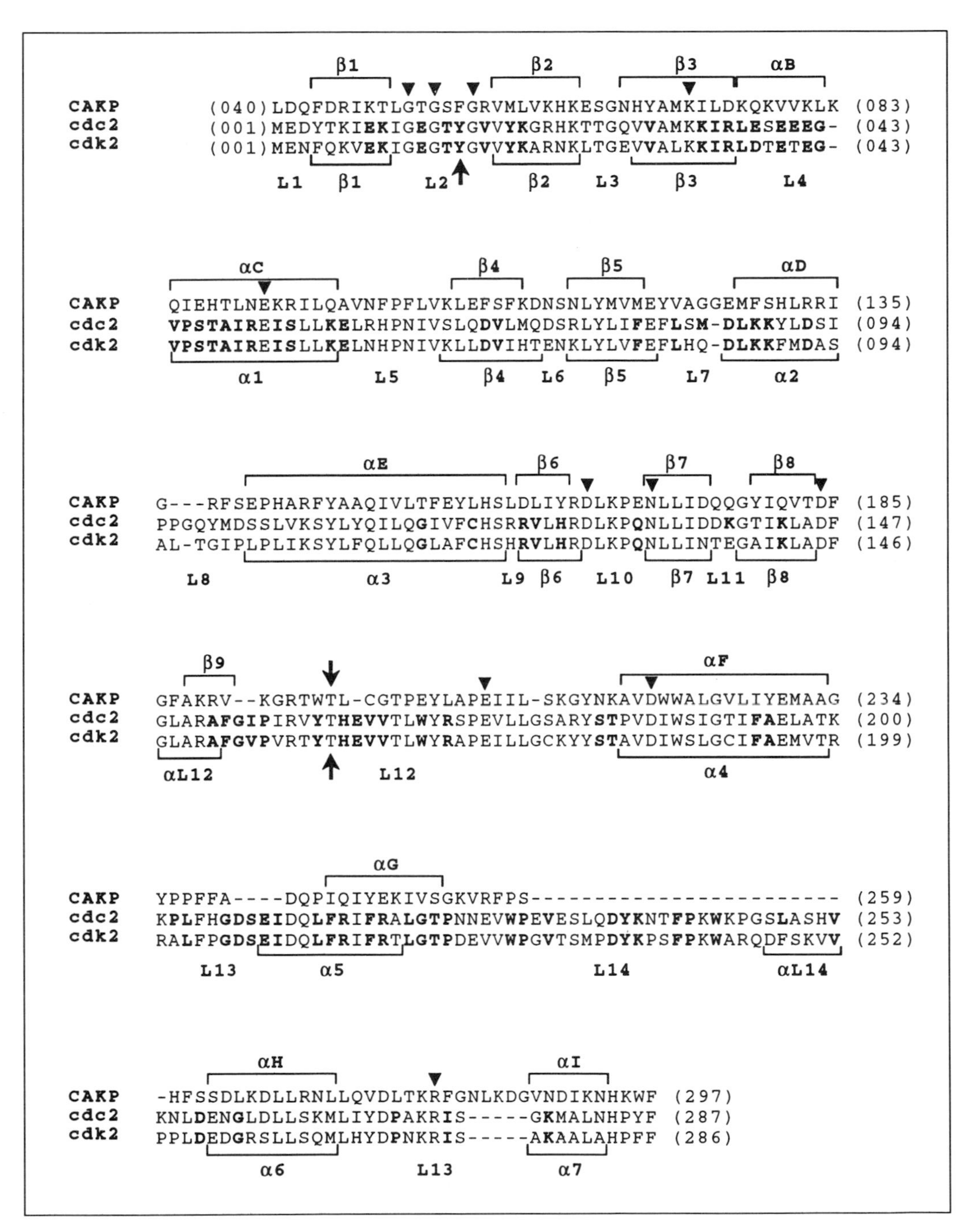

Fig. 6 Sequence alignment of hscdc2 with hscdk2 and with the catalytic core of the cyclic AMP-dependent protein kinase. The Tyr15 and T161 phosphorylation sites are indicated by arrows. Residues that are highly conserved throughout the protein kinase family are indicated by small triangles. Residues that lie in helices and beta-strands (from refs 134 and 135) are bracketed. Residues specific for the cdc2 kinase family, i.e. those which are conserved in the cdc2 kinase family (Fig. 5) but are not common to other serine/threonine kinases are bold-faced. Many of these residues are also conserved in cdk2.

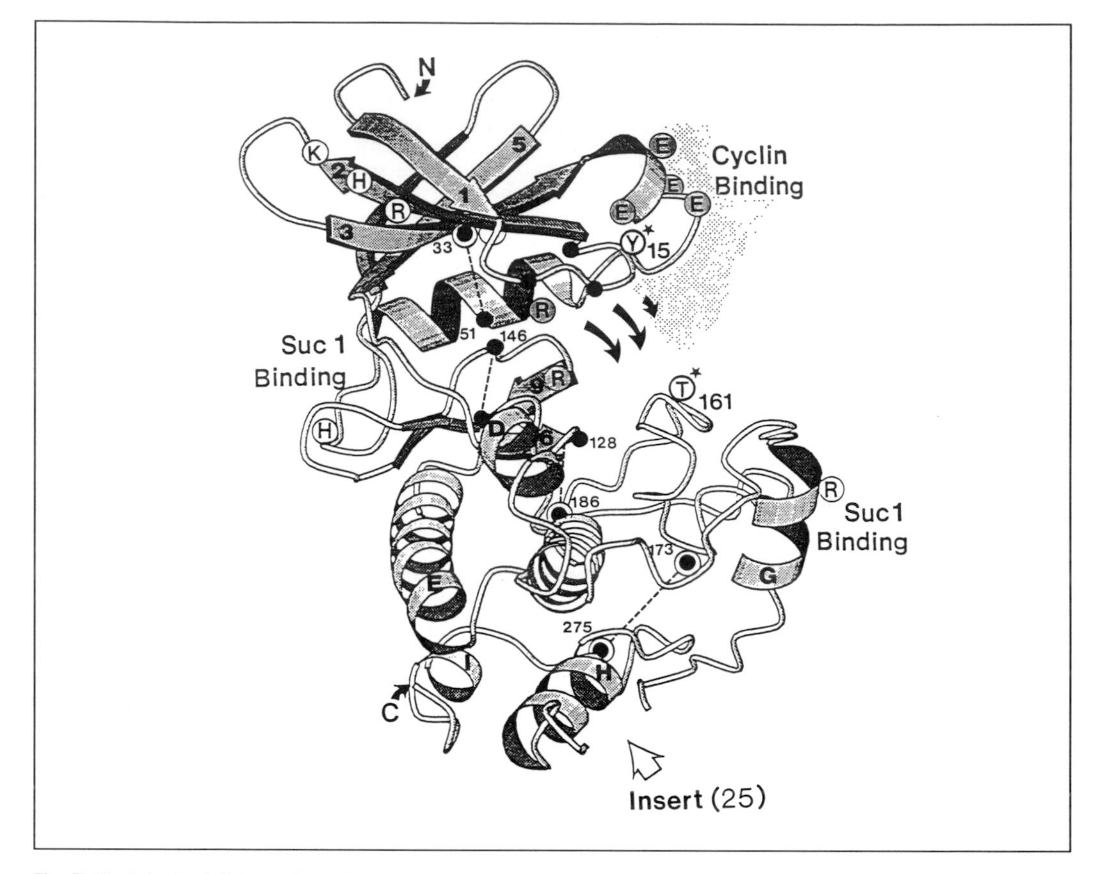

Fig. 7 Model of cdc2 based on the structure of the cAMP-dependent protein kinase. Ribbon structure of the cdc2 kinase. Essential residues that are conserved throughout the protein kinase family are indicated as solid circles, and those that interact are connected by dashed lines. Several of these essential residues are numbered. The site where a major insert occurs between residues 229 and 254 is shown as a gap and indicated by the arrow. The two major sites of phosphorylation, Tyr15 and Thr161, are designated with a star. Putative cyclin and suc1 binding sites are indicated. The arrows at the cleft interface indicate the general location of the 'PSTAIR' sequence that faces towards the surface containing Thr161. The N- and C-terminus are indicated as N and C. (Courtesy of Dr S. Taylor.)

Phosphorylation of cdc2 on Thr161 and binding to cyclins are both required for the activation of cdc2. The loop region L12, also known as T-loop, contains the Thr161 phosphorylation site. This region is located in the substrate binding cleft and it is thought to inhibit the binding of the substrate (135). As a result of the binding of cyclins, the T-loop would be removed from the cleft. Once the T-loop becomes exposed, Thr161 phosphorylation could take place and this would lead to an activation of the kinase. In agreement with this model, phosphorylation of cdc2 on Thr161 is cyclin-dependent *in vitro* (69). Another prediction of the model is that cyclin binding and phosphorylation of cdc2 on Thr161 would induce a re-orientation of the β- and γ-phosphates of ATP, such that phosphotransfer is allowed.

The Thr14, Tyr15, and Thr161 phosphorylation sites are near the cyclin binding domain. Phosphorylation of Thr14 is predicted to affect the orientation of the ATP γ-phosphate (135). Tyr15, on the other hand, is more distant from the ATP molecule and its phosphorylation does not seem to inhibit the catalytic activity of cdc2 by interfering with the binding of ATP (137). The phosphorylated Thr161 and another residue of the big lobe (Arg151) are thought to interact with cyclins and with the 'PSTAIR' region. This interaction is predicted to play an important role in the activation of the cdc2 kinase (134). Phosphorylation of Tyr15 would displace the 'PSTAIR' region preventing it from interacting with the Thr161 phosphate and the Arg151. This is one possible explanation for the inhibitory effect of Tyr15 phosphorylation.

3.2 Substrates of the mitotic cdc2 kinase

One of the most difficult tasks in the study of protein kinases is the identification of their physiological substrates. A genetic approach to the problem is lacking. Genetics can be very helpful in identifying the upstream regulators of a kinase, but it is of very limited use for the identification of downstream targets. On the other hand, a biochemical approach can be useful to find putative substrates of a kinase, but it does not help to establish which, among them, are the physiological substrates. Usually, if a substrate is phosphorylated *in vitro* by a kinase, the sites of phosphorylation *in vitro* are compared with the pattern of phosphorylation *in vivo*. If the *in vitro* and *in vivo* data are compatible, the substrate is considered a putative physiological target for that kinase. It is not unusual to find that two different kinases can phosphorylate the same substrate at identical sites. For example, Tyr15 of cdc2 can be phosphorylated by wee1, but also by the src kinase (58, 128). In this case how is it possible to discriminate the physiological kinase? Several criteria have been suggested, which may help to make a distinction (138). First, the kinase and the substrate should have a similar subcellular distribution. Second, phosphorylation of the substrate should induce a physiological event which is compatible with the expected function of the kinase. Third, if temperature- or cold-sensitive kinase mutant cell lines are available, in these cell lines phosphorylation of the substrate should be temperature- or cold-sensitive. Fourth, if the kinase is known to be cell cycle regulated, as it is for cdc2, the timing of activation of the kinase should be parallel to the timing of phosphorylation of the substrate. These criteria, however, are not always sufficient to identify with certainty a physiological substrate for cdc2. The substrates of the mitogen activated protein (MAP) kinases are phosphorylated at a consensus site (L/P-X-S/T-P), which overlaps with that of cdc2 (139). MAP kinases are activated during re-entry in the cell cycle at the G0/G1 transition, but also during M phase. Therefore, some of the mitotic substrates of cdc2 are potential targets for the MAP kinases, and indeed common substrates for cdc2 and MAP kinases have been identified (see for example ref. 140). The unambiguous identification of physiological substrates for cdc2 requires all these considerations to be taken into account.

Cdc2 is a serine/threonine protein kinase and its consensus sequence, derived

Table 1 Mitotic substrates of cdc2

Substrate	Phosphorylation *in vitro*	Corresponding sites are phosphorylated *in vivo*	Phosphorylation sites	References
Cytoskeletal proteins				
Lamins	+	+	LSPTR	(143–146)
			PSPTS	
Vimentin	+	+	SSPGG	(148)
Caldesmon	+	+	?	(149, 150)
LC-20 (Myosin Regulatory Chain)	+	+	(*SS*RKA)	(151)
MAP-p220	+	+	?	(140)
RMSA-1	+	?	?	(153)
Protein kinases				
p60^c-src^	+	+	OTPNK	(154, 156)
			RTPSR	
			TSPQR	
p150^c-abl^	+	+	DTPCI	(157, 158)
p85^gag-mos^	+	+	LSPSV	(159, 160)
CKII α-subunit	+	+	?	(161)
CKII β-subunit	+	+	KSPVK	(161, 162)
DNA-binding proteins				
Histone H1	+	+	K/R S/T P X K/R	(163, 164)
HMG I, Y, P1	+	+	PTPKR	(167, 168)
			TTPGB	
			PSPVK	
Nucleolin	+	+	ATPAKK	(170, 171)
Myb	+	+	?	(169)
Other proteins				
NO38	+	+	?	(170)
APP	+	+	VTPEE	(172)
rab4	+	+	RSPRB	(173, 174)
rab1	+	?	?	(173)
cyclin B	+	?	PSPVP	(175)
		?	TSPDV	
cdc25	+	+	?	(84, 122)

from studies using peptides substrates is S/T P X K/R (141, 142). This consensus matches the phosphorylation site(s) of most of the substrates identified so far (see Table 1). The requirement for a basic residue in position+3 is probably not essential, since it is not present in all putative substrates. The phosphorylation site of myosin-II regulatory light chain (SSRKA) represents an exception, in that the proline in position+1 is missing. It might be relevant to note, however, that in this case phosphorylation occurs at the N-terminus of the molecule. It has become evident that cyclins play an important role in the recognition of the substrates. Additional structural features (other than the consensus sequence) in the substrate might be essential in directing phosphorylation by a specific cdc2 complex (see Section 2.2.2).

Potential substrates of the cdc2 kinase at mitosis include cytoskeletal proteins, tyrosine and serine/threonine protein kinases, DNA-binding proteins, and a number of other proteins with different functions (Table 1).

Lamins are proteins which constitute the nuclear lamina (a multi-protein layer furnishing structural support to the inner nuclear membrane) and become phosphorylated at mitosis, when the lamina is disassembled. Lamina disassembly usually occurs prior to nuclear membrane breakdown. All three known types of lamins (A, B, and C) can be phosphorylated by the mitotic cdc2 kinase at sites which are important for lamina disassembly (143–145). Mutation of these sites confers resistance to lamina disassembly both *in vitro* and *in vivo*, indicating that phosphorylation of lamins by the cdc2 kinase triggers a relevant physiological event (146, 147).

Other candidate substrates, which may play an important role in the cytoskeletal rearrangements which take place at mitosis, are vimentin, a member of the cytoplasmic intermediate filaments family (148), and caldesmon, a regulator of actin filaments (149, 150).

The Myosin-II regulatory light chain (LC-20) is regulated by mitotic phosphorylation, and can be phosphorylated by the cdc2 kinase *in vitro* (151). During mitosis, a contractile ring of myosin-II and actin filaments is formed around the cell at the site where cytokinesis is going to occur. Phosphorylation of LC-20 by cdc2–cyclin B might inhibit the myosin ATPase activity essential for contraction at the cleavage ring, thereby delaying cytokinesis until MPF is inactivated.

Studies in *Xenopus* egg-extracts have indicated that the dynamics of the spindle microtubules can be regulated by the cdc2 kinase (152). A microtubule associated protein of 220 kDa (MAP-p220), which is specifically phosphorylated at mitosis, has been identified in *Xenopus* eggs (140). Unphosphorylated p220 isolated from interphase extracts binds to microtubules and can stimulate tubulin polymerization, an activity that is lost upon phosphorylation of p220 at mitosis. Therefore, phosphorylation of p220 might play an important role in the regulation of microtubule assembly. *In vitro*, both MPF and MAP kinase can phosphorylate p220 at sites which are phosphorylated *in vivo*. This is an example where it is difficult to distinguish which kinase is responsible for the phosphorylation of the substrate *in vivo*. The possibility exists that both kinases contribute to the phosphorylation of p220 at mitosis.

The chromosomal protein RMSA-1 (regulator of mitotic spindle assembly-1) has been shown to be required for the formation of the mitotic spindle (153). The protein is specifically phosphorylated in M phase, and is a substrate of cdc2 *in vitro*. It has been proposed that phosphorylation of RMSA-1 by cdc2 at mitosis plays an important role in the assembly of the spindle.

Another putative substrate of cdc2, which could have an important role for the reorganization of the cytoskeleton, is $p60^{src}$ (154, 155). Phosphorylation of $p60^{src}$ at mitosis coincides with an increase in its kinase activity. Mutation of the sites phosphorylated by cdc2 has been shown to cause a decrease in the mitotic activation of the $p60^{src}$ kinase (156).

$p150^{c\text{-}abl}$ contains a DNA binding domain in the carboxyl terminal part of the molecule, which is specifically phosphorylated at mitosis and which can be phosphorylated by cdc2 *in vitro* (157). *In vivo* or *in vitro* phosphorylation of this region does not affect the kinase activity of $p150^{c\text{-}abl}$, but abolishes its DNA-binding activity (158). Phosphorylation of $p150^{c\text{-}abl}$ by cdc2 might therefore play an important role in regulating the subcellular localization and the biological activity of the tyrosine kinase.

The $p85^{gag\text{-}mos}$ viral oncoprotein, which is activated at mitosis by phosphorylation, is also an *in vitro* substrate for cdc2 (159, 160). Phosphorylation of the α (catalytic) and β (regulatory) subunits of casein kinase-II at mitosis has also been reported. *In vitro*, both subunits are substrates for the cdc2 kinase (161, 162). However, M-phase specific phosphorylation does not correlate with any change in casein kinase-II activity. As for $p150^{c\text{-}abl}$, phosphorylation of the α and β casein kinase-II subunits might affect the subcellular distribution of the kinase rather than its biochemical activity.

Histone H1 is the substrate generally used for assaying the cdc2 kinase activity *in vitro*. M-phase specific phosphorylation of histone H1 *in vivo* has been known for a long time (31) and it is very likely to be catalysed by the cdc2 kinase (163, 164). So far, the functional significance of histone H1 phosphorylation at mitosis has not been determined. It has been thought to play a role in chromatin condensation by inducing changes in the nucleosome packing (31). The consensus motif for the cdc2 kinase is frequently found in nuclear proteins involved in transcriptional regulation, and the formation of a β-turn structure which can bind DNA directly has been proposed. Phosphorylation of this motif would induce a structural change and abolish the binding of the protein to DNA. The removal of transcription factors from DNA would then allow the chromatin to condense (165, 166). Although this idea is very speculative, there is experimental evidence that the cdc2 kinase can phosphorylate chromatin-associated proteins. Members of the HMG (high mobility group) non-histone proteins have been identified as potential substrates of the mitotic cdc2 kinase (167, 168). The product of the proto-oncogene c-Myb can be phosphorylated *in vitro* by cdc2 at sites which are specifically phosphorylated at mitosis (169). This phosphorylation has been shown to correlate with a decrease in DNA-binding affinity of c-Myb . Nucleolin, an abundant nucleolar protein which is known to bind ribosomal DNA, is also phosphorylated by the cdc2 kinase at specific mitotic sites (170, 171). Phosphorylation of this protein might play a role in disassembly of the nucleolus at mitosis.

Other potential substrates of cdc2 are the nucleolar protein NO38 (170), the amyloid precursor protein APP (involved in the etiology of Alzheimer's disease) (172), and the small GTP-binding proteins rab1 and rab4 (173, 174). It is not clear what the physiological consequences of the phosphorylation of NO38 and of APP by cdc2 are. rab1 and rab4 play a role in endomembrane traffic, and their phosphorylation by cdc2 might be related to the inhibition of vesicle fusion which is observed at mitosis.

Cyclin-B is also phosphorylated by the active cdc2 kinase (175). Although this phosphorylation almost certainly occurs *in vivo*, it probably does not reflect a regulatory event, since it is not required for cyclin degradation and it does not affect the activation of the cdc2 kinase at mitosis (175).

cdc2 is also a regulator of its own activity. The rapid activation and inactivation of MPF during mitosis is regulated by the existence of positive and negative feedback loops primed by the cdc2 kinase at the onset of M phase (see below). Phosphorylation of the cdc25C phosphatase by cdc2 is part of the positive feedback loop required for the full activation of cdc2 at entry into mitosis (see Section 4.2).

4. Feedback mechanisms regulating the timing of mitosis

4.1 Checkpoint controls: the dependence of mitosis on S phase

In eukaryotes the regular succession of different cell cycle phases is guaranteed by control mechanisms called checkpoints (176). Checkpoints are molecular controls which act at specific points of the cell cycle and do not allow entry into the subsequent phases of the cell cycle if the necessary preceding events have not been properly completed. Entry into mitosis is controlled by a checkpoint which monitors completion of S phase. If cells contain unreplicated or damaged DNA, entry into mitosis is delayed until DNA replication or repair is completed. Evidence in support of such a mechanism of control comes from the discovery of mutants or drugs which specifically abrogate the dependency of mitosis on completion of S phase (see Chapter 1). In mammals, for example, caffeine treatment of cells exposed to DNA synthesis inhibitors or to ionizing radiation induces entry into mitosis, but it has no observable effect on normal cycling cells (177). Very little is known about the molecular mechanisms which monitor the presence of unreplicated or damaged DNA, and how the signal is transmitted to the cell cycle control machinery. Experiments in *S. pombe* and in *Xenopus* extracts have suggested that the tyrosine-phosphorylation state of cdc2, which plays a major role in regulating the timing of mitosis, is regulated by the checkpoint coupling mitosis to the completion of DNA replication (178). In vertebrates, entry into mitosis is determined by the phosphorylation state of cdc2 on both Tyr15 and Thr14 (63). Similar to what is observed in *S. pombe* by overexpressing cdc25 (179), addition of bacterially-expressed cdc25 to a *Xenopus* extract can induce entry into mitosis in the presence of unreplicated DNA (180). The activity of the cdc25 C phosphatase in such an extract is low in S phase and elevated at mitosis (121), while the opposite is true for the activity of the wee1 kinase (94). The question that arises from these findings is what regulates the cdc25 C and wee1 activities? Phosphorylation events seem to play an important role in cell cycle regulation of both activities (see Section 2.3). There is evidence that the activity of the cdc25 C phosphatase is regulated by the cdc2–cyclin B kinase via a positive feedback loop started during G2/M (84).

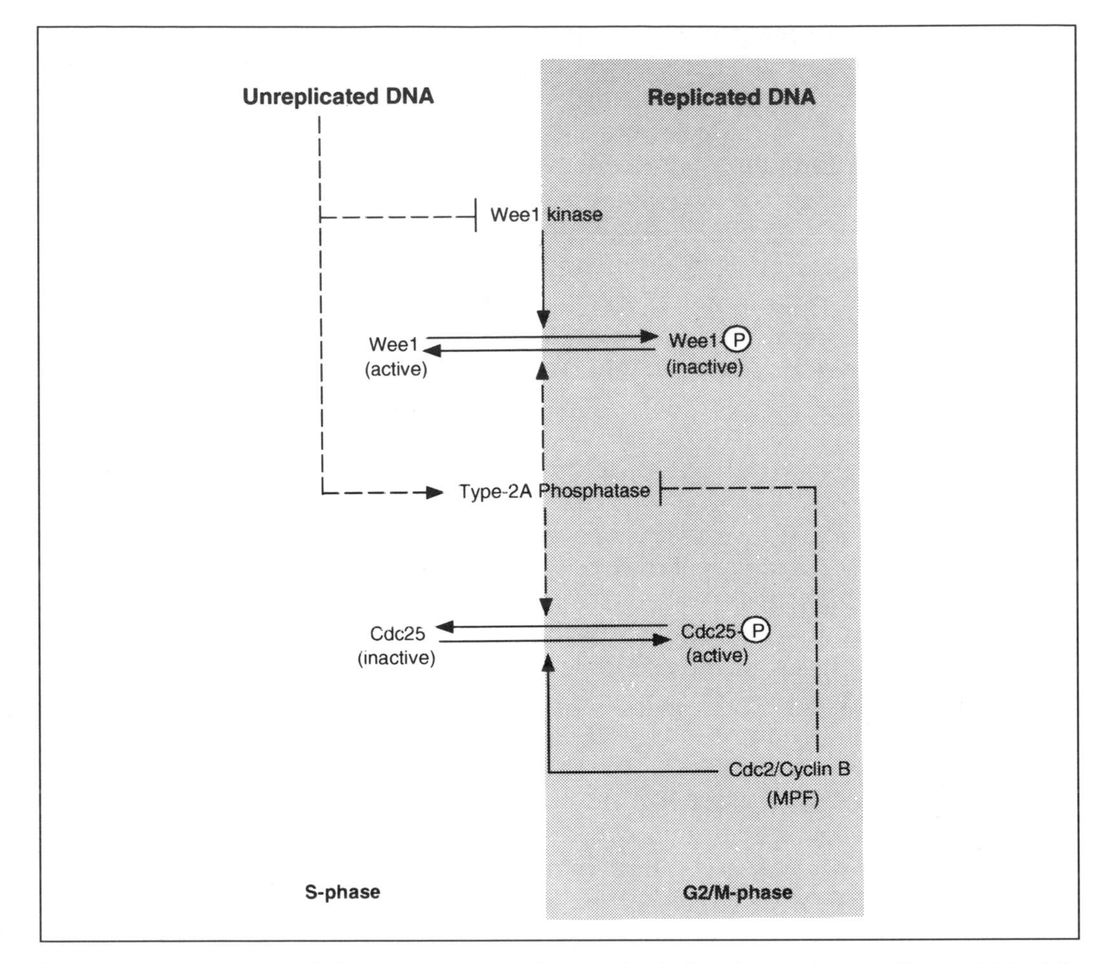

Fig. 8 Coupling of mitosis to DNA-replication: a working hypothesis. Putative regulatory pathways which might be important for the checkpoint control coupling M to S phase are shown. The dashed lines indicate that the regulatory feedback might be indirect, since the pathways are likely to involve additional control elements which have not yet been identified. (—|: inhibitory feedback. →: activatory feedback).

Phosphorylation and activation of cdc25 C are necessary but not sufficient for entry into mitosis. Active, phosphorylated cdc25 C cannot induce maturation when injected in *Xenopus* oocytes, owing to the high cdc25 C dephosphorylation rate. Thiophosphorylated cdc25 C, on the other hand, is resistant to phosphatase hydrolysis and can induce oocyte maturation under the same experimental conditions (84). These results suggest that cdc25 C dephosphorylation plays an important role in regulating the activity of this enzyme. It is proposed that after completion of DNA replication, cdc25 is activated, and that a negative feedback loop starts which inhibits the wee1 kinase (see Fig. 3). The primary event initiating such regulatory loops could be the inactivation of a type-2A phosphatase, which keeps the cdc25 phosphatase inactive and the wee1 kinase active during S phase. If such a hypothesis is correct, the scenario would be the following (Fig. 8). In a normal cell cycle, a type-2A phosphatase, regulating the cdc25 and the wee1 activities, is gradually inactivated during the G2 phase. Owing to the balance between the activity of kinases and phosphatases, concomitantly to the inhibition of the type-2A phosphatase, wee1 is inactivated and cdc25 becomes active. Activation of cdc25 C induces cdc2 tyrosine dephosphorylation, which brings about activation of preMPF and entry into mitosis. If DNA synthesis is inhibited, a feedback control mechanism blocks inactivation of the type-2A phosphatase until DNA replication is completed. Mechanisms other than a feedback control of the type-2A phosphatase are also possible. For instance, the checkpoint could control the activity of wee1 by regulating the kinase which phosphorylates and inactivates wee1 (Fig. 8). Alternatively, the checkpoint could operate through a mechanism independent of the regulation of cdc2 on tyrosine. It has been suggested, for example, that PCNA and $p21^{CIP1/WAF1/SDI1}$ could function as signalling molecules detecting DNA damage (49). However, in transformed cells where S and M phases are still coupled, PCNA and p21 are not associated with cdc2–cyclin B, which seems to exclude this hypothesis. Further experimental evidence will be required to establish which mechanism couples MPF activation to completion of S phase.

4.2 Feedback loops: the activation and inactivation of MPF

The activation of MPF at mitosis is an all-or-nothing event reminiscent of the changes in membrane potential (action potential) observed during the stimulation of a nerve axon. The action potential of a nerve axon is primed by a stimulus whose intensity reaches or exceeds a certain threshold level. Once the stimulus has reached the threshold, feedback loop mechanisms are activated that trigger the rapid increase and decrease of the signal. In a nerve axon, two major feedback loops (one positive and one negative) regulate the changes of membrane polarization produced after the stimulus (181). The similarity between the profile of MPF activity and the action potential of a nerve axon strongly suggests the existence of feedback loop mechanisms which regulate the activity of MPF at mitosis. Indeed, early studies in *Xenopus* oocytes suggested the existence of an auto-amplificatory mechanism of MPF activation, which was shown not to be dependent on protein synthesis (57, 13). The molecular mechanisms regulating the transient spike of MPF activation at mitosis have been understood only recently. Two feedback loops, one positive and one negative, have been shown to modulate MPF activity (Fig. 9). The positive feedback loop operates through the phosphorylation and activation of the cdc25 C phosphatase by the active cdc2–cyclin B complex (84) (see also Fig. 4). The negative feedback loop is induced by MPF through the activation of the cyclin degradation machinery (182). The molecular mechanism triggering the negative loop is unknown, but it is likely to involve the activation of ubiquitin-conjugating enzymes (79). In order to explain the kinetics of MPF activation, the positive loop must be faster than the negative one. This can be explained by the

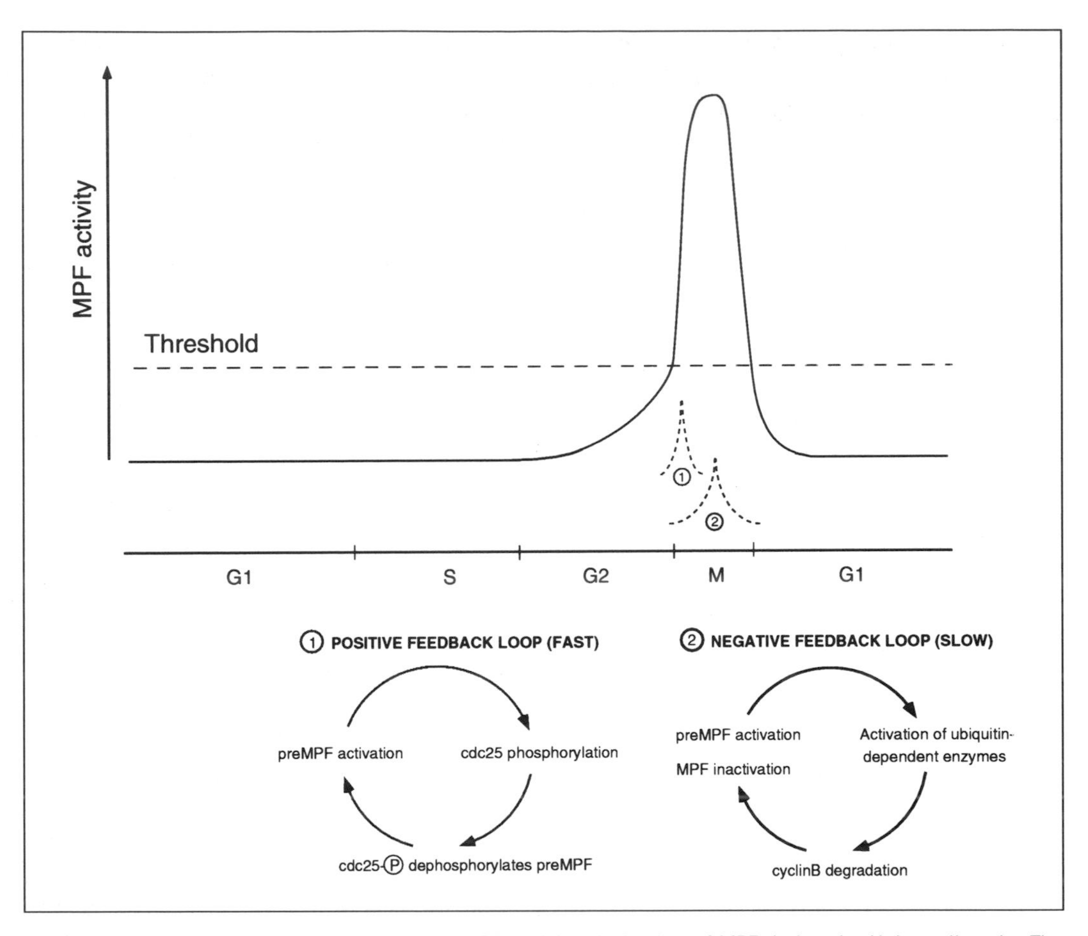

Fig. 9 Feedback loop mechanisms regulating MPF activity. Activation of MPF during the HeLa cell cycle. The relative timing of the different cell cycle phases is approximately to scale. Two feedback loops responsible for the rapid activation and inactivation of MPF are shown. (1) Positive feedback loop triggered by the phosphorylation of cdc25C (fast). (2) Negative feedback loop triggered by the activation of ubiquitin-conjugating enzymes (slow). The dashed lines indicate, respectively, cdc25C activity and cyclin B ubiquitination (not to scale). Threshold = level of MPF activity required for the entry into mitosis. MPF activity = histone H1 kinase activity of the cdc2–cyclin B complex during the HeLa cell cycle.

fact that ubiquitination and degradation of cyclins is a multi-step pathway, which is clearly more complex than the event of phosphorylation of cdc25. Other feedback loop mechanisms may also be important in the regulation of MPF; for example, inactivation of wee1 (see Fig. 4), and activation of the Thr161 phosphatase are probably also controlled by feedback loop mechanisms. However, the simplest model for MPF activation predicts that after completion of DNA replication, a progressive inactivation of wee1 and a concomitant activation of cdc25 would induce a gradual increase in the cdc2 kinase activity; once a certain threshold level has been reached, the cdc2 kinase would rapidly phosphorylate and activate the cdc25 phosphatase. The active cdc25 would then dephosphorylate inactive preMPF, inducing a positive feedback loop at the onset of mitosis (Fig. 9, raising phase). Concomitantly with the activation of the cdc25 phosphatase, a slow negative feedback loop is started. This loop, which involves activation of cyclin-ubiquitinating enzymes, would induce cyclin degradation and MPF inactivation, which are required for the exit from mitosis (Fig. 9, falling phase).

5. Conclusions and unanswered questions

Since the initial identification of a factor promoting M phase in *Physarum polycephalum* (1), enormous progress has been made in the characterization of the molecular components and regulators of MPF. The activity of the cdc2 kinase, the catalytic subunit of MPF, is regulated at mitosis by its association with B-type cyclins, and by multiple events of phosphorylation and dephosphorylation on conserved residues. In addition to cyclin binding, phosphorylation of cdc2 on Thr161 is required for activation of the kinase. Entry into mitosis is triggered by dephosphorylation of cdc2 on Thr14 and Tyr15. The phosphorylation state of the Thr14 and Tyr15 residues is tightly regulated, and probably plays a role in coupling entry into mitosis to completion of S phase.

The finding that in non-transformed cells, the inhibitor protein $p21^{CIP1/WAF1/SDI1}$ is associated with the cdc2–cyclin B complex suggests the existence of an additional level of control which is lost during cellular transformation. Since $p21^{CIP1/WAF1/SDI1}$ is also found associated with other cdk–cyclin complexes, it remains to be clarified whether *in vivo* $p21^{CIP1/WAF1/SDI1}$ acts as a general inhibitor of cyclin-dependent kinases or whether it regulates specific cdk–cyclin complexes preferentially.

The role played by cyclins in targeting the associated kinase to the proper subcellular localization and in the recognition of the substrate is not fully understood. Two domains have been identified in cyclins: the cyclin box and the destruction box. The cyclin box is known to be necessary and sufficient for the activation of the cdc2 kinase. Although A- and B-type cyclins have a similar destruction box, they are degraded at different times in the cell cycle (cyclin A at the beginning of M phase, cyclin B at the metaphase/anaphase transition). It is possible that degradation of the various cyclins is triggered by different ubiquitin-conjugating enzymes which recognize only a specific type of cyclin.

The crystal structure of cdk2 and the three-dimensional modelling of cdc2 on the crystal structure of cAPK have revealed some important structural features that are likely to be conserved in other members of the cdk family. In order to understand how the activation of cdc2 takes place at the molecular level we will have to await for more structural information on the cyclins.

Understanding the functional role of MPF will require a better knowledge of the molecular targets of the cdc2 kinase at mitosis. The substrates identified so far indicate that cdc2 can directly phosphorylate some important structural proteins. There is evidence that some of the relevant substrates are kinases or enzymes controlling multiple regulatory pathways and amplifying the input signal received by the cdc2 kinase. Phosphorylation of cdc25 C by cdc2, for example, is part of a positive feedback loop ensuring the rapid activation of the cdc2 kinase at the onset of mitosis.

The mechanism of regulation of cdc2–cyclin B at mitosis should be considered as a paradigm for understanding the regulation of other cdk–cyclin complexes during the cell cycle. Since the different cdk–cyclin complexes become active not only at the G1/S and G2/M transition, but also during the G1, S and G2 phases, it is probably an oversimplification to think of SPF and MPF as the unique driving forces of the cell cycle. It seems likely that other factors exist which regulate progression through the G1, S, and G2 phases of the cell cycle. The scenario will undoubtedly become more complete (and complex) in the near future, and understanding of the network of controls which regulate the cell cycle will be a great challenge for all of us.

Acknowledgements

The authors would like to thank Paul Clarke, Ingrid Hoffmann, Christian Kambach, Liz Moynihan, Tim Humphrey, Kristi Chrispell Forbes, and Tamar Enoch for comments and critical reading of the manuscript.

References

1. Rusch, H. P., Sachsenmaier, W., Behrens, K., and Gruter, V. (1966) Synchronization of mitosis by the fusion of the plasmodia of *Physarum polycephalum*. *J. Cell Biol.*, **31,** 204.
2. Johnson, R. T. and Rao, P. N. (1970) Mammalian cell fusion: induction of premature chromosome condensation in interphase nuclei. *Nature,* **226,** 717.
3. Rao, P. N. and Johnson, R. T. (1970) Mammalian cell fusion: studies on the regulation of DNA synthesis and mitosis. *Nature,* **225,** 159.
4. Sherr, C. J. (1993) Mammalian G1 cyclins. *Cell,* **73,** 1059.
5. Pardee, A. B. (1989) G1 events and regulation of cell proliferation. *Science,* **246,** 603.
6. Masui, Y. and Markert, C. L. (1971) Cytoplasmic control of nuclear behavior during meiotic maturation of frog oocytes. *J. Exp. Zool.*, **177,** 129.
7. Smith, L. D. and Ecker, R. E. (1971) The interaction of steroids with *Rana pipiens* oocytes in the induction of maturation. *Dev. Biol.*, **25,** 233.
8. Wasserman, W. J. and Smith, L. D. (1978) The cyclic behaviour of a cytoplasmic factor controlling nuclear membrane breakdown. *J. Cell. Biol.*, **78,** R15.
9. Sunkara, P. S., Wright, D. A., and Rao, P. N. (1979) Mitotic factors from mammalian cells induce germinal vesicle breakdown and chromosome condensation in amphibian oocytes. *Proc. Natl. Acad. Sci. USA,* **76,** 2799.
10. Nelkin, B., Nichols, C., and Vogelstein, B. (1980) Protein factor(s) from mitotic CHO cells induce meiotic maturation in *Xenopus laevis* oocytes. *FEBBS Lett.*, **109,** 233.
11. Kishimoto, T., Kuriyama, R., Kondo, H., and Kanatani, H. (1982). Generality of the action of various maturation-promoting factors. *Exp. Cell Res.*, **137,** 121.
12. Weintraub, H., Buscaglia, M., Ferrez, M., Weiller, S., Boulet, A., Fabre, F., and Baulieu, E. E. (1982) Mise en evidence d'une activite 'MPF' chez *Saccharomyces cerevisiae*. *C. R. Acad. Sci. (Paris)* III, **295,** 787.

13. Gerhart, J., Wu, M., and Kirschner, M. (1984) Cell cycle dynamics of an M-phase-specific cytoplasmic factor in *Xenopus laevis* oocytes and eggs. *J. Cell Biol.*, **98,** 1247.
14. Hartwell, L. H. (1974) *Saccharomyces cerevisiae* cell cycle. *Bacteriol. Rev.*, **38,** 164.
15. Nurse, P., Thuriaux, P., and Nasmyth, K. (1976) Genetic control of the cell division cycle in the fission yeast *Schizosaccharomyces pombe*. *Mol. Gen. Genet.*, **146,** 167.
16. Nurse, P. and Bissett, Y. (1981) Genes required for G1 for commitment to cell cycle and in G2 for control of mitosis in fission yeast. *Nature*, **292,** 558.
17. Nurse, P. and Thuriaux, P. (1980) Regulatory genes controlling mitosis in the fission yeast *Schizosaccharomyces pombe*. *Genetics*, **96,** 627.
18. Beach, D., Durkacz, B., and Nurse, P. (1982) Functionally homologous cell cycle control genes in budding and fission yeast. *Nature*, **300,** 706.
19. Lee, M. G. and Nurse, P. (1987) Complementation used to clone a human homologue of the fission yeast cell cycle control gene *cdc2*. *Nature*, **327,** 31.
20. Russell, P. and Nurse, P. (1986) $cdc25^{+}$ functions as an inducer in the mitotic control of fission yeast. *Cell*, **45,** 145.
21. Russell, P. and Nurse, P. (1987) Negative regulation of mitosis by $wee1^{+}$, a gene encoding a protein kinase homolog. *Cell*, **49,** 559.
22. Hayles, J., Beach, D., Durkacz, B., and Nurse, P. (1986) The fission yeast cell cycle control gene *cdc2*: isolation of a sequence *suc1* that suppresses cdc2 mutant function. *Mol. Gen. Genet.*, **202,** 291.
23. Booher, R. and Beach, D. (1987) Interaction between $cdc13^{+}$ and $cdc2^{+}$ in the control of mitosis in fission yeast; dissociation of the G1 and G2 roles of the $cdc2^{+}$. *EMBO J.*, **6,** 3441.
24. Brizuela, L., Draetta, G., and Beach, D. (1987) $p13^{suc1}$ acts in the fission yeast cell division cycle as a component of the $p34^{cdc2}$ protein kinase. *EMBO. J.*, **6,** 3507.
25. Booher, R. N., Alfa, C. E., Hyams, J. S., and Beach, D. H. (1989) The fission yeast cdc2/cdc13/suc1 protein kinase: regulation of catalytic activity and nuclear localization. *Cell*, **58,** 485.
26. Moreno, S., Hayles, J., and Nurse, P. (1989) Regulation of $p34^{cdc2}$ protein kinase during mitosis. *Cell*, **58,** 361.
27. Fisher, P. (1987) Disassembly and reassembly of nuclei in cell-free systems. *Cell*, **48,** 175.
28. Lohka, M. J., Hayes, M. K., and Maller, J. L. (1988) Purification of maturation-promoting factor, an intracellular regulator of early mitotic events. *Proc. Natl. Acad. Sci. USA*, **85,** 3009.
29. Gautier, J., Norbury, C., Lohka, M., Nurse, P., and Maller, J. (1988) Purified maturation-promoting factor contains the product of a *Xenopus* homolog of the fission yeast cell cycle control gene $cdc2^{+}$. *Cell*, **54,** 433.
30. Dunphy, W. G., Brizuela, L., Beach, D., and Newport, J. (1988) The *Xenopus* cdc2 protein is a component of MPF, a cytoplasmic regulator of mitosis. *Cell*, **54,** 423.
31. Bradbury, E. M., Inglis, R. J., and Matthews, H. R. (1974) Control of cell division by very lysine-rich histone (F1) phosphorylation. *Nature*, **247,** 257.
32. Labbé, J. C., Lee, M. G., Nurse, P., Picard, A., and Dorée, M. (1988) Activation at M-phase of a protein kinase encoded by a starfish homologue of the cell cycle control gene $cdc2^{+}$. *Nature*, **335,** 251.
33. Arion, D., Meijer, L., Brizuela, L., and Beach, D. (1988) cdc2 is a component of the M phase-specific histone H1 kinase: evidence for identity with MPF. *Cell*, **55,** 371.
34. Simanis, V. and Nurse, P. (1986) The cell cycle control gene $cdc2^{+}$ of fission yeast encodes a protein kinase potentially regulated by phosphorylation. *Cell*, **45,** 261.

35. Evans, T., Hunt, T., and Youngblom, J. (1982) On the role of maternal mRNA in sea urchins: studies of a protein which appears to be destroyed at a particular point during each division cycle. *Biol. Bull.*, **163,** 372.
36. Evans, T., Rosenthal, E. T., Youngblom, J., Distel, D., and Hunt, T. (1983) Cyclin: a protein specified by maternal mRNA in sea urchin eggs that is destroyed at each cleavage division. *Cell*, **33,** 389.
37. Wilt, F. H., Sakai, H., and Mazia, D. (1967) Old and new protein in the formation of the mitotic apparatus in cleaving sea urchin eggs. *J. Mol. Biol.*, **27,** 1.
38. Swenson, K., Farrell, K. M., and Ruderman, J. V. (1986) The clam embryo protein cyclin A induces entry into M-phase and the resumption of meiosis in *Xenopus* oocytes. *Cell*, **47,** 861.
39. Pines, J. and Hunt, T. (1987) Molecular cloning and characterization of the mRNA for cyclin from sea urchin eggs. *EMBO J.*, **6,** 2987.
40. Solomon, M., Booher, R., Kirschner, M., and Beach, D. (1988) Cyclin in fission yeast. *Cell*, **54,** 738.
41. Goebl, M. and Byers, B. (1988) Cyclin in fission yeast. *Cell*, **54,** 739.
42. Draetta, G., Luca, F., Westendorf, J., Brizuela, L., Ruderman, J., and Beach, D. (1989) cdc2 protein kinase is complexed with both cyclin A and B: evidence for proteolytic inactivation of MPF. *Cell*, **56,** 829.
43. Murray, A. W. and Kirschner, M. W. (1989) Cyclin synthesis drives the early embryonic cell cycle. *Nature*, **339,** 275.
44. Minshull, J., Blow, J. J., and Hunt, T. (1989) Translation of cyclin mRNA is necessary for extracts of activated *Xenopus* eggs to enter mitosis. *Cell*, **56,** 947.
45. Murray, A. W., Solomon, M. J., and Kirschner, M. W. (1989) The role of cyclin synthesis and degradation in the control of maturation promoting factor activity. *Nature*, **339,** 280.
46. Labbé, J. C., Capony, J. P., Caput, D., Cavadore, J. C., Derancourt, J., Kaghdad, M., Lelias, J. M., Picard, A., and Dorée, M. (1989) MPF from starfish oocytes at first meiotic metaphase is a heterodimer containing one molecule of cdc2 and one molecule of cyclin B. *EMBO J.*, **8,** 3053.
47. Gautier, J., Minshull, J., Lohka, M., Glotzer, M., Hunt, T., and Maller, J. L. (1990) Cyclin is a component of MPF from *Xenopus*. *Cell*, **60,** 487.
48. Draetta, G. and Beach, D. (1988) Activation of cdc2 protein kinase during mitosis in human cells: cell cycle-dependent phosphorylation and subunit rearrangement. *Cell*, **54,** 17.
49. Xiong, Y., Zhang, H., and Beach, D. (1993) Subunit rearrangement of the cyclin-dependent kinases is associated with cellular transformation. *Genes Dev.*, **7,** 1572.
50. Riabowol, K., Draetta, G., Brizuela, L., Vandre, D., and Beach, D. (1989) The cdc2 kinase is a nuclear protein that is essential for mitosis in mammalian cells. *Cell*, **57,** 393.
51. Th'ng, J. P. H., Wright, P. S., Hamaguchi, J., Norbury, C. J., Nurse, P., and Bradbury, E. M. (1990) The FT210 cell line is a mouse G2 phase mutant with a temperature-sensitive *cdc2* gene product. *Cell*, **63,** 313.
52. Giordano, A., Whyte, P., Harlow, E., Franza, B. R., Beach, D., and Draetta, G. (1989) A 60 kd cdc2-associated polypeptide complexes with the E1A proteins in adenovirus-infected cells. *Cell*, **58,** 981.
53. Pagano, M., Pepperkok, R., Verde, F., Ansorge, W., and Draetta, G. (1992) Cyclin A is required at two points in the human cell cycle. *EMBO J.*, **11,** 761.
54. Pines, J. (1993) Cyclins and cyclin-dependent kinases: take your partners. *TIBS*, **18,** 195.

55. Pagano, M., Pepperkok, R., Lukas, J., Baldin, V., Ansorge, W., Bartek, J., and Draetta, G. (1993) Regulation of the human cell cycle by the cdk2 protein kinase. *J. Cell Biol.*, **121,** 101.
56. Van den Heuvel, S. and Harlow, E. (1993) Distinct roles for cyclin-dependent kinases in cell cycle control. *Science*, **262,** 2050.
57. Wasserman, W. J. and Masui, Y. (1975) Effects of cycloheximide on a cytoplasmic factor initiating meiotic maturation in *Xenopus* oocytes. *Exp. Cell Res.*, **91,** 381.
58. Draetta, G., Piwnica, W. H., Morrison, D., Druker, B., Roberts, T., and Beach, D. (1988) Human cdc2 protein kinase is a major cell-cycle regulated tyrosine kinase substrate. *Nature*, **336,** 738.
59. Morla, A., Draetta, G., Beach, D., and Wang, J. Y. J. (1989) Reversible tyrosine phosphorylation of cdc2: dephosphorylation accompanies activation during entrance into mitosis. *Cell*, **58,** 193.
60. Dunphy, W. and Newport, J. (1989) Fission yeast p13 blocks mitotic activation and tyrosine dephosphorylation of the *Xenopus* cdc2 protein kinase. *Cell*, **58,** 181.
61. Gould, K. L. and Nurse, P. (1989) Tyrosine phosphorylation of the fission yeast $cdc2^+$ protein kinase regulates entry into mitosis. *Nature*, **342,** 39.
62. Krek, W. and Nigg, E. (1991) Differential phosphorylation of vertebrate $p34^{cdc2}$ kinase at the G1/S and G2/M transitions of the cell cycle: identification of major phosphorylation sites. *EMBO J.*, **10,** 305.
63. Norbury, C., Blow, J., and Nurse, P. (1991) Regulatory phosphorylation of the $p34^{cdc2}$ protein kinase in vertebrates. *EMBO J.*, **10,** 3321.
64. Krek, W. and Nigg, E. (1991) Mutations of $p34^{cdc2}$ phosphorylation sites induce premature mitotic events in HeLa cells: evidence for a double block to $p34^{cdc2}$ kinase activation in vertebrates. *EMBO J.*, **10,** 3331.
65. Gould, K. L., Moreno, S., Owen, D. J., Sazer, S., and Nurse, P. (1991) Phosphorylation at Thr167 is required for *Schizosaccharomyces pombe* $p34^{cdc2}$ function. *EMBO J.*, **10,** 3297.
66. Solomon, N., Lee, T., and Kirschner, M. W. (1992) Role of phosphorylation in $p34^{cdc2}$ activation: identification of an activating kinase. *Mol. Biol. Cell*, **3,** 13.
67. Ducommun, B., Brambilla, P., Felix, M.-A., Franza, B. R. J., Karsenti, E., and Draetta, G. (1991) cdc2 phosphorylation is required for its interaction with cyclin. *EMBO J.*, **10,** 3311.
68. Shuttleworth, J., Godfrey, R., and Colman, A. (1990) p40MO15, a cdc2-related protein kinase involved in negative regulation of meiotic maturation of *Xenopus* oocytes. *EMBO J.*, **9,** 3233.
69. Fesquet, D., Labbe, J. C., Derancourt, J., Capony, J. P., Galas, S., Girard, F., Lorce, T., Shuttleworth, J., Doree, M., and Cavadore, J. C. (1993) The MO15 gene encodes the catalytic subunit of a protein kinase that activates cdc2 and other cyclin-dependent kinases (CDKs) through phosphorylation of threonine 161 and its homologs. *EMBO J.*, **12,** 3111.
70. Solomon, M., Harper, W., and Shuttleworth, J. (1993) CAK, the $p34^{cdc2}$ activating kinase contains a protein related to MO15. *EMBO J.*, **12,** 3133.
71. Poon, R. Y. C., Yamashita, K., Adamkzewski, J., Hunt, T., and Shuttleworth, J. (1993) The cdc2-related protein $p40^{MO15}$ is a component of a $p33^{cdk2}$-activating kinase. *EMBO J.*, **12,** 3123.
72. Lorca, T., Labbé, J. C., Devault, A., Fesquet, D., Capony, J. P., Cavadore, J. C., and Dorée, M. (1992) Dephosphorylation of cdc2 on threonine 161 is required for cdc2 kinase inactivation and normal anaphase. *EMBO J.*, **11,** 2381.

73. Krek, N., Marks, J., Schmitz, N., Nigg, E. A., and Simanis, V. (1992) Vertebrate $p34^{cdc2}$ phosphorylation site mutants: effect upon cell cycle progression in the fission yeast *Schizosaccharomyces pombe*. *J. Cell. Sci.*, **102,** 42.
74. Pines, J. and Hunter, T. (1989) Isolation of a human cyclin cDNA: evidence for cyclin mRNA and protein regulation in the cell cycle and for interaction with $p34^{cdc2}$. *Cell*, **58,** 833.
75. Girard, F., Strausfeld, U., Fernandez, A., and Lamb, N. (1991) Cyclin A is required for the onset of DNA replication in mammalian fibroblasts. *Cell*, **67,** 1169.
76. Zindy, F., Lamas, E., Chenivesse, X., Sobczak, J., Wang, J., Fesquet, D., Henglein, B., and Brechot, C. (1992) Cyclin A is required in S phase in normal epithelial cells. *Biochem. Biophys. Res. Commun.*, **182,** 1144.
77. Kobayashi, H., Stewart, E., Poon, R., Adamczewski, J. P., Gannon, P., and Hunt, T. (1992) Identification of the domains in cyclin A required for binding to, and activation of, $p34^{cdc2}$ and $p32^{cdk2}$ protein kinase subunits. *Mol. Biol. Cell*, **3,** 1279.
78. Lees, E. and Harlow, E. (1993) Sequences within the conserved cyclin box of human cyclin A are sufficient for binding and activation of cdc2 kinase. *Mol. Cell. Biol.*, **13,** 1194.
79. Glotzer, M., Murray, A., and Kirschner, M. (1990) Cyclin is degraded by the ubiquitin pathway. *Nature*, **349,** 132.
80. Gallant, P. and Nigg, E. A. (1992) Cyclin B2 undergoes cell-cycle dependent nuclear translocation and, when expressed as non-destructible mutant, causes mitotic arrest in HeLa cells. *J. Cell. Biol.*, **117,** 213.
81. Tyers, M., Tokiwa, G., Nash, R., and Futcher, B. (1992) The CLN3–CDC28 kinase complex is regulated by proteolysis and phosphorylation. *EMBO J.*, **11,** 1773.
82. Pines, J. and Hunter, T. (1991) Human cyclins A and B1 are differentially located in the cell and undergo cell cycle-dependent nuclear transport. *J. Cell Biol.*, **115,** 1.
83. Cardoso, M. C., Leonhardt, H., and Nadal-Ginard, B. (1993) Reversal of terminal differentiation and control of DNA replication: cyclin A and cdk2 specifically localize at subnuclear sites of DNA replication. *Cell*, **74,** 979.
84. Hoffmann, I., Clarke, P. R., Marcote, M. J., Karsenti, E., and Draetta, G. (1993) Phosphorylation and activation of human cdc25-C by cdc2–cyclin B and its involvement in the self-amplification of MPF at mitosis. *EMBO J.*, **12,** 53.
85. Peeper, D. S., Parker, L. L., Ewen, M. E., Tobes, M., Hall, F. L., Xu, M., Zantema, A., van der Eb, A. J., and Piwnica-Worms, H. (1993) A- and B-type cyclins differentially modulate substrate specificity of cyclin–cdk complexes. *EMBO J.*, **12,** 1947.
86. Richardson, H., Lew, D. J., Henze, M., Sugimoto, K., and Reed, S. (1992) Cyclin-B homologs in *Saccharomyces cerevisiae* function in S phase and in G2. *Genes Dev.*, **6,** 2021.
87. Hunter, T. and Pines, J. (1991) Cyclins and cancer. *Cell*, **66,** 1071.
88. Hadwiger, J. A., Wittenberg, C., Mendenhall, M. D., and Reed, S. I. (1989) The *Saccharomyces cerevisiae* CKS1 gene, a homolog of the *Schizosaccharomyces pombe* $suc1^{+}$ gene, encodes a subunit of the Cdc28 protein kinase complex. *Mol. Cell Biol.*, **9,** 2034.
89. Richardson, H. E., Stueland, C. S., Thomas, J., Russell, P., and Reed, S. I. (1990) Human cDNAs encoding homologs of the small $p34^{CDC28/cdc2}$-associated protein in *Saccharomyces cerevisiae* and *Schizosaccharomyces pombe*. *Genes Dev.*, **4,** 1332.
90. Ducommun, B. and Draetta, G. Unpublished data.
91. Brizuela, L., Draetta, G., and Beach, D. (1989) Activation of human CDC2 protein as a histone H1 kinase is associated with complex formation with the p62 subunit. *Proc. Natl. Acad. Sci. USA*, **86,** 4362.

92. Hayles, J., Aves, S., and Nurse, P. (1986) *suc1* is an essential gene involved in both the cell cycle and growth in fission yeast. *EMBO J.*, **5,** 3373.
93. Basi, G. (1993) Regulation of the cell cycle by p13^{suc1} in *Shizosaccharomyces pombe*. Ph.D. thesis, EMBL, Heidelberg.
94. Tang, Y. and Reed, S. I. (1993) The Cdk-associated protein Cks1 functions both in G1 and G2 in *Saccharomyces cerevisiae*. *Genes Dev.*, **7,** 822.
95. Meyerson, M., Enders, G. H., Wu, C.-L., Su, L.-K., Gorka, C., Nelson, C., Harlow, E., and Tsai, L.-H. (1992) A family of human cdc2-related protein kinases. *EMBO J.*, **11,** 2909.
96. Parge, H. E., Arvai, A. S., Murtari, D. J., Reed, S. I., and Tainer, J. A. (1993) Human ckshs2 atomic structure: a role of its hexameric assembly in cell cycle control. *Science*, **262,** 387.
97. Hunter, T. (1993) Braking the cycle. *Cell*, **75,** 839.
98. Harper, J. W., Adami, G. R., Wei, N., Keyomarsi, K., and Elledge, S. J. (1993) The p21 Cdk-interacting protein Cip1 is a potent inhibitor of G1 cyclin-dependent kinases. *Cell*, **75,** 805.
99. El-Deiry, W. S., Tokino, T., Velculescu, V. E., Levy, D. B., Parson, R., Trent, J. M., Lin, D., Mercer, W. E., Kinzler, K. W., and Vogelstein, B. (1993) WAF1, a potential mediator of p53 tumor suppression. *Cell*, **75,** 817.
100. Noda, A., Ning, Y., Venable, S. F., Pereira-Smith, O. M., and Smith, J. R. (1994) Cloning of senescent cell-derived inhibitors of DNA synthesis using an expression screen. *Exp. Cell. Res.*, **211,** 90.
101. Xiong, Y., Hannon, G., Zhang, H., Casso, D., Kobayashi, R., and Beach, D. (1993) p21 is an universal inhibitor of cyclin kinases. *Nature*, **366,** 701.
102. Iragashi, M., Nagata, A., Jinno, S., Suto, K., and Okayama, H. (1991) Wee1^{+}-like gene in human cells. *Nature*, **353,** 80.
103. McGowan, C. H. and Russell, P. (1993) Human wee1 kinase inhibits cell division by phosphorylating p34cdc2 exclusively on Tyr15. *EMBO J.*, **12,** 75.
104. Featherstone, C. and Russell, P. (1991) Fission yeast p107^{wee1} mitotic inhibitor is a tyrosine/serine kinase. *Nature*, **349,** 808.
105. Parker, L. and Piwnica-Worms, H. (1992) Inactivation of the p34cdc2–cyclin B complex by the human wee1 tyrosine kinase. *Science*, **257,** 1955.
106. Parker, L., Atherton-Fessler, S., Lee, M., Ogg, S., Falk, J., Swenson, K., and Piwnica-Worms, H. (1991) Cyclin promotes the tyrosine phosphorylation of p34^{cdc2} in a wee1^{+} dependent manner. *EMBO J.*, **10,** 1255.
107. Meijer, L., Azzi, L., and Wang, J. (1991) Cyclin B targets p34^{cdc2} for tyrosine phosphorylation. *EMBO J.*, **10,** 1545.
108. Parker, L., Atherton-Fessler, S., and Piwnica-Worms, H. (1992) p107^{wee1} is a dual-specific kinase that phosphorylates p34^{cdc2} on tyrosine 15. *Proc. Natl. Acad. Sci. USA*, **89,** 2917.
109. Solomon, M. J., Glotzer, M., Lee, T. L., Philippe, M., and Kirschner, M. W. (1990) Cyclin activation of p34^{cdc2}. *Cell*, **63,** 1013.
110. Lundgren, K., Walworth, N., Booher, R., Dembski, M., Kirschner, M., and Beach, D. (1991) *mik1* and *wee1* cooperate in the inhibitory tyrosine phosphorylation of cdc2. *Cell*, **64,** 1111.
111. Russell, P. and Nurse, P. (1987) The mitotic inducer *nim1*$^{+}$ functions in a regulatory network of protein kinase homologs controlling the initiation of mitosis. *Cell*, **49,** 569.

112. Coleman, T. R., Tang, Z., and Dunphy, W. G. (1993) Negative regulation of the wee1 protein kinase by direct action of the nim1/cdr1 mitotic inducer. *Cell*, **72**, 919.
113. Félix, M. A., Cohen, P., and Karsenti, E. (1990) Cdc2 H1 kinase is negatively regulated by a type 2A phosphatase in the *Xenopus* early embryonic cell cycle: evidence from the effects of okadaic acid. *EMBO J.*, **9**, 675.
114. Lee, T. H., Solomon, M. J., Mumby, M. C., and Kirschner, M. W. (1991) INH, a negative regulator of MPF, is a form of protein phosphatase 2A. *Cell*, **64**, 415.
115. Clarke, P., Hoffmann, I., Draetta, G., and Karsenti, E. (1993) Dephosphorylation of cdc25-C by a type-2A protein phosphatase: specific regulation during the cell cycle in *Xenopus* egg extracts. *Mol. Biol. Cell*, **4**, 397.
116. Kinoshita, N., Ohkura, H., and Yanagida, M. (1990) Distinct, essential roles of type 1 and type 2A protein phosphatase in the control of the fission yeast cell division cycle. *Cell*, **63**, 405.
117. Kinoshita, N., Yamano, H., Niwa, W., Yoshida, T., and Yanagida, M. (1993) Negative regulation of mitosis by the fission yeast protein phosphatase ppa2. *Genes Dev.*, **7**, 1059.
118. Russell, P., Moreno, S., and Reed, S. I. (1989) Conservation of mitotic controls in fission and budding yeasts. *Cell*, **57**, 295.
119. Edgar, B. A. and O'Farrell, P. H. (1989) Genetic control of cell division patterns in the *Drosophila* embryo. *Cell*, **57**, 177.
120. Alphey, L., Jimenez, J., White-Cooper, H., Dawson, I., Nurse, P., and Glover, D. (1992) *Twine*, a cdc25 homolog that functions in the male and female germline of *Drosophila*. *Cell*, **69**, 977.
121. Kumagai, A. and Dunphy, W. G. (1992) Regulation of the cdc25 protein during the cell cycle in *Xenopus* extracts. *Cell*, **70**, 139.
122. Izumi, T., Walker, D., and Maller, J. (1992) Periodic changes in phosphorylation of the *Xenopus* cdc25 phosphatase regulates its activity. *Mol. Biol. Cell*, **3**, 927.
123. Kakizuka, A., Sebastian, B., Borgmeyer, U., Hermans-Borgmeyer, I., Bolado, J., Hunter, T., Hoekstra, M., and Evans, R. (1992) A mouse cdc25 homolog is differentially and developmentally expressed. *Genes Dev.*, **6**, 578.
124. Sadhu, K., Reed, S. I., Richardson, H., and Russell, P. (1990) Human homolog of fission yeast *cdc25* is predominantly expressed in G1. *Proc. Natl. Acad. Sci. USA*, **87**, 5139–5143.
125. Galactionov, K. and Beach, D. (1991) Specific activation of cdc25 tyrosine phosphatase by B-type cyclins: evidence for multiple roles of mitotic cyclins. *Cell*, **67**, 1181.
126. Nagata, A., Igarashi, M., Jinno, S., Suto, K., and Okayama, H. (1991) An additional homolog of the fission yeast *cdc25* gene occurs in humans and is highly expressed in some cancer cells. *New Biol.*, **3**, 959.
127. Dunphy, W. G. and Kumagai, A. (1991) The cdc25 protein contains an intrinsic phosphatase activity. *Cell*, **67**, 189.
128. Gautier, J., Solomon, M. J., Booher, R. N., Bazan, J. F., and Kirschner, M. W. (1991) cdc25 is a specific tyrosine phosphatase that directly activates p34cdc2. *Cell*, **67**, 197.
129. Sebastian, B., Kakizuka, A., and Hunter, T. (1993) Cdc25M2 activation of cyclin-dependent kinases by dephosphorylation of threonine-14 and tyrosine-15. *Proc. Natl. Acad. Sci. USA*, **90**, 3521.
130. Millar, J. B. A., Lenaers, G., and Russell, P. (1992) *Pyp*3 PTPase acts as a mitotic inducer in fission yeast. *EMBO J.*, **11**, 4933.
131. Hoffmann, I., Clarke, P., Draetta, G., and Karsenti, E. (1993) Regulation of cyclin-

dependent kinases and tyrosine phosphatases during the cell cycle. *Eur. J. Cell Biol.*, **61** (Suppl. 38), 6.

132. Millar, J. B. A., Blevitt, J., Gerace, L., Sadhu, K., Featherstone, C., and Russell, P. (1991) p55^{cdc25} is a nuclear protein required for initiation of mitosis in human cells. *Proc. Natl. Acad. Sci. USA*, **88,** 10500.
133. Ducommun, B., Brambilla, P., and Draetta, G. (1991) Mutations at sites involved in suc1 binding inactivate cdc2. *Mol. Cell. Biol.*, **11,** 6177.
134. Marcote, M. J., Knighton, D. R., Basi, G., Sowadski, J. M., Brambilla, P., Draetta, G., and Taylor, S. S. (1993) A three-dimensional model of the cdc2 protein kinase: identification of cyclin and suc1 binding regions. *Mol. Cell. Biol.*, **13,** 5122.
135. De Bondt, H. L., Rosenblatt, J., Jancarik, J., Jones, H. D., Morgan, D. O., and Kim, S.-H. (1993) Crystal-structure of cyclin-dependent kinase 2. *Nature*, **363,** 595.
136. Knighton, D., Zheng, J., Ten Eyck, L. F., Ashford, V., Xuong, N., Taylor, S. S., and Sowadski, J. (1991) Crystal structure of the catalytic subunit of cyclin adenosine monophosphate-dependent protein kinase. *Science*, **253,** 407.
137. Atherton-Fessler, S., Parker, L., Geahlen, R., and Piwnica-Worms, H. (1993) Mechanisms of p34^{cdc2} regulation. *Mol. Cell. Biol.*, **13,** 1675.
138. Nigg, E. (1991) The substrates of the cdc2 protein kinase. In *Cyclin-dependent kinases. Seminars in cell biology*, Vol. 2, G. Draetta (ed.). Saunders, p. 261.
139. Nigg, E. (1993) Cellular substrates of p34^{cdc2} and its companion cyclin-dependent kinases. *TICB*, **3,** 296.
140. Shiina, N., Moriguchi, T., Ohta, K., Gotoh, Y., and Nishida, E. (1992) Regulation of a major microtubule associated protein by MPF and MAP kinase. *EMBO J.*, **11,** 3977.
141. Kamijo, M., Yasuda, H., Yau, P. M., Yamashita, M., Nagahama, Y., and Ohba, Y. (1992) Preference of human cdc2 kinase for peptide substrate. *Pept. Res.*, **5,** 281.
142. Marin, O., Meggio, F., Draetta, G., and Pinna, L. (1992) The consensus sequences for cdc2 kinase and for casein kinase-II are mutually incompatible. *FEBS Lett.*, **301,** 111.
143. Peter, M., Nakagawa, J., Dorée, M., Labbé, J. C., and Nigg, E. A. (1990) *In vitro* disassembly of the nuclear lamina and M phase-specific phosphorylation of lamins by cdc2 kinase. *Cell*, **61,** 591.
144. Ward, G. E. and Kirschner, M. W. (1990) Identification of cell cycle-regulated phosphorylation sites on nuclear lamin C. *Cell*, **61,** 561.
145. Dessev, G., Iovcheva-Dessev, C., Bischoff, J. R., Beach, D., and Goldman, R. (1991) A complex containing p34^{cdc2} and cyclin B phosphorylates the nuclear lamina and disassembles nuclei of clam oocytes *in vitro*. *J. Cell Biol.*, **112,** 523.
146. Peter, M., Heitlinger, E., Haener, M., Aebi, U., and Nigg, E. A. (1991) Disassembly of *in vitro* formed lamin head-to-tail polymers by cdc2 kinase. *EMBO J.*, **10,** 1535.
147. Heald, R., and McKeon, F. (1990) Mutations of phosphorylation sites in lamin A that prevent nuclear lamina disassembly in mitosis. *Cell*, **61,** 579.
148. Chou, Y.-H., Bischoff, J. R., Beach, D., and Goldman, R. D. (1990) Intermediate filament reorganization during mitosis is mediated by p34^{cdc2} phosphorylation of vimentin. *Cell*, **62,** 1063.
149. Yamashiro, S., Yamakita, Y., Hosoya, H., and Matsumura, F. (1991) Phosphorylation of non-muscle caldesmon by p34^{cdc2} kinase during mitosis. *Nature*, **349,** 169.
150. Mak, A., Watson, M., Litwin, C., and Wang, J. (1991) Phosphorylation of caldesmon by cdc2 kinase. *J. Biol. Chem.*, **266,** 6678.
151. Satterwhite, L. L., Lohka, M. J., Wilson, K. L., Scherson, T. Y., Cisek, L. J., Corden,

J. L., and Pollard, T. D. (1992) Phosphorylation of myosin-II regulatory light chain by cyclin-p34^{cdc2}: a mechanism for the timing of cytokinesis. *J. Cell. Biol.*, **118,** 595.

152. Verde, F., Labbé, J. C., Dorée, M., and Karsenti, E. (1990) Regulation of microtubule dynamics by cdc2 protein kinase in cell-free extracts of *Xenopus* eggs. *Nature*, **343,** 233.
153. Yeo, J.-P., Alderuccio, F., and Toh, B.-H. (1994) A new chromosomal protein essential for mitotic spindle assembly. *Nature*, **367,** 288.
154. Shenoy, S., Choi, J., Bagrodia, S., Copeland, T. D., Maller, J. L., and Shalloway, D. (1989) Purified maturation promoting factor phosphorylates pp60c-src at the sites phosphorylated during fibroblast mitosis. *Cell*, **57,** 763.
155. Morgan, D. O., Kaplan, J. M., Bishop, J. M., and Varmus, H. E. (1989) Mitosis-specific phosphorylation of p60$^{c\text{-}src}$ by p34^{cdc2}-associated protein kinase. *Cell*, **57,** 775.
156. Shenoy, S., Chackalaparampil, I., Bagrodia, S., Lin, P.-H., and Shalloway, D. (1992) Role of p34^{cdc2} mediated phosphorylations in two step activation of p60$^{c\text{-}src}$ during mitosis. *Proc. Natl. Acad. Sci. USA*, **89,** 7237.
157. Kipreos, E. T. and Wang, J. Y. (1990) Differential phosphorylation of c-Abl in cell cycle determined by cdc2 kinase and phosphatase activity. *Science*, **248,** 217.
158. Kipreos, E. T. and Wang, J. Y. J. (1992) Cell cycle regulated binding of c-abl tyrosine kinase to DNA. *Science*, **256,** 382.
159. Liu, J., Singh, B., Wlodek, D., and Arlinghaus, R. B. (1990) Cell cycle-mediated structural and functional alteration of p85$^{gag\text{-}mos}$ protein kinase activity. *Oncogene*, **5,** 171.
160. Bai, W. L., Singh, B., Karshin, W. L., Shonk, R. A., and Arlinghaus, R. B. (1991) Phosphorylation of v-mos Ser 47 by the mitotic form of p34cdc2. *Oncogene*, **6,** 1715.
161. Litchfield, D. W., Lozeman, F. J., Cicirelli, M. F., Harrylock, M., Ericsson, L. H., Piening, C. J., and Krebs, E. G. (1991) Phosphorylation of the beta subunit of casein kinase II in human A431 cells. Identification of the autophosphorylation site and a site phosphorylated by p34cdc2. *J. Biol. Chem.*, **266,** 20380.
162. Litchfield, D. W., Luescher, B., Lozeman, F. J., Eisenman, R. N., and Krebs, E. G. (1992) Phosphorylation of casein kinase II *in vitro* and at mitosis. *J. Biol. Chem.*, **267,** 13943.
163. Langan, T. A., Zeilig, C. E., and Leichtling, B. (1980) Analysis of multiple sites of phosphorylation of H1 histone. In *Protein Phosphorylation and Bioregulation*. Thomas, G., Podesta, E. J. and Gordon, J. (ed.). Karger, Basel, p. 1.
164. Langan, T. A., Gautier, J., Lohka, M., Hollingsworth, R., Moreno, S., Nurse, P., Maller, J., and Sclafani, R. A. (1989) Mammalian growth-associated H1 histone kinase: a homolog of *cdc2*$^{+}$/*CDC28* protein kinase controlling mitotic entry in yeast and frog cells. *Mol. Cell. Biol.*, **9,** 3860.
165. Moreno, S. and Nurse, P. (1990) Substrates for p34^{cdc2}: *in vivo veritas*? *Cell*, **61,** 549.
166. Reeves, R. (1992) Chromatin changes during the cell cycle. *Curr. Opin. Cell. Biol.*, **4,** 413.
167. Meijer, L., Ostvold, A.-C., Walaas, S. I., Lund, T., and Laland, S. G. (1991) High-mobility-group proteins P1, I, Y as substrates of the M-phase-specific p34^{cdc2}/cyclin^{cdc13} kinase. *Eur. J. Biochem*, **196,** 557.
168. Nissen, M. S., Langan, T. A., and Reeves, R. (1991) Phosphorylation by cdc2 kinase modulates DNA binding activity of high mobility group I nonhistone chromatin protein. *J. Biol. Chem.*, **266,** 19945.
169. Luescher, B. and Eisenman, R. N. (1992) Mitotis-specific phosphorylation of the nuclear oncoproteins myc and myb. *J. Cell Biol.*, **118,** 775.

170. Peter, M., Nakagawa, J., Dorée, M., Labbé, J.-C., and Nigg, E. A. (1990) Identification of major nucleolar proteins as candidate mitotic substrates of cdc2 kinase. *Cell*, **60,** 791.
171. Belenguer, P., Caizergues-Ferrer, M., Labbé, J.-C., Dorée, M., and Amalric, F. (1990) Mitosis specific phosphorylation of nucleolin by $p34^{cdc2}$ kinase. *Mol. Cell. Biol.*, **10,** 3607.
172. Suzuki, T., Oishi, M., Marshak, D. R., Czernik, A. R., Nairn, A. C., and Greengard, P. (1994) Cell-cycle dependent regulation of the phosphorylation and metabolism of the Alzheimer amyloid precursor protein. *EMBO J.*, **13,** 1114.
173. Bailly, E., McCaffrey, M., Touchot, N., Zaharaoui, A., Goud, B., and Bornens, M. (1991) Phosphorylation of two small GTB-binding proteins of the Rab family by $p34^{cdc2}$. *Nature*, **350,** 715.
174. Van der Sluijis, P., Hull, P., Huber, L. A., Male, P., Goud, B., and Mellman, I. (1992) Reversibile phosphorylation–dephosphorylation determines the localization of rab4 during the cell cycle. *EMBO J.*, **11,** 4379.
175. Izumi, T. and Maller, J. L. (1991) Phosphorylation of *Xenopus* cyclins B1 and B2 is not required for cell cycle transitions. *Mol. Cell. Biol.*, **11,** 3860.
176. Hartwell, L. H. and Weinert, T. A. (1989) Checkpoints: controls that ensure the order of cell cycle events. *Science*, **246,** 629.
177. Schlegel, R. and Pardee, A. B. (1986) Caffeine-induced uncoupling of mitosis from the completion of DNA replication in mammalian cells. *Science*, **232,** 1264.
178. Enoch, T. and Nurse, P. (1991) Coupling M phase and S phase: controls maintaining the dependence of mitosis on chromosome replication. *Cell*, **65,** 921.
179. Enoch, T. and Nurse, P. (1990) Mutation of fission yeast cell cycle control genes abolishes dependence of mitosis on DNA replication. *Cell*, **60,** 665.
180. Kumagai, A. and Dunphy, W. (1991) The cdc25 protein controls tyrosine dephosphorylation of the cdc2 protein in a cell-free system. *Cell*, **64,** 903.
181. Hodgkin, A. L. and Huxley, A. F. (1952) A quantitative description of membrane current and its application to conduction and excitation in nerve. *J. Physiol.*, **117,** 500.
182. Félix, M. A., Labbé, J. C., Dorée, M., Hunt, T., and Karsenti, E. (1990) Cdc2-kinase triggers cyclin degradation in interphase extracts of amphibian eggs. *Nature*, **346,** 379.
183. Basi, G. and Draetta, G. (1995) $p13^{suc1}$ of fission yeast regulates two distinct forms of the mitotic cdc2 kinase. *Mol. Cell. Biol.*, in press.
184. Solomon, M. J. (1994) The function(s) of the $p34^{cdc2}$-activating kinase. *Trends in Biochemical Science*, **19,** 496.
185. Guan, K.-L., Jenkins, C. W., Li Y., Nichols, M. J., Wu, X., O'Keefe, C. L., Matera, A. G. and Xiong, Y. (1994) Growth suppression by p18, a $p16^{INK4/MTS1}$ and $p14^{INK4B/MTS2}$-related CDK6 inhibitor, correlates with wild-type Rb function. *Genes Dev.*, **8,** 2939.

6 Cyclin-dependent kinases: an embarrassment of riches?

JONATHON PINES and TONY HUNTER

1. Introduction

One of the most exciting advances in our understanding of the cell cycle has been the identification in multicellular organisms of a family of protein kinases that are involved in the regulation of different checkpoints. These protein kinases are related both in their primary structure, and because they bind and are activated by the cyclin family of proteins. The latter property has led to this family of protein kinases being called the 'cyclin-dependent kinases', or cdks. The cyclins are a family of structurally related proteins that typically vary in amount during the cell cycle in a programmed manner (the cyclins are reviewed in Chapter 5). The first two cyclins to be isolated, cyclins A and B, were called mitotic cyclins, because they were specifically destroyed in mitosis, implicating them in the regulation of the G2/M transition. Subsequently, another class of cyclins was discovered in yeast that have roles in the G1/S transition, and are called G1 cyclins. More than ten different cyclins have been identified in humans and in budding yeast, and the different cyclin/cdk complexes are involved in regulating several different checkpoints throughout the cell cycle. Furthermore, it now appears that some cyclin–cdk complexes regulate processes quite separate from the cell cycle. The paradigm for the cyclin–cdk interaction is that between cdc2 and cyclin B (reviewed in Chapter 5). Therefore, in this article we will only outline the salient points.

2. The cdc2 gene family

2.1 The cyclin B–cdc2 paradigm

2.1.1 Conserved regions of cdc2 kinase

In brief, the cdc2 gene product is a 34 kDa protein with all the hallmarks of a protein kinase as defined by Hanks *et al.* (1). It is highly conserved through evolution, such that the fission yeast *cdc2* gene can be replaced by a human cdc2 cDNA, and the yeast cells will still properly regulate their cell cycle (2). This is reflected in the conservation of protein sequence; yeast and human cdc2 proteins are 74 per cent identical. One of the most conserved regions is a stretch of 16 amino acids just C-terminal to the ATP binding domain. This is the PSTAIRE region, named for the partial sequence of the conserved amino acids, which at one time was taken as the hallmark of cdc2 itself. However, we now find this sequence is well conserved in most of the cdks identified, probably because this region is important for binding to cyclins, as mutational studies have shown (3). The crystal structure of monomeric cdk2 shows that most of the residues necessary for cyclin binding are clustered at the top of the ATP-binding lobe of the enzyme (4).

In contrast to the cyclins, cdc2 and the other cdks are present during the cell cycle at relatively constant levels. However, monomeric cdc2 is only weakly active as a protein kinase (5). Its activity is greatly enhanced when it is bound by a cyclin, primarily a B-type cyclin, which is thought to have both an activating and a targeting role in the cyclin–cdk complex. The cyclin B–cdc2 is itself regulated through both inhibitory and activatory phosphorylation events (described by Basi and Draetta in Chapter 5). Thus the cyclin B–cdc2 complex can accumulate in an inactive form until it is required at the transition from interphase into mitosis. The protein kinases involved in the inhibition of cyclin B–cdc2 have been identified as the products of the *wee1* and *mik1* genes in yeast (6–9), and a potential mik1 homologue has been found in animal cells (10–12). Similarly, the protein phosphatase responsible for the Y15 (and T14) dephosphorylation has also been conserved through evolution. It is the product of the *cdc25* gene in fission yeast (13, 14), and multiple homologues have been found in frogs (15, 16) and in humans (17, 18). In the activation of cyclin B–cdc2 at mitosis, cdc25 and cyclin B–cdc2 interact in a positive-feedback loop, in which cdc25 is phosphorylated and activated by cyclin B–cdc2 (19, 20) (see Chapter 5).

2.1.2 cdc2 in yeast acts at two points in the cell cycle

In fission yeast, *cdc2* conditional mutants have the unusual property compared to other *cdc* mutants of arresting under restrictive conditions at one of two distinct points in the cell cycle—either at START, where the cell commits itself to another round of DNA replication (rather than to the alternative pathways of sporulation or mating), or at the G2/M transition (21). Thus, the same protein kinase molecule is capable of fulfilling two different regulatory roles in the cell cycle. This is because cdc2 is associated with the G1 cyclins for its function at START, but with the mitotic cyclins at the G2/M transition (reviewed in Chapter 4). The two types of cyclin–cdc2 complexes have either a different substrate specificity, or different sets of substrates are available to them through temporal or spatial regulation in the cell. Based on this seminal work in yeast, it was assumed that the same principles would apply to the cell cycle in multicellular organisms. However, it soon became apparent that in animal cells there are several protein kinases related in structure to cdc2, and that these regulate separate cell cycle checkpoints.

2.1.3 There is more than one cell cycle kinase in animal cells

There are several lines of evidence that the original 'cdc2', isolated from human cells by Lee and Nurse (2) through its ability to replace a defective fission yeast *cdc2* gene, functions primarily in the regulation of the entry into mitosis, and not in the initiation of DNA replication. The mouse tissue culture cell line FT210, which was selected as a *cdc* conditional mutant, has a temperature-sensitive cdc2 (22, 23). At the restrictive temperature these cells only arrest at the G2/M transition and have no apparent defect in the initiation of DNA replication, suggesting that cdc2 is only required at mitosis (24). Similarly, when tissue culture cells are micro-injected with anti-cdc2 antibodies they are still able to replicate their DNA but cannot enter mitosis (25, 26). The alternative explanation to these results, that the mutant or antibody-bound cdc2 is unable to bind mitotic cyclins but is still able to bind G1 cyclins, is made less likely by complementary evidence from studies on the introduction of a dominant-negative cdc2 mutant into tissue culture cells, and from frog egg cell-free extracts. The dominant-negative cdc2 stops cells entering mitosis, but has no effect on DNA replication (27). When frog egg cell-free extracts are immunodepleted of cdc2 they are unable to enter mitosis, but, as in mammalian cells, demonstrate no defect in DNA replication (28). However, as a caveat to concluding that cdc2 has no role in DNA replication in vertebrates, there is a report that anti-sense cdc2 oligonucleotides prevent T cells from replicating their DNA after antigen stimulation, presumably by blocking accumulation of cdc2 protein which is present at very low levels in resting T cells (29). Thus, there may be cell-type differences in which protein kinase controls each cell cycle checkpoint, and some facets of cell cycle regulation may differ in cells other than fibroblasts with which most studies have been performed. Nevertheless, cdc2 has not been found as an active protein kinase in association with any of the candidate vertebrate G1 cyclins. Instead, some of the G1 cyclins were found to co-immunoprecipitate with ~ 34 kDa proteins distinct from cdc2 that are recognized by an antibody made to the PSTAIRE epitope. These studies suggested that there are other protein kinases related in sequence to cdc2 involved in the regulation of DNA synthesis in animal cells. Several candidate protein kinases have now been isolated.

2.2 A family of cdc2-related proteins identified by PCR

The most extensive set of cdc2-related cDNAs has been identified through the use of the polymerase chain reaction (PCR). Several groups have used PCR to identify cdc2-like proteins in human, *Xenopus*, and *Drosophila* cDNA libraries (Fig. 1 and Table 1). In all, twelve cdc2-related cDNAs were isolated from human cells (30), two from *Drosophila* (31), and two from *Xenopus* (32, 33). These cDNAs were assayed for their ability to complement *cdc2* in fission yeast, or to complement the *cdc2* homologue, *CDC28*, in budding yeast. Only one of the *Xenopus* and one of the *Drosophila* cDNAs was able to replace *cdc2* or *CDC28* in yeast cells, and is therefore defined as the *cdc2* homologue (cdk1) in these organisms. The function

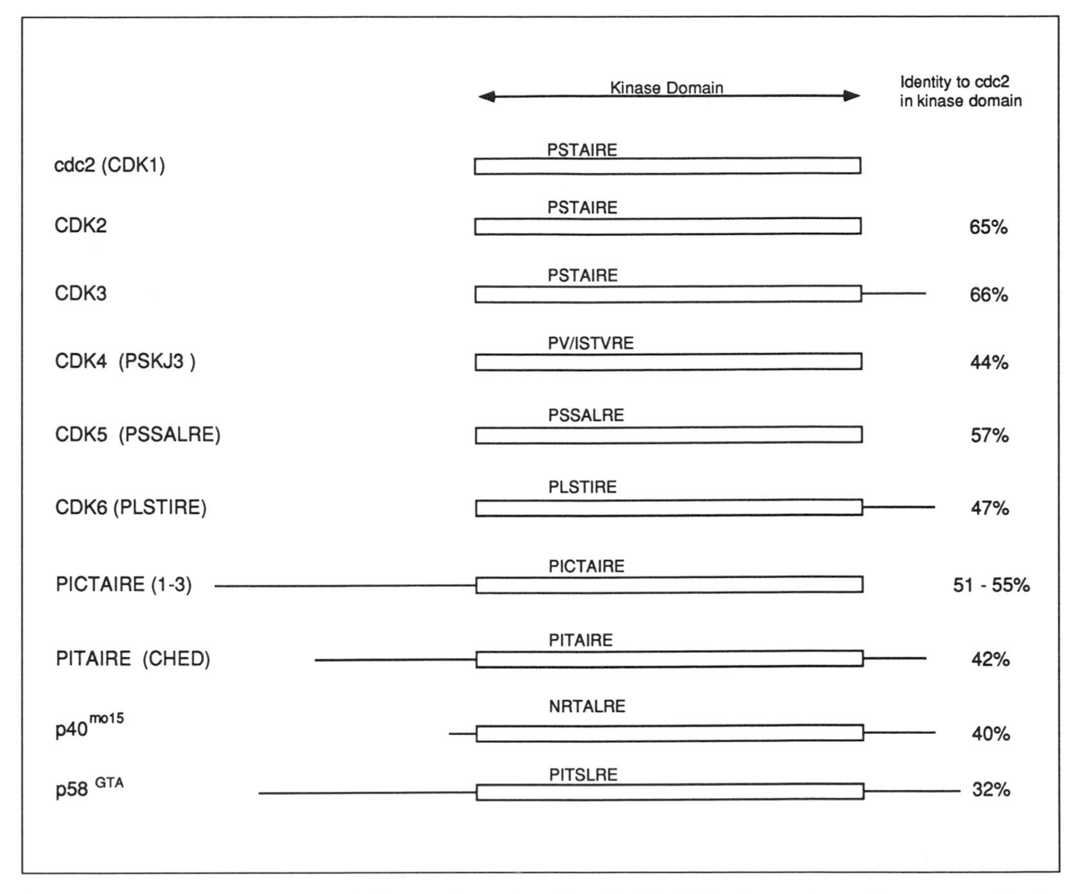

Fig. 1 Schematic of the various cdc2-related proteins identified by PCR. The amino acid sequences in the conserved PSTAIRE motif and the overall identity to $p34^{cdc2}$ in the kinase domain are shown.

of the non-complementing cdc2-related protein, CDC2c, in *Drosophila* is becoming clearer. It apparently binds with cyclin E, and thus may be involved in the regulation of E2F-mediated transcription in G1 phase (P. O'Farrell, personal communication). We shall return to this point in Section 4.4. The non-complementing *Xenopus* cDNA, now called cdk2, but originally called Eg1 (33), appears to be involved in DNA replication. The evidence for this comes from the converse experiment to that described for *Xenopus* cdc2 above. When frog egg cell-free extracts are immuno-depleted of cdk2 they are unable to replicate DNA, but are still able to enter mitosis (28). The human homologue to frog cdk2 is 90 per cent identical in sequence, but surprisingly, human cdk2 is able to complement in yeast, whereas frog cdk2 cannot do so. Nevertheless, given the high degree of homology to frog cdk2, human cdk2 is a likely candidate for the protein kinase involved in the regulation of DNA replication, and the introduction of a dominant negative form of human cdk2 suggests that it is also required for the G1 to S transition (27) (see Chapter 7).

Table 1 Cdks and their activating cyclins, and other cdc2-related cDNAs

CDK	PSTAIRE epitope	Associated cyclin	Cell cycle stage	% Identity to CDK1 (human)	Reference
cdc2Hs (CDK1)	EGV**PSTAIRE**ISLLKE	A, B	G2 → M,	100	2
cdc2X1 *Xenopus*	EGV**PSTAIRE**ISLLKE	A, B	G2 → M,	87	32
cdc2Dm *Drosophila*	EGV**PSTAIRE**ISLLKE	A, b	G2 → M,	72	31
CDK2Hs	EGV**PSTAIRE**ISLLKE	A, D, E	G1 → S, S	65	34–36, 38
CDK2 *Xenopus*	EGV**PSTAIRE**ISLLKE	E	G1 → S, S	64	33
Cdc2c *Drosophila*	EGV**PSTAIRE**ISLLKE	E	G1 → S, S	57	31
CDK3	EGV**PSTAIRE**ISLLKE	N/D	N/D	66	30
CDK4 (PSKJ3)	GGL**PV/ISTVRE**VALLRR	D	G1 → S, S (?)	44	37
CDK5 (PSSLARE)	EGV**PSSALRE**ICLLKE	D	G1 → S (?)	57	30, 38
CDK6 (PLISTRE)	EGM**PLISTRE**VAVLRH	D	N/D	47	39, 41
PCTAIRE 1–3	EGA**PCTAIRE**VSLLKD	N/D	N/D	51–55	30
CHED (PITAIRE)	EGF**PITAIRE**IKILRQ	N/D	N/D	42	30
$p40^{mo15}$	DGI**NRTALRE**IKLLQE	H	N/D	40	42–44
p58-GTA	EGF**PITSLRE**INTILK	N/D	N/D	32	64

Of the twelve cdc2-related human cDNAs that have been isolated, three were able to complement a defective budding yeast *CDC28* gene (30). One of these was the cDNA originally isolated by Lee and Nurse, and the other two cDNAs were about 70 per cent identical to it in predicted amino acid sequence. Because these proteins were able to function properly in the yeast cell cycle, they probably play a role in the human cell cycle and associate with cyclins. Thus, these three cDNAs have been designated cdk1 (the original Lee and Nurse *cdc2* protein), cdk2 (34–36), and cdk3 (30). The role of cdk3 in the cell cycle is still not clear. No cyclin partner has yet been found, but a dominant negative form of cdk3 will delay cells in G1 phase, suggesting that cdk3, like cdk2, is involved in the G1–S transition (27).

Although they were unable to complement *cdc2* or *CDC28* in yeast, three of the other cdc2-related cDNAs have now been shown to be associated with cyclins, and have therefore been designated cdk4 (37), cdk5 (38) (Table 1), and cdk6 (39). All three cdks bind the D-type cyclins, but the complexes comprising cdk4 and cdk6 are the most active cyclin D-directed protein kinases in G1 phase (39–41).

At the time of writing, a new cyclin and CDK interaction has just been identified. The $p40^{MO15}$ protein is known to be the cdc2-activating kinase (CAK) responsible for phosphorylating p34cdc2 on T161 (42–44), and it has now been shown to have a cyclin-like partner, called cyclin H (45, 46). So $p40^{MO15}$ has itself been renamed CDK7. Cyclin H is most similar (20 per cent identical) to the yeast cyclins mcs2 (47) and ccl1 (48), and its closest mammalian relative is cyclin C, although cyclin C itself is unable to activate human $p40^{MO15}$ (45). There is little change in CAK activity through the cell cycle, although there is a decrease in G0 (46, 45). In addition CAK is exclusively in the nucleus (46) which raises the issue of how T161 is phosphorylated in the cytoplasmic cyclin B1/Cdc2 (see below).

3. cdk partners: cyclins A to *n* (where *n* is a large number)

Thus there are at least six true cdks in animal cells. What is known of their activating cyclin partners? To date there are eight types of cyclin, from A to H that all share an ~ 150 amino acid region of homology called the 'cyclin box' (49) (Fig. 2). From deletion and point mutation studies it is clear that the cyclin box region is the part of the protein that binds to the cdk (50, 51). Cyclins can be roughly divided into two subfamilies; the G1 cyclins (C, D, and E), and the original mitotic cyclins (A and B). Too little is known of cyclins F and G (52) to assign them to either of the two groups as yet. The G1 cyclins differ from the mitotic cyclins in their overall primary structure, and this has implications for their stability in the cell cycle. The G1 cyclins are short-lived proteins that are rapidly turned over throughout the cell cycle, and thus the levels of G1 cyclins are determined by the rate of transcription of their mRNAs (reviewed in ref. 53).

Compared with the mitotic cyclins, the G1 cyclins have a longer C-terminal

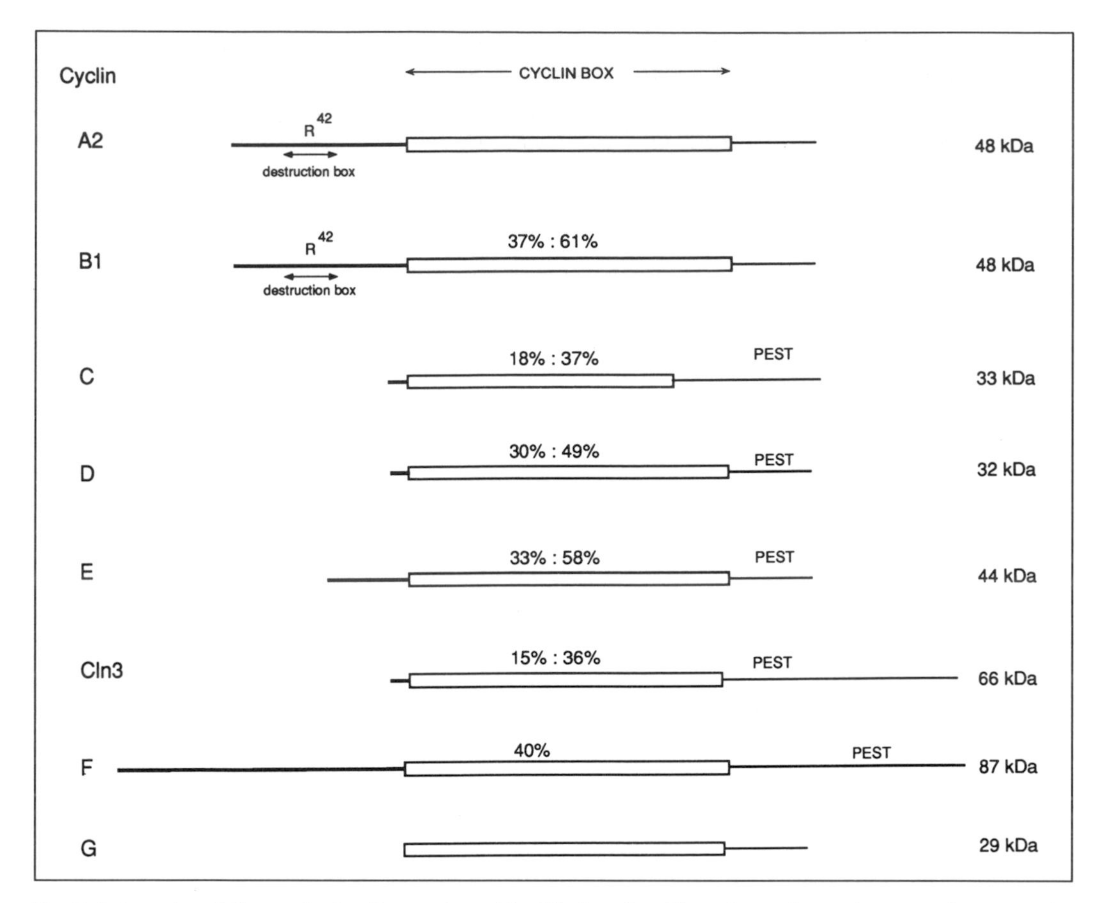

Fig. 2 Schematic of the cyclin family members identified so far. The schematics and scores shown are for human cyclins, with the exception of the budding yeast Cln3. The per cent identity: per cent similarity to human cyclin A in the cyclin box, and the relative positions of the mitotic destruction box and G1 cyclin PEST sequences are shown. (PEST sequences have been correlated with rapidly turned over proteins.)

sequence after the cyclin box, and it is this part of the protein that seems to confer instability on the G1 cyclins (Fig. 2). In contrast, the mitotic cyclins, A and B, are completely stable throughout interphase, but are rapidly proteolysed in mitosis by an ubiquitin-dependent pathway (54). Cyclin A and B transcription is also cell-cycle dependent (55–57), and so the levels of cyclins A and B are determined both by their transcription, and by proteolysis (58–60). Cyclin A and B destruction is in part regulated by a conserved region—the destruction box, containing a conserved arginine residue—that is approximately mid-way from the N-terminus to the cyclin box, and becomes ubiquitinated in mitosis (54). The destruction box region differs between the A and B cyclins, which probably accounts for the finding that in mitosis, cyclin A is degraded before cyclin B. Significantly, the cyclin box in the G1 cyclins is at the N-terminus, and thus G1 cyclins lack the mitotic cyclin destruction box.

Cyclins A, B, D, and E have been paired with cdks (Table 1), and we shall deal with them in detail below. There are no published protein data on cyclins F and G, and thus no indication as yet with which cdk they are associated (although cyclin F is reported to bind to one of the PCTAIRE proteins *in vitro*—M. Meyerson and E. Harlow, personal communication). Cyclin C has an associated protein kinase activity when immunopurified from *Drosophila* embryos (61) but the protein kinase responsible has not been identified, although it does bind to $p13^{suc1}$, a protein known to bind cdc2-related proteins (62). It is also not clear what role, if any, the cyclin C–cdk complex plays in the cell cycle, although cyclin C is highly conserved between man and *Drosophila* (71 per cent identity), which indicates that it may be important in cell regulation. In serum-stimulated cells, cyclin C mRNA increases in early G1, before those of the other cyclins (63). This has been taken as an indication that cyclin C may be involved in regulating the transition from the quiescent (G0) to the cycling state, but as yet there are no other clues to its function in multicellular organisms. Recently, a novel cyclin (*mcs2*) has been isolated from fission yeast that is most closely related in sequence to the C-type cyclins (47). *msc2* was first isolated as a suppressor of mitotic catastrophe in a mutant *cdc2* strain, which indicates that *msc2* may act in the induction of mitosis. The mcs2 cyclin does not bind to cdc2, but from genetic evidence it functionally interacts with, and so may associate with, another protein kinase, cyk1, that has ~ 30–33 per cent identity with the cdc2 family (47). Although *msc2* and cyclin C may not be homologues, they might perform analogous roles in the cell cycle.

4. The roles of cdk1 and cdk2

4.1 Cyclin B–cdk1: the M phase kinase

The B-type cyclins (B1 and B2) only associate with cdk1, and these complexes are the mitosis-specific kinases. Although the level of cdk1 protein is constant through the cell cycle, its mRNA is transcribed in a cell cycle-dependent fashion (65–67). cdk1 mRNA transcription and synthesis increase as cells enter S phase, and decline in early G1 phase (in a similar manner to cyclin A2 mRNA). Thus the rate of cdk1 proteolysis must increase in parallel with the increase in cdk1 mRNA transcription and synthesis to maintain a constant level of cdk1 (67). This has led to the suggestion that the two pools of cdk1 in the cell, newly synthesized, and 'old' cdk1, may have different properties (65). Such a model ties in with evidence from studies on fission yeast showing that if the pool of cdc2 in G2 cells is degraded, newly synthesized cdc2 will drive the cell into another round of DNA replication rather than into mitosis (68). Similarly, there may be two pools of cdk4, given the finding that cyclin D begins to associate with cdk4 once cdk4-synthesis increases in late G1 phase (37). Therefore, one of the roles of the different cyclin–cdk complexes may be to act as markers by which the cell can judge its progress through the cell cycle.

The subcellular location of the cyclin B1– and B2–cdk1 complexes is also cell cycle regulated. During S and G2 phases the complexes accumulate in the cytoplasm, becoming associated with the centrosomes late in interphase (69–72), but once cells reach mitosis the complexes move into the nucleus (73–75) (Fig. 3). The cyclin B–cdk1 complex enters the nucleus at the beginning of prophase, before nuclear envelope breakdown, and may well be the protein kinase responsible for the phosphorylation and consequent dissolution of the nuclear lamina (76). The relocalization of cyclin B–cdk1 is conserved from starfish to man, and may be another method by which the cell prevents its premature activation. Once it has entered the nucleus, cyclin B1–cdk1 becomes associated with the spindle, especially the spindle caps and main spindle fibres. There are data to suggest that the active cyclin B–cdk1 complexes can have profound effects on microtubules (72), but their exact role in organizing the spindle has yet to be defined (reviewed in ref. 77).

As mentioned above, the activity of the cyclin B–cdk1 complexes is negatively regulated by phosphorylation on the threonine 14 (T14) and tyrosine 15 (Y15) residues in the ATP-binding region of cdk1, promoted by the wee1 and mik1 protein kinases. In order for cells to enter mitosis, the cyclin B–cdk1 complexes are activated by the cdc25 phosphatase (reviewed in more detail in Chapter 5). The highly regulated activation of the cyclin B–cdk1 complexes prompts the question as to whether the other cyclin–cdk complexes are also regulated by phosphorylation and dephosphorylation of T14 and Y15.

4.2 Cyclin A–cdk2 and cdk1: the S and M cyclin?

Cyclin A has been implicated in the regulation of both DNA replication (78), and of mitosis (56, 78). Although there are two types of cyclin A, cyclin A1 and A2, at the moment it appears that cyclin A1 is restricted to a role in meiosis (59). Like the D-type and E-type cyclins, the overproduction of A-type cyclins has been correlated with oncogenesis. The human cyclin A2 gene was identified as the site of integration of a fragment of hepatitis B virus (HBV) in a hepatocellular carcinoma (79). The integration resulted in the expression of a stabilized, chimeric cyclin A2 protein, in which the first ~ 200 amino acids (which includes the destruction box) of cyclin A2 are replaced with the PreS protein of the HBV, thus preventing its destruction at mitosis (80).

Cyclin A2 binds to both cdk1 and cdk2 in a cell cycle-dependent manner. Cyclin A2 synthesis begins as cells enter S phase and at this point it is exclusively associated with cdk2 (36, 56). There is fairly good evidence that the cyclin A–cdk2 complex is required for DNA replication, though not for the G1 to S transition itself (see also Chapter 7). Three groups have reported that anti-cyclin A2 antibodies will inhibit DNA replication when microinjected into cells, as judged by failure to incorporate BrdU (78, 81, 82). However, there is a recent report that, as judged by the incorporation of ^{3}H-thymidine, cells microinjected with anti-cyclin A2 antibodies are able to replicate DNA, but synthesize less than 10 per cent of the DNA compared with control cells (E. Harlow, personal communication). This suggests that cyclin A2 is not required to initiate DNA replication, but is necessary for its continued synthesis. Thus a compromise view of the relative roles of the cyclins would be that the cyclin E–cdk2 complex is responsible for the G1–S transition and the initiation of DNA replication, whereas cyclin A–cdk2 is required for progression through S phase. (Though one should not exclude the possibility that the cyclin D–cdk complexes also have a role in this process.) As mentioned above, the substrates for cyclin–cdk complexes in S phase remain obscure; one of the candidates is the RF-A protein identified as part of the DNA replication complex (reviewed in Chapter 7).

Recent data suggest that one of the roles of the cyclin A2–cdk2 complex in cell proliferation is connected with cell adhesion (83). It appears that a stable cell line of NRK cells expressing ectopic cyclin A2 is able to grow in suspension, whereas the parental cell line is anchorage-dependent. This effect is specific to cyclin A2; ectopic expression of cyclin D1 or E does not allow the cells to grow in suspension. Thus it is suggested that one of the events required for cyclin A transcription is a signal from surface adhesion molecules, in either late G1 or very early S phase (83).

Once cells progress into late G2 phase, cyclin A2 becomes associated with cdk1 (78). The exact roles of the cyclin A–cdk1 complex versus the cyclin A–cdk2 complex have not been defined. Indeed, in certain HeLa cell clones, cyclin A2 only forms a complex with cdk2 (56), not with cdk1, suggesting that cyclin A–cdk1 is not essential for cell cycle regulation. However, the caveat to this is that HeLa cells have been transformed by the E7 protein which may change the behaviour of cyclin A. One potential role for cyclin A–cdk1/cdk2 in late G2 phase is in the reorganization of the cytoskeleton in preparation for mitosis (72).

4.3 Cyclin E–cdk2: the G1/S kinase?

Both the D-type and the E-type cyclin cdk complexes have been implicated in the regulation of the G1/S transition. Of all the putative G1 cyclins, cyclin E is the most closely related in sequence to the original mitotic cyclins, A and B (63, 84). Cyclin E primarily binds to cdk2 *in vivo*, although it also forms a complex with cdk1 (84, 85). However, the cyclin E–cdk1 complex does not have detectable protein kinase activity when isolated from asynchronously growing cells, although it may only be active at one specific point in the cell cycle. Cyclin E mRNA and protein levels, and the activity of the cyclin E–cdk2 complex all peak at the G1/S transition, sharply declining as cells progress through mid and late S phase (85). Thus, the behaviour of the cyclin E–cdk2 complex correlates with a role in the initiation of DNA replication, and there is good evidence to support this. The most compelling data are from the studies on a cyclin E mutation in *Drosophila*. In these flies, once the maternal store of cyclin E mRNA is degraded at cycle 15, all cell division ceases and cells arrest in G1 phase (86). Thus cyclin E is essential for the cells to enter S phase. By contrast, flies that are deficient in cyclin A are able to go through S phase, but all the cells arrest in G2 phase (87). When anti-cyclin E antibodies are microinjected into mammalian cells they partially prevent cells from replicating their DNA, as judged by a reduction in new DNA synthesis estimated by BrdU or ^{3}H-thymidine labelling (E. Harlow, personal communication). Additionally, when cyclin E is overexpressed in cells, the cells progress through G1 and into S phase at a faster rate (88). Such cells also have a diminished requirement for growth factors, indicating that cyclin E may overlap with the D-type cyclins in integrating growth factor signal transduction into the cell cycle. The amplification of cyclin E is also seen in a significant percentage of breast cancer tissue (89).

TGFβ1 causes cells to arrest in G1 with Rb in its unphosphorylated state, and mediates this effect primarily through the cyclin E–cdk2 complex; although there is a report that it also down-regulates cdk4 (90), and thus depresses the formation of the cyclin D kinase (see Section 5.1). TGFβ-treatment inhibits the appearance of cyclin E–cdk2 protein kinase activity both by preventing the cyclin E–cdk2 complex from assembling, and by preventing the activation of any newly formed complex (91). The activation of the cyclin E–cdk2 complex also appears to be the point at which radiation damage in G1 phase can delay the cell cycle (92). These effects in part are mediated through small inhibitor proteins (CDI) that bind and inhibit cyclin–cdk complexes, and may or may not also explain the defect in cyclin–cdk complex assembly. CDIs will be dealt with in more detail in Section 6.2.

4.4 Cyclin E and cyclin A interact with transcription factors

Cyclin E shares some characteristics with cyclin A, which regulates cells once they have entered S phase. Both cyclin E and cyclin A in conjunction with cdk2 associate with the Rb-related protein p107, and with the transcription factor E2F (93–96) (see Fig. 4). There are now convincing data to suggest that the E2F family of factors are responsible for stimulating the transcription of many genes required for DNA replication, such as ribonucleotide reductase and DNA polymerase α. It thus appears that the E2F family may be the human homologue to SBF and MBF in budding yeast. In *S. cerevisiae* there is an intricate interplay between the transcription factors SBF (a complex of the Swi4 and Swi6 proteins) and MBF (a complex of Swi4 and $p120^{MBP1}$ proteins), and the different G1 cyclin (Cln) and mitotic cyclin (Clb)–Cdc28 protein kinase complexes (reviewed in Chapter 3). Suffice it to say here that the Cln–Cdc28 kinases and the Swi4/Swi6 transcription factor complex are involved in a positive feedback loop at START (97–99). Later in the cell cycle, the Clb–Cdc28 complexes inhibit the activity of SBF and may activate MBF (100).

It is not clear exactly what effect, stimulatory or inhibitory, cyclin A or cyclin E association with E2F has on the transcription factor. At present, all data suggest that E2F is most active when freed from any associating proteins, but it is possible the cyclin–cdk complexes modulate E2F activity rather than completely repress it. E2F apparently acts at the minimum as a heterodimer when it activates transcription, and two of the components involved have been cloned. The protein common to all forms of E2F is called DP-1, and the other is (confusingly) called E2F. An additional complication is that at least three different types of E2F—E2F1, 2, and 3—have been cloned, but the roles in transcription of the different E2F forms have

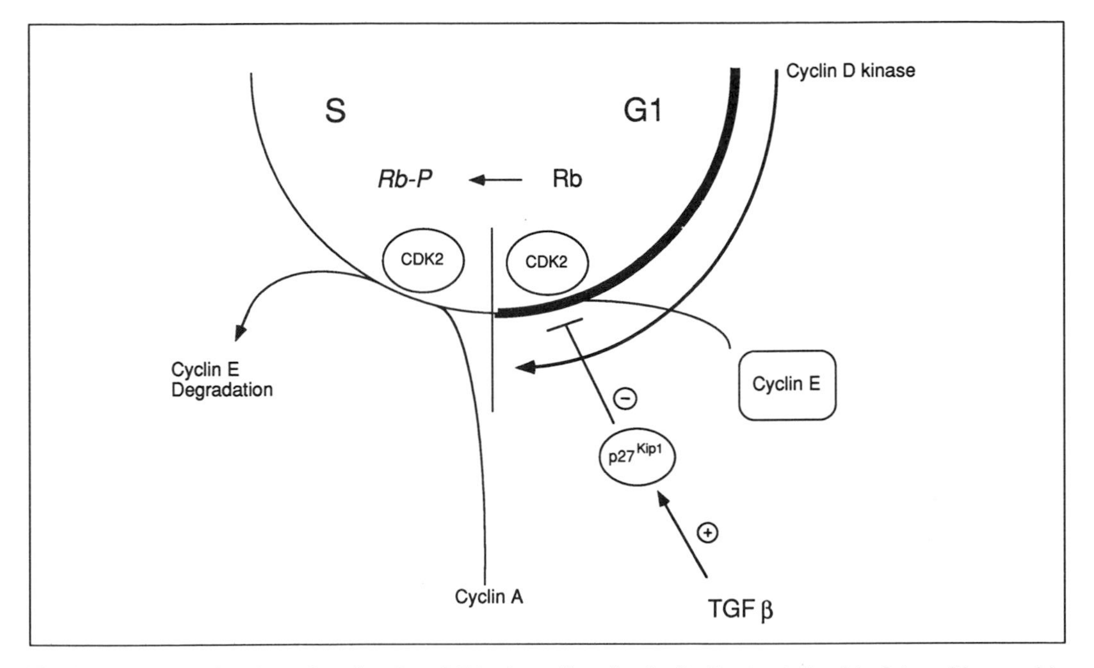

Fig. 3 The points of action of cyclins A and E in the cell cycle. Cyclin E acts at the G1–S transition, and is negatively regulated by TGFβ via a CDK inhibitor, $p27^{Kip1}$, that inhibits the kinase activity of the cyclin E–CDK2 complex. Cyclin E is degraded early in S phase and is replaced by cyclin A, whose synthesis begins once cells have entered S phase.

not yet been demarcated. Like Rb, p107 inhibits SAOS-2 cell growth in G1, and the inactivation of p107 could be one reason for the formation of the cyclin E–p107 complex. Both Rb and p107 inhibit E2F transcriptional activation (101), and therefore one might expect the cyclin E–p107–E2F complex to activate transcription. Another possibility is that while bound to DNA via E2F, the associated cyclin/cdk can phosphorylate an adjacent DNA-bound protein. Interestingly, in human cells the CCG1 protein, which causes a G1 arrest when mutated (102), has recently been shown to be TAF250, a component of the basic transcription machinery, and to have homology to Swi4. CCG1 associates with the TATA binding protein (TBP) in the TFIID complex, forming part of the RNA polymerase II initiation complex (102–104). This makes CCG1 a likely candidate to be a part of the machinery involved in the regulation of cell cycle transcription, and thus a potential target for the cyclin–cdk complexes. Alternatively, the association between cyclin–cdks and transcription factors in human cells may be connected with emerging evidence that sites of transcription are used as origins of replication (105, 106).

Whatever its function, the cyclin E–p107–E2F complex accumulates in late G1 and peaks in early S phase (96), in parallel with the synthesis of cyclin E. This complex disappears with the destruction of cyclin E in mid to late S phase, and is replaced by one containing p107, E2F, and cyclin A2 which persists through the rest of the cell cycle. Cyclin A2 is able to bind to either p107 or to E2F1 independ-

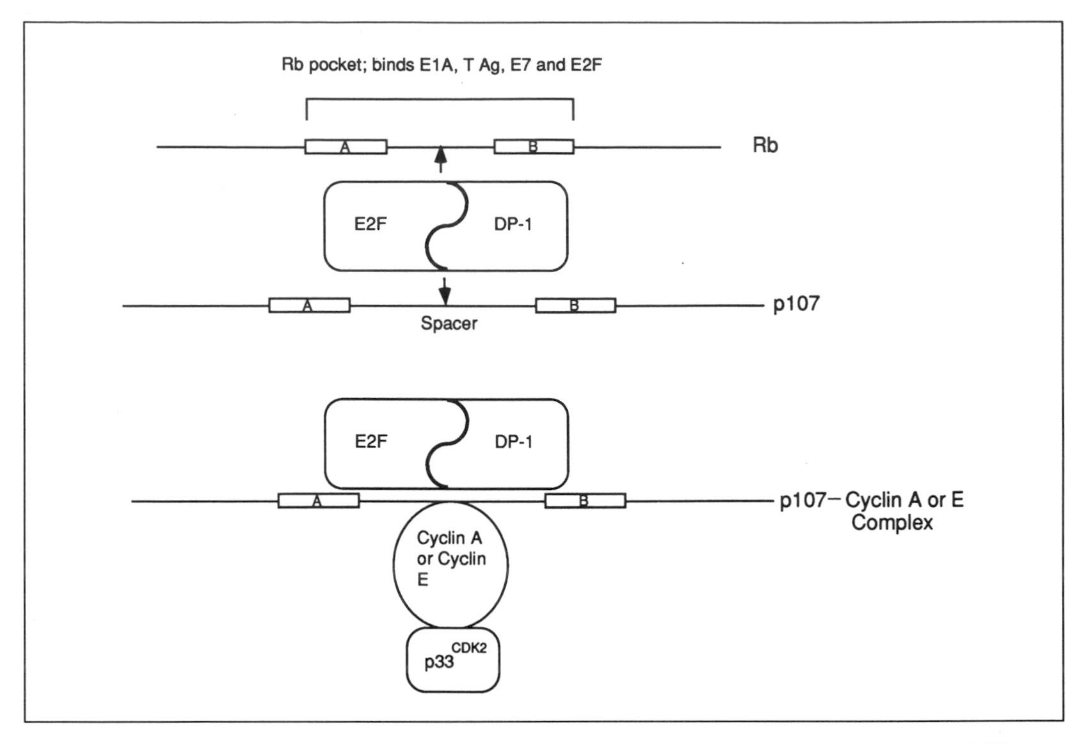

Fig. 4 The interaction between Rb-related proteins, the transcription factor E2F, and the cyclin–CDK complexes are shown. The Rb pocket is defined as two conserved regions—the A and B boxes—separated by a spacer region. E2F is thought to bind to p107 as a heterodimer, one component of which is DP-1 and the other is an as yet unidentified member of the E2F family.

ently. The sites of interaction between cyclin A2 and p107 and E2F1 have been mapped. Cyclin A2 binds to the first 51 residues of E2F1 (E. Harlow, personal communication), and to the 'spacer' region of p107 (107). The spacer is the non-conserved region between p107 and Rb, which lies between the two regions that are homologous between the two proteins—the 'A' and 'B' boxes (see Fig. 4). The A and B boxes of Rb and p107 are the sites of interaction with various viral oncoproteins such as SV40 T antigen, adenovirus E1A and the papilloma virus E7 protein (see Fig. 4). Thus when these viral proteins bind to p107, they form a ternary complex with cyclin A–cdk2. Interestingly, the papilloma E7 protein is also able to bind directly to cyclin A2 (108), and may therefore be targeting cyclin A2 for an additional purpose compared with adenovirus E1A or SV40 T antigen.

5. The roles of cdk2, 4, and 5

5.1 Signal transduction and all that

There are three types of cyclin D, which vary in amounts in different tissues and cell types (63, 109, 110). The three D-type cyclins are not redundant; cyclin D1

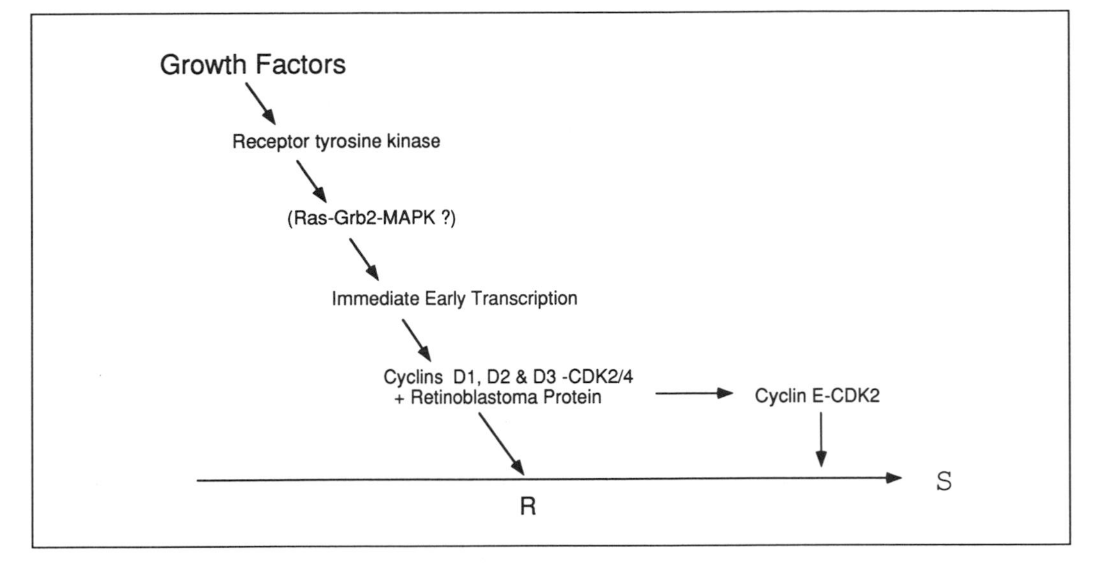

Fig. 5 Signal transduction and the potential roles of the D-type cyclins. The synthesis of the D-type cyclins is absolutely dependent on the presence of serum, although the cyclins act around the restriction point (R). There are some indications that the D2 and D3 cyclins act upstream of cyclin D1. The D-type cyclins are thought to interact with the retinoblastoma protein, and to be upstream of cyclin E.

plays a different role in differentiation, and has a different affinity for the retinoblastoma protein (Rb) *in vitro*, compared with cyclins D2 and D3. Protein kinase partners for the D-type cyclins have also been isolated. They are cdk2 (38), cdk4 (previously isolated as PSKJ3) (37), cdk5 (previously known as PSSALRE) (38), and cdk6 (PLISTRE) (39, 41). Of these, the most important in proliferating cells appears to be cdk4 (although there may be some overlap with cdk6). In addition, D-type cyclins co-immunoprecipitate a 21 kDa protein (p21), and PCNA (38) (see below).

D-type cyclins are the most closely linked with signal transduction (Fig. 4). D-type cyclin mRNA and protein synthesis are dependent on growth factors. When quiescent BAC1 mouse macrophages are stimulated by CSF-1, cyclin D1 mRNA and protein first appear 4 to 6 h later (109), at about the time that the cells become committed to a round of DNA synthesis, and remain elevated through the rest of the cell cycle. (Although some experiments suggest that cyclin D1 does vary in level through the cell cycle (111). The anomaly may reflect a difference between cell types.) If growth factors are withdrawn during the cell cycle, cyclin D transcription stops and the mRNAs disappear. Because D-type cyclins are short-lived proteins this results in the rapid disappearance of the proteins too. Similarly, cyclin D1 transcription is induced after growth factor stimulation of quiescent normal human fibroblasts; cyclin D1 and D3 mRNAs appear at the same time as the level of *fos* mRNA declines, and reach a maximum prior to S phase (112). Microinjection experiments suggest that cyclin D1 acts before S phase. When anti-cyclin D1 antibodies are microinjected into serum-starved fibroblasts they block cells from replicating their DNA. Anti-cyclin D1 antibodies prevent DNA replication when the antibodies are microinjected up to 8 h after serum stimulation, at which time cells are in mid-G1 phase, but not when microinjected at 16 h when cells are beginning to enter S phase (113). This suggests that cyclin D1 acts in mid to late G1 phase, but not in S phase itself, although it is possible that cyclin D1 is modified or becomes part of a multimeric complex in late G1 phase and is no longer recognized by the antibodies. Data suggest that the D-type cyclins may be rate-limiting for the G1–S transition. Moderate overexpression (five-fold over normal) has been found to diminish the cell's requirement for growth factors and to shorten G1 phase, thus speeding up the cell cycle as a whole (114). Cyclin D1 becomes associated with cdk4 just before the start of DNA synthesis, and the complex accumulates to a peak in early S phase, before declining in late S and G2 phases (37). The changes in the amount of cyclin D1 associated with cdk4 parallel the changes in cdk4 mRNA levels during the cell cycle, although the amount of cdk4 protein remains constant. This suggests that cyclin D1 may preferentially associate with newly synthesized cdk4—a theme we have already encountered in the association of cdk1 with cyclin B.

5.2 D-type cyclins are implicated in oncogenesis

The level of cyclin D in the cell is probably very carefully regulated, because the overproduction of cyclin D had been linked with cell transformation (dealt with in more detail in Chapter 9). Candidate oncogenes overexpressed in certain parathyroid tumours and B cell lymphomas, PRAD1 (115) and *bcl1* (116, 117) respectively, have both been identified as cyclin D1. Similarly, the *Vin1* proto-oncogene has been shown to be cyclin D2 (118). Cyclin D1 or D2 are also overexpressed in many tumour types in most cases as a result of gene amplification (119, 120). Thus the overexpression of cyclin D has been correlated with transformation, and indeed causes tumours when cells overexpressing cyclin D1 are injected into nude mice (121).

5.3 D-type cyclins interact with the retinoblastoma protein

The carcinogenic potential of the D-type cyclins must be related to their regulatory role in the cell cycle, but as yet there are few indications as to their substrates. A clue to the substrates of cyclin D1–cdk4 comes from the observation that the cell cycle is not inhibited by microinjected anti-cyclin D1 antibodies in cells that lack a functional Rb, or that are transformed by SV40 T antigen (122). This implies that the only important substrates of cyclin D1–cdk4 are Rb and perhaps other targets of T antigen. Rb is thus far the only *in vitro* substrate for the D-type cyclin–cdk complexes (123), and is phosphorylated on the same sites *in vitro* by cyclin D1–cdk4 as are phosphorylated at the G1/S transition *in vivo*. Immunoprecipitates of cyclin D1 from synchronized cells have protein kinase activity from late G1 phase onwards, and these immunoprecipitates will phosphorylate Rb. Cyclin D1 is able to overcome the growth-inhibitory action of Rb when transfected into SAOS-2 cells, although Rb does not become noticeably more phosphorylated in these cells (124). In contrast, in another study in which cyclin D2 was found to reverse Rb-mediated inhibition of SAOS-2 cell growth there was concomitant Rb hyperphosphorylation (125, 126). This difference is consistent with the fact that Rb will form a complex with cyclins D2 or D3 but not cyclin D1 *in vitro*, and in the presence of cdk4 this results in the phosphorylation of Rb and the dissociation of the cyclin D2 or D3–Rb complex (123). That the phosphorylation of Rb probably causes the complex to dissociate is shown by an experiment in which the replacement of cdk4 with an inactive mutant results in the formation of a ternary complex between cyclin D2 or D3, Rb and cdk4 (123). Although they appear to have somewhat different properties, all three D-type cyclin–cdk complexes are strongly implicated in the control of the G1/S transition through inactivation of Rb by phosphorylation.

D-type cyclins also form a complex with PCNA, the auxiliary subunit of DNA polymerase δ, and this can be immunoprecipitated from growing cells as what appears to be a tertiary complex containing cyclin D, a cdk, PCNA, and p21 (38). Thus it is possible that the D-type cyclins may be involved in DNA replication, although cyclin D substrates in S phase have not been identified. There are also contrasting data from immunofluorescence studies which suggest that cyclin D1 is a nuclear protein in G1 phase, but moves into the cytoplasm in S phase, leading to the proposal that the D-type cyclins may act as S phase inhibitors (113).

6. Regulatory versatility

6.1 Regulation of cyclin–cdk complexes by phosphorylation

In vertebrates there are multiple members of the cdc25 phosphatase family, which raises the possibility that different types of cdc25 (cdc25A, B and C) dephosphorylate separate cyclin–CDK complexes (17). There are some data to support this proposal. For example, recent data have shown that the Cdc25A phosphatase is essential for the G1/S transition (127), and is activated by cyclin E-CDK2 complex at the end of G1 phase, suggesting that these molecules may form a positive feedback loop at the initiation of DNA replication in an analogous fashion to Cdc25C and cyclin B-cdc2 (128). In addition, when antibodies specific for cdc25A or cdc25C are microinjected into cells, they cause cells to arrest at different points in the cell cycle. Anti-cdc25A antibodies cause cells to arrest in mitosis, whereas anti-cdc25C antibodies block cells at the G2/M transition, suggesting that the 2 types of phosphatase act on different cell cycle regulators (17, 18).

However, an additional level of complexity is that not all of the population of any one type of cyclin–cdk complex is regulated in the same manner. Some of the cdk2 in the cyclin A–cdk2 complex is phosphorylated on T14/Y15, and these residues can be dephosphorylated by cdc25C or cdc25B *in vitro*, resulting in an increase in the protein kinase activity of the complex (129–131). But at some stages in the cell cycle, only a fraction of the cdk2 associated with cyclin A2 is phosphorylated at these residues, indicating that some cyclin A–cdk2 complexes in the cell are active and others are not (129). Similar characteristics have been observed for the cyclin E–cdk2 complexes (85, 131). One possible explanation is that there may be differential regulation of cyclin A/E–cdk2 complexes localized to different parts of the cell. For example, cyclin A/E–cdk2 associated with p107 and/or E2F may be regulated separately from those complexes not bound to transcription factors. Phosphorylation by $p40^{MO15}$/CAK is another potential variable, because even if a cyclin-bound cdk molecule is not phosphorylated on T14 and Y15, it will not be active unless it is phosphorylated by $p40^{MO15}$/CAK at T161.

There are no data concerning the phosphorylation state of cdk4, cdk5, or cdk6, but it is interesting to note that cdk4 and cdk6 differ from the other cdks in that they only have a tyrosine residue in the ATP binding site, the adjacent residue is not threonine but alanine. Thus, if the cyclin D–cdk4 or cyclin D–cdk6 complexes are regulated by phosphorylation, they can only be regulated by a tyrosine kinase and tyrosine phosphatase.

The tyrosine kinases responsible for cdc2 phosphorylation are the products of the *wee1* and *mik1* genes in *S. pombe*. A putative human wee1 homologue, upon isolation of a full length clone, now seems to be more closely related to mik1, and will phosphorylate cdk1 on Y15 (10). A true human wee1 gene has not yet been isolated, and nor has a T14-specific protein kinase, though these may be one and the same gene. There is evidence that the T14 kinase activity is physically separable from the Y15 kinase activity in frog egg extracts (132). Given the precedent of multiple cdc25 family members, it is possible that there may be more than one human wee1 or mik1 homologue, each specific for a different cyclin–cdk complex. In yeast and frogs, wee1 is itself negatively regulated by the nim1 protein kinase (133, 134), and by an unidentified protein kinase at mitosis (135). Thus each human wee1 homologue could be regulated by a specific nim1 homologue.

6.2 Negative regulation of cyclin–cdk complexes by CDIs; a second link to oncogenesis

The latest motif to appear in the regulation of the cell cycle is inhibition by proteins that bind directly to the cyclin–cdk complexes (Fig. 6). The first example of this is the FAR1 protein in budding yeast. FAR1 is a 120 kDa protein that is required for mating factor to arrest the cell cycle at START (136). Mating factor binds to a serpentine membrane receptor that activates a MAP kinase pathway, at the end of which is the MAP kinase homologue FUS3. FUS3 phosphorylates and activates FAR1 (137, 138), which binds to and inactivates the CLN2–CDC28 kinase that is required for START.

Several recent papers have presented evidence for a series of proteins (CDIs) in human cells that bind and inactivate various cyclin–cdk complexes, although unlike FAR1, the inhibitors involved are all small proteins (20–27 kDa) (reviewed in refs 139–141). The first of these is a 21 kDa protein that was identified in several

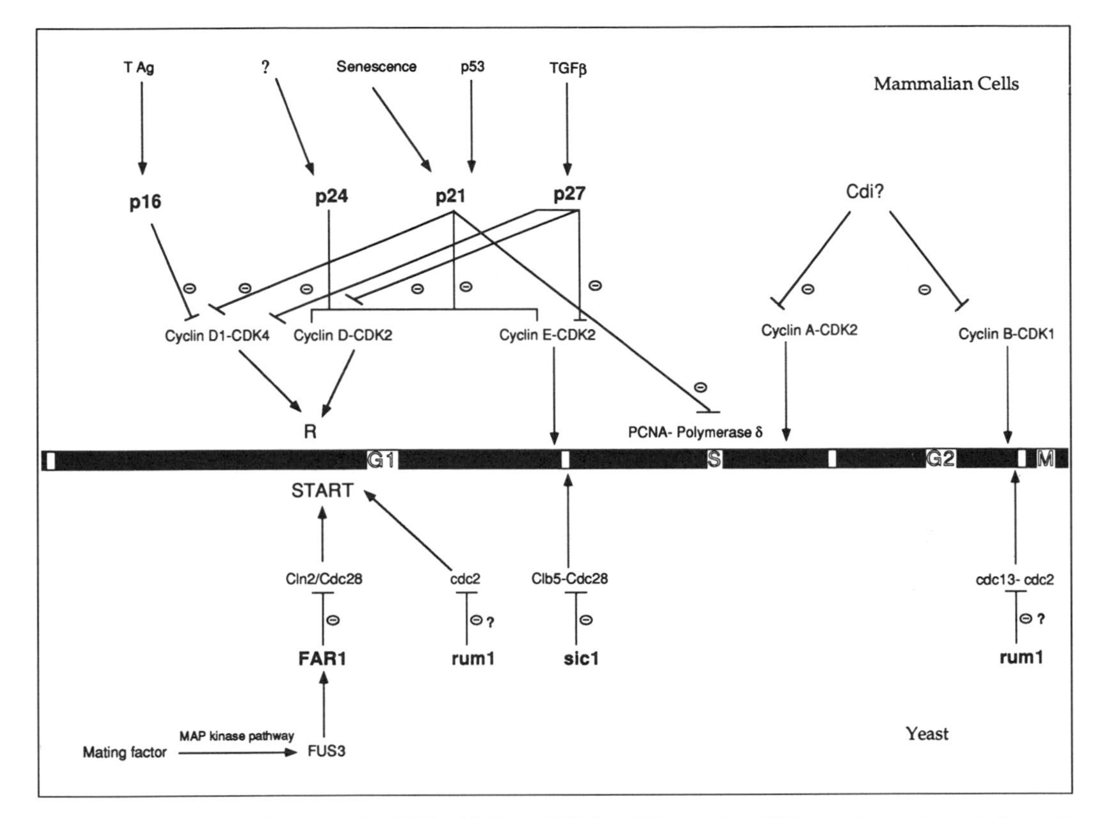

Fig. 6 The interactions between the CDK inhibitors (CDIs) and the cyclin–CDK complexes through the cell cycle. The CDIs isolated so far from mammalian cells are shown above the line, and those from budding yeast (Far1 and Sic1) and rum1 from fission yeast, below the line. Note that for most of the mammalian CDIs their inhibitory action has only been shown *in vitro*, so their *in vivo* roles are still uncertain.

different ways: in the yeast 2 hybrid system as a protein that interacts with cdc2 (Cip1) (142, 143), as an inhibitor derived from senescent cells (sdi1) (144), as a protein induced by wild-type p53 (waf1) (145), and as a protein that binds tightly to cyclin–cdk2 complexes (CAP20) (146). It has been proposed that this p21 protein be called pic1 (p53-regulated inhibitor of cdks) in future (139). p21 forms a ternary complex with several cdk2 complexes, those involving cyclin A, cyclin D1, and cyclin E, and it also binds more weakly to cdk1 and cdk3 (142). The exact mechanism by which p21 inhibits cyclin–cdk kinase activity is not clear, although it does not appear to inhibit by affecting the phosphorylation state of the cdk. An analysis of p21 mRNA in growing, quiescent, and senescent cells suggests a role for p21 in negatively regulating the entry into S phase. p21 mRNA is up-regulated as cells become quiescent, and after serum stimulation of quiescent cells, dropping away as cells enter S phase. p21 mRNA is also increased 10- to 20-fold at cellular senescence (144). The gene encoding p21 maps to chromosome 6p21.2, and the promoter region has a p53 binding site that will confer p53 inducibility on a reporter gene, and p21 is induced by wild-type but not mutant p53 (145). p21 is induced when G1 cells are irradiated in G1 phase, but not in cells that are mutant in p53 (92). Overexpression of wild-type p53 arrests cells in G1 phase, and this could be through the induction of pic1 which in turn inhibits G1 cyclin–cdk complexes. Conversely, overexpressing p21 suppress growth in various human tumour cell lines, and inhibits DNA synthesis in non-transformed cells (142). The negative effect of p21 on DNA replication in normal cells can be overcome by overexpressing SV40 T antigen, and in T antigen transformed cells p21 is absent from cyclin–cdk complexes. Thus p21 may provide one link between the response to DNA damage—induction of p53—and inhibition of the cell cycle, and its displacement from cyclin–cdk complexes may also be involved in cellular transformation by viral oncoproteins.

A recent striking observation is that p21 is able to inhibit DNA replication directly (147), without the mediation of cyclin–cdk complexes, by binding to PCNA, the auxillary subunit of DNA polymerase δ. Thus if DNA is damaged in S phase it would stop DNA replication immediately through the induction of p21. What remains unclear is how the cell overcomes p21 inhibition after DNA is repaired; whether p21 is post-translationally inactivated, or specifically degraded, or whether the other cell cycle components accumulate until they titrate it out.

The association between p21 and the DNA replication apparatus has potential parallels with two yeast CDIs that have been recently cloned and characterized. The first of these is the product of the *rum1*$^{+}$ gene in *S. pombe* (148). *rum1* is required for the proper coordination of DNA replication and mitosis in fission yeast ('rum' stands for replication uncoupled from mitosis). Cells deleted for *rum 1* are apparently incapable of judging whether the DNA is unreplicated, or whether replication is complete, and thus the cells initiate mitosis inappropriately from G1 phase. Conversely, cells that overproduce rum1 remain in a permanent G1 state and so continuously replicate their DNA without undergoing mitosis. *rum1* encodes a small protein that is able to bind and inhibit cdc2 complexes *in vitro*; therefore it is probably involved in defining fission yeast cdc2 complexes in the G1 rather than G2 state.

A budding yeast CDI, p40^{SIC1} (149) also regulates the G1 state of the cell through binding and inhibiting the S phase Clb5–CDC28 complex. Once the cell has passed START and is ready to enter S phase, p40SIC1 is degraded by the ubiquitin pathway, and the Clb5–CDC28 complex is activated to initiate DNA replication (170).

In mammalian cells, p27^{KIP1}, a CDI related in structure to p21, is responsible for blocking cells in G1 phase in response to TGFβ treatment, mostly through inhibiting cyclin E–cdk2. p27^{KIP1} and p21 are 42 per cent identical in a 60 amino acid region of the N-terminus (150, 151), suggesting that they may bind and inhibit cyclin–cdk complexes in a similar fashion. p27^{KIP1} is a more potent inhibitor of cyclin E–cdk2 than cyclin A–cdk2, and there are also data to suggest that p27^{KIP1} can inhibit the phosphorylation of T160 in cyclin E–cdk2 by p40^{MO15} (150, 152). p21 and p27^{KIP1} differ in their affinity for cdks; p21 binds most tightly to cdk2, whereas p27^{KIP1} binds more to cdk4 (150). p27 is also a more potent inhibitor of cyclin

D–cdk4 protein kinase activity (and of cyclin B–cdc2 activity) than p21. In addition, $p27^{KIP1}$ seems to be able to bind the D-type cyclins themselves—which might increase its affinity for the cyclin D–cdk4 complex even further. Indeed, more than 50 per cent of the cyclin D1–cdk4 in untreated Swiss 3T3 cells appears to be associated with $p27^{KIP1}$ in anti-p27 immunoprecipitates, whereas none of the other cyclins or cdks can be detected (151). Unlike p21, which is primarily regulated at the level of transcription, the level of $p27^{KIP1}$ mRNA and protein remains the same in quiescent and proliferating cells, and throughout the cell cycle (151). $p27^{KIP1}$ is regulated through binding a heat-labile 'masking' factor, and is unmasked when cells are treated with TGFβ, or become contact-inhibited (152). $p27^{KIP1}$ can be sequestered by cyclin D2–cdk4 complexes, which has led to the suggestion that cyclin E–cdk2 can only be activated after a sufficient amount of cyclin D–cdk4 complex has accumulated to sequester $p27^{KIP1}$ (152).

The yeast 2 hybrid system identified a second small protein that interacts with cdk1. This protein, $p24^{Cip2}$ (139) or $p24^{Cdi1}$ (153), has significant homology to VH1-type protein tyrosine phosphatases, and has phosphotyrosine phosphatase activity (but apparently not phosphoserine or phosphothreonine phosphatase activity) *in vitro* (153). Mutating the conserved cysteine residue to serine eliminates phosphatase activity, as is true for other members of this phosphatase family. p24 mRNA peaks in late G1 cells, and overexpressing the wild-type p24 will delay S phase in both budding yeast and HeLa cells. This effect is lost when the essential cysteine is mutated (153). However, the substrates for the protein are not known. p24 does not dephosphorylate Y15 in cdk1, which is reassuring because this would be expected to activate the kinase!

Two small inhibitors, $p16^{INK4}$ and $p15^{INK4B}$ have been found associated specifically with the D-type cyclin–CDK complexes. In nontransformed cells, the cyclin D–CDK4 complex is associated with p21Pic1 and PCNA. By contrast, in SV40 T antigen-transformed human cells, CDK4 is primarily bound to $p16^{INK4}$ alone (154). The sequence of $p16^{INK4}$ has 4 ankyrin repeats (155); motifs that are thought to mediate protein–protein interaction. *In vitro* p16 will bind CDK4 and displace cyclin D1, thus inhibiting cyclin D–CDK4 protein kinase activity (155). The displacement of cyclin D1 by p16 provides an explanation for the absence of cyclin D1, p21, and PCNA from CDK4 complexes in transformed cells.

p16 maps to 9p21 and has just been identified as a potential tumour suppressor gene, because this locus is rearranged, deleted or mutated in a majority of human tumour cell lines (156, 157). This has led to the suggestion that p16 may correspond to the tumour suppressor gene that maps to this region of chromosome 9 (*MTS1*), and whose inactivation has been implicated in tumourigenesis in a wide variety of cancers, and in one form of familial melanoma. However, it has subsequently been found that p16 is mutated in a much smaller fraction of primary tumours (158), implying that the deletion/modification of p16 expression in cell lines may be a secondary event during the establishment of cell lines. Thus at the time of writing, the status of p16 as a tumour suppressor is probable but unproven.

The p16 gene is adjacent to a gene encoding a very similar protein, now called p15^{INK4B} (159). Both proteins have 4 ankyrin motifs and both bind and inhibit only CDK4 and CDK6. The sequences of p16 and p15 are 44 per cent identical in the first 50 amino acids and 97 per cent identical in the following 81 amino acids. The levels of p15 mRNA and protein are induced more than 30 fold after TGFβ treatment of HaCaT keratinocytes, indicating that p15 and p27^{Kip1} are responsible for arresting cells in G1 phase in response to TGFβ.

In T antigen transformed cells, another novel protein, p19, is associated with CDK4 (154), but it has not yet been cloned and thus its effect on the kinase complex is not known.

Because the CDIs were only identified in 1993, there are many unanswered questions regarding them. How many CDIs are there? What is their role in the cell cycle? Are different CDIs involved at different checkpoints, or are they more specifically involved in signal transduction? How are they regulated in non-transformed versus transformed cells? How do they inhibit the CDKs? (Is it by acting as pseudosubstrates?) This area of the cell cycle will be one of very active research in the future.

6.3 Multiple cyclin–cdk complexes facilitate complex signal integration

The presence of multiple cyclins and multiple cdks, and the potential to regulate each of the cyclin–cdk complexes at different points in the cell cycle (Fig. 7) by

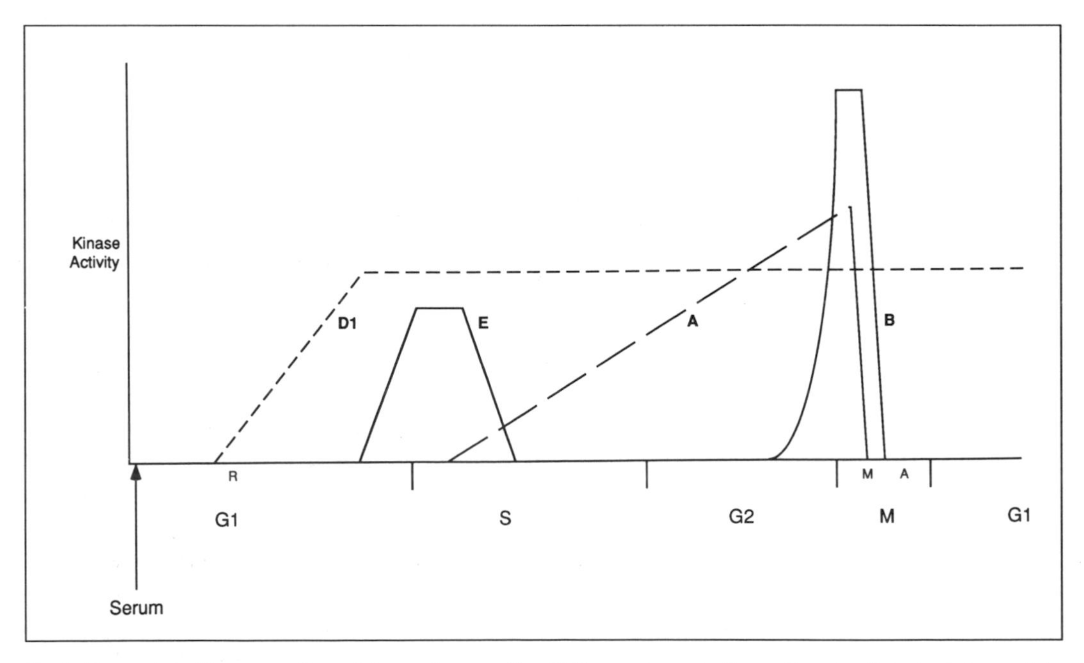

Fig. 7 The protein kinase profiles of the various cyclin–CDK complexes in the cell cycle. Note that the protein kinase activity of cyclin D refers to the D1 cyclin; D2 and D3 activity may differ from this profile.

transcription, proteolysis, steric inhibition, subcellular localization, and phosphorylation all vastly increase the potential to integrate the many positive and negative signals for cell proliferation impinging on a cell in a multicellular organism. Thus, growth factors stimulate D-type cyclin synthesis in order to proceed through the restriction (R) point (the point at which progression through the cell cycle becomes independent of growth factors). TGFβ prevents the activation of cycline–cdk2 to stop cells entering S phase through the production of p27^{Kip1}. DNA damaged by radiation in G1 prevents the activation of cyclin E–cdk2 through the production of pic1, whereas DNA damaged by radiation in G2 prevents the dephosphorylation of cyclin B–cdk1, and also down-regulates the synthesis of cyclin B mRNA and protein to delay entry into mitosis (160, 161). This may explain why the cdk family has expanded from the single cdk associating with multiple cyclins as found in yeast cells. The situation is made more complicated in that human cdk2 is able to associate with at least three different cyclins, A, D, and E, and conversely, cyclin A can associate with cdk1 or cdk2. Furthermore, D-type cyclins are able to associate with at least four different cdks; cdk2, cdk4, cdk5, and cdk6. The relative roles, substrate specificities, and degree of redundancy among all these different complexes have yet to be determined, as has the manner in which the cyclins and cdks select their partners. Indeed, a critical issue is the substrate specificities of the different cyclin–cdk complexes in the cell. The substrate specificity of the cdks is influenced by the cyclin to which it is bound, although with the exception of the D-type cyclin complexes, most cyclin–cdk complexes have similar, but not identical, specificities *in vitro*. However, if any complex is found to phosphorylate a unique substrate *in vivo* this would imply that it has a non-redundant role in the cell cycle.

It is also interesting to note that cellular senescence may be determined more by the down-regulation of cdk subunits than of the G1 cyclin partners as primary cells pass the crisis point in culture and begin to senesce (~30 passages). Senescent human fibroblasts increase expression of Pic1 (144) and the levels of cdk2 mRNA and protein drop sharply, but cyclin D1 and E mRNAs appear to be deregulated and are present at high levels (although the corresponding protein levels have not been measured) (168). The levels of cdk4, 5, and 6 in senescent cells have not yet been assayed.

Another unanswered question is whether different cell types employ different cyclin–cdk complexes for the same checkpoint control. It is already clear that a different subset of the D-type cyclins are expressed in macrophages compared with T cells, and ectopic expression of cyclin D1 has different effects on the differentiation of granulocytes compared with cyclin D2 or D3 (162), but the degree of redundancy between cyclins D2 and D3 is not known. There may be a similar difference in which cdk partner is present in individual cell types. Some evidence in support of this is the finding that although cdk5 is ubiquitously expressed in tissues, it is most abundant and seems to have significant protein kinase activity only in terminally differentiated neuronal cells (163, 164). It has also been found that cdk5 is associated with a 35 kDa protein partner that is specific to brain cells, which has limited homology to the cyclin family (165, 166). We are thus left with the intriguing finding that the cyclin–cdk complex motif has diverged to regulate processes quite distinct from the cell cycle, such as the phosphorylation of tau and neurofilament proteins which are phosphorylated on cdk consensus sites *in vivo* (164, 167). Recent data concerning the control of the phosphate pathway in budding yeast lend even more weight to this view. Here, the Pho85 protein, a close structural homologue to Cdc28, is activated by the Pho80 protein, a close relative of the Hcs26 and OrfD cyclins. The active Pho80–Pho85 complex phosphorylates the Pho4 transcription factor to prevent it from binding to DNA (169). This pathway is highly analogous to the interactions between the Cln–Cdc28 protein kinase complex and the Swi4–Swi6 transcription factor complex. Furthermore, Pho81, a small protein with several ankyrin repeats, has recently been shown to bind and inhibit the Pho80–Pho85 complex, in striking parallel with the recently cloned CDI, Sic1.

In conclusion, we are left with the model for cyclin–cdk regulation of the cell cycle shown in Fig. 8. In this model, distinct cyclin–cdk complexes are shown as regulating specific checkpoints in the mammalian cell cycle, but this is not meant

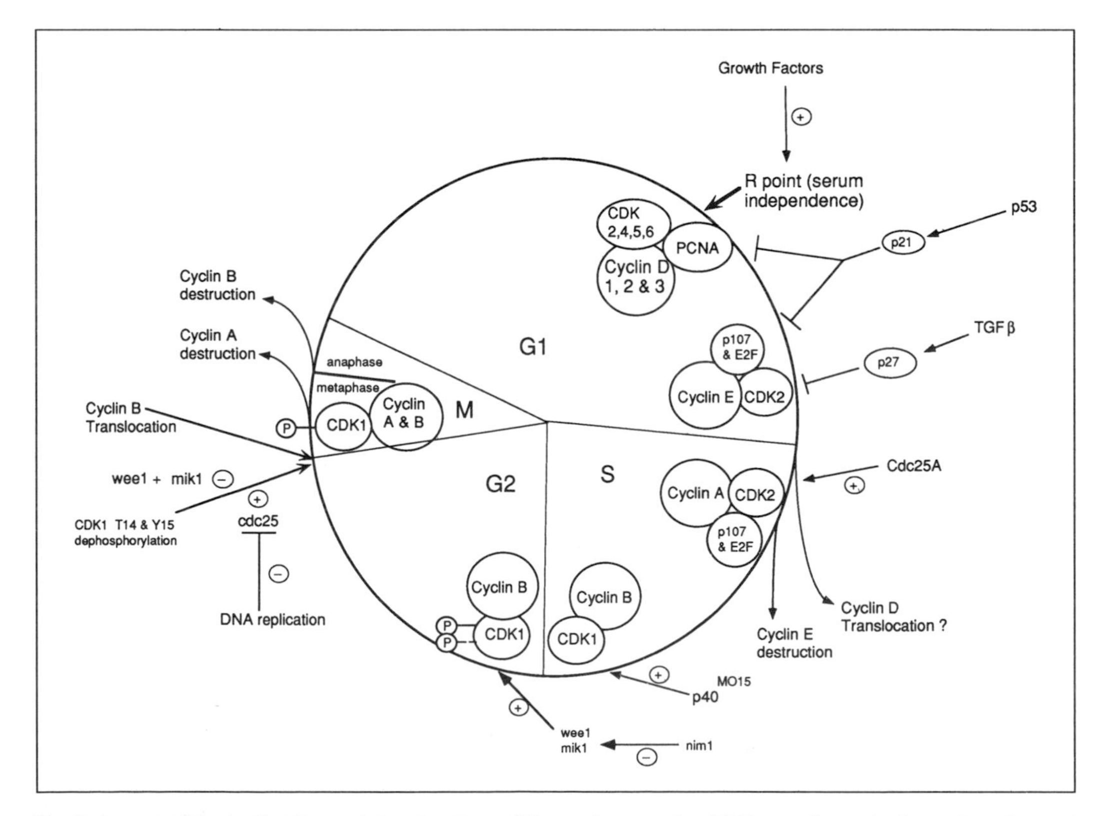

Fig. 8 A model illustrating the points of action of the various cyclin–CDK complexes in the cell cycle, and their modulation by growth factors and the state of the DNA. The protein kinases which regulate the cyclin B–CDK1 complex are shown; whether these also regulate other cyclin–CDK complexes *in vivo* has yet to be shown. Also shown are two of the CDIs whose points of action in the cell cycle have been partially defined.

to exclude the possibility that *in vivo* the cyclin–cdk complexes interact or indeed have overlapping functions at different cell cycle transitions.

References

1. Hanks, S., Quinn, A., and Hunter, T. (1988) The protein kinase family: conserved features and deduced phylogeny of the catalytic domains. *Science*, **241**, 42.
2. Lee, M. G. and Nurse, P. (1987) Complementation used to clone a human homologue of the fission yeast cell cycle control gene cdc2. *Nature*, **327**, 31.
3. Ducommun, B., Brambilla, P., and Draetta, G. (1991) Mutations at sites involved in Suc1 binding inactivate Cdc2. *Mol. Cell. Biol.*, **11**, 6177.
4. De Bondt, H. L., Rosenblatt, J., Jancarik, J., Jones, H. D., Morgan, D. O., and Kim, S.-H. (1993) Crystal structure of human cdk2: implications for the regulation of cyclin-dependent kinases by phosphorylation and cyclin binding. *Nature*, **363**, 595.
5. Draetta, G. and Beach, D. (1988) Activation of cdc2 protein kinase during mitosis in human cells: cell cycle-dependent phosphorylation and subunit rearrangement. *Cell*, **54**, 17.
6. Russell, P. and Nurse, P. (1987) Negative regulation of mitosis by *wee1*$^+$, a gene encoding a protein kinase homolog. *Cell*, **49**, 559.
7. Featherstone, C. and Russell, P. (1991) Fission yeast p107wee1 mitotic inhibitor is a tyrosine/serine kinase. *Nature*, **349**, 808.
8. Lundgren, K., Walworth, N., Booher, R., Dembski, M. M. K., and Beach, D. (1991) mik1 and wee1 cooperate in the inhibitory tyrosine phosphorylation of cdc2. *Cell*, **64**, 1111.
9. Parker, L. L., Atherton, F. S., Lee, M. S., Ogg, S., Falk, J. L., Swenson, K. I., and Piwnica-Worms, H. (1991) Cyclin promotes the tyrosine phosphorylation of cdc2 in a *wee1*$^+$ dependent manner. *EMBO J.*, **10**, 1255.
10. Parker, L. L. and Piwnica-Worms, H. (1992) Inactivation of the cdc2–cyclin B complex by the human WEE1 tyrosine kinase. *Science*, **257**, 1955.
11. McGowan, C. H. and Russell, P. (1993) Human Wee1 kinase inhibits cell division by phosphorylating cdc2 exclusively on Tyr15. *EMBO J.*, **12**, 75.
12. Heald, R., McLoughlin, M., and McKeon, F. (1993) Human Wee1 maintains mitotic timing by protecting the nucleus from cytoplasmically activated Cdc2 kinase. *Cell*, **74**, 463.
13. Russell, P. and Nurse, P. (1986) *cdc25*$^+$ functions as an inducer in the mitotic control of fission yeast. *Cell*, **45**, 145.
14. Millar, J. B. A., McGowan, C. H., Lenaers, G., Jones, R., and Russell, P. (1991) p80^{cdc25} mitotic inducer is the tyrosine phosphatase that activates cdc2 kinase in fission yeast. *EMBO J.*, **10**, 4301.
15. Dunphy, W. G. and Kumagai, A. (1991) The cdc25 protein contains an intrinsic phosphatase activity. *Cell*, **67**, 189.
16. Kumagai, A. and Dunphy, W. G. (1991) Molecular mechanism of the final steps in the activation of MPF. *Cold Spring Harbor Symp. Quant. Biol.*, **56**, 585.
17. Galaktionov, K. and Beach, D. (1991) Specific activation of cdc25 tyrosine phosphatases by B-type cyclins: evidence for multiple roles of mitotic cyclins. *Cell*, **67**, 1181.
18. Millar, J. B., Blevitt, J., Gerace, L., Sadhu, K., Featherstone, C., and Russell, P. (1991) p55CDC25 is a nuclear protein required for the initiation of mitosis in human cells. *Proc. Natl. Acad. Sci. USA*, **88**, 10500.

19. Kumagai, A. and Dunphy, W. G. (1992) Regulation of the cdc25 protein during the cell cycle in *Xenopus* extracts. *Cell*, **70,** 139.
20. Hoffmann, I., Clarke, P. R., Marcote, M. J., Karsenti, E., and Draetta, G. (1993) Phosphorylation and activation of human cdc25-C by cdc2–cyclin B and its involvement in the self-amplification of MPF at mitosis. *EMBO J.*, **12,** 53.
21. Nurse, P. and Bisset, Y. (1981) Gene required in G1 for commitment to cell cycle and in G2 for control of mitosis in fission yeast. *Nature*, **292,** 558.
22. Th'ng, J. P., Wright, P. S., Hamaguchi, J., Lee, M. G., Norbury, C. J., Nurse, P., and Bradbury, E. M. (1990) The FT210 cell line is a mouse G2 phase mutant with a temperature-sensitive cdc2 gene product. *Cell*, **63,** 313.
23. Yasuda, H., Kamijo, M., Honda, R., Nakamura, M., Hanaoka, F., and Ohba, Y. (1991) A point mutation in C-terminal region of cdc2 kinase causes a G2-phase arrest in a mouse temperature-sensitive FM3A cell mutant. *Cell Struct. Func.*, **16,** 105.
24. Hamaguchi, J. R., Tobey, R. A., Pines, J., Crissman, H. A., Hunter, T., and Bradbury, E. M. (1992) Requirement for cdc2 kinase is restricted to mitosis in the mammalian cdc2 mutant FT210. *J. Cell Biol.*, **117,** 1041.
25. Riabowol, K., Draetta, G., Brizuela, L., Vandre, D., and Beach, D. (1989) The cdc2 kinase is a nuclear protein that is essential for mitosis in mammalian cells. *Cell*, **57,** 393.
26. Pagano, M., Pepperkok, R., Lukas, J., Baldin, V., Ansorge, W., Bartek, J., and Draetta, G. (1993) Regulation of the cell cycle by the cdk2 protein kinase in cultured human fibroblasts. *J. Cell Biol.*, **121,** 101.
27. Van den Heuvel, S. and Harlow, E. (1994) Distinct roles for cyclin-dependent kinases in cell cycle control. *Science*, **262,** 2050.
28. Fang, F. and Newport, J. W. (1991) Evidence that the G1–S and G2–M transitions are controlled by different cdc2 proteins in higher eukaryotes. *Cell*, **66,** 731.
29. Furukawa, Y., Piwnica-Worms, H., Ernst, T. J., Kanakura, Y., and Griffin, J. D. (1990) cdc2 gene expression at the G1 to S transition in human T lymphocytes. *Science*, **250,** 805.
30. Meyerson, M., Enders, G. H., Wu, C. L., Su, L. K., Gorka, C., Nelson, C., Harlow, E., and Tsai, L. H. (1992) A family of human cdc2-related protein kinases. *EMBO J.*, **11,** 2909.
31. Lehner, C. F. and O'Farrell, P. H. (1990) *Drosophila* cdc2 homologs: a functional homolog is coexpressed with a cognate variant. *EMBO J.*, **9,** 3573.
32. Milarski, K. L., Dunphy, W. G., Russell, P., Gould, S. J., and Newport, J. W. (1991) Cloning and characterization of *Xenopus* cdc2, a component of MPF. *Cold Spring Harbor Symp. Quant. Biol.*, **56,** 377.
33. Paris, J., Le, G. R., Couturier, A., Le, G. K., Omilli, F., Camonis, J., MacNeill, S., and Philippe, M. (1991) Cloning by differential screening of a *Xenopus* cDNA coding for a protein highly homologous to cdc2. *Proc. Natl. Acad. Sci. USA*, **88,** 1039.
34. Elledge, S. J. and Spottswood, M. R. (1991) A new human p34 protein kinase, cdk2, identified by complementation of a cdc28 mutation in *Saccharomyces cerevisiae*, is a homolog of *Xenopus* Eg1. *EMBO J.*, **10,** 2653.
35. Ninomiya, T. J., Nomoto, S., Yasuda, H., Reed, S. I., and Matsumoto, K. (1991) Cloning of a human cDNA encoding a CDC2-related kinase by complementation of a budding yeast cdc28 mutation. *Proc. Natl. Acad. Sci. USA*, **88,** 9006.
36. Tsai, L.-H., Harlow, E., and Meyerson, M. (1991) Isolation of the human cdk2 gene that encodes the cyclin A and adenovirus E1A-associated p33 kinase. *Nature*, **353,** 174.
37. Matsushime, H., Ewen, M. E., Strom, D. K., Kato, J. Y., Hanks, S. K., Roussel,

M. F., and Sherr, C. J. (1992) Identification and properties of an atypical catalytic subunit (p34PSK-J3/cdk4) for mammalian D-type G1 cyclins. *Cell*, **71**, 323.

38. Xiong, Y., Zhang, H., and Beach, D. (1992) D-type cyclins associate with multiple protein kinases and the DNA replication and repair factor PCNA. *Cell*, **71**, 504.
39. Meyerson, M. and Harlow, E. (1994) Identification of G1 kinase activity for cdk6, a novel cyclin D partner. *Mol. Cell. Biol.*, **14**, 2077.
40. Matsushime, H., Quelle, D. E., Shurtleff, S. A., Shibuya, M., Sherr, C. J., and Kato, J. (1994) D-type cyclin-dependent kinase activity in mammalian cells. *Mol. Cell. Biol.*, **14**, 2066.
41. Bates, S., Bonetta, L., MacAllan, D., Parry, D., Holder, A., Dickson, C., and Peters, G. (1994) cdk6 (PLSTIRE) and cdk4 (PSK-J3) are a distinct subset of the cyclin-dependent kinases that associate with cyclin D1. *Oncogene*, **9**, 71.
42. Fesquet, D., Labbé, J. C., Derancourt, J., Capony, J. P., Galas, S., Girard, F., Lorca, T., Shuttleworth, J., Dorée, M., and Cavadore, J. C. (1993) The MO15 gene encodes the catalytic subunit of a protein kinase that activates cdc2 and other cyclin-dependent kinases (cdks) through phosphorylation of Thr161 and its homologues. *EMBO J.*, **12**, 3111.
43. Poon, R. Y. C., Yamashita, K., Adaczewski, J., Hunt, T., and Shuttleworth, J. (1993) The cdc2-related protein $p40^{MO15}$ is the catalytic subunit of a protein kinase that can activate $p33^{cdk2}$ and cdc2. *EMBO J.*, **12**, 3123.
44. Solomon, M. J., Harper, J. W., and Shuttleworth, J. (1993) CAK, the cdc2 activating kinase, contains a protein identical or closely related to $p40^{MO15}$. *EMBO J.*, **12**, 3133.
45. Fisher, R. P. and Morgan, D. O. (1994) A novel cyclin associates with MO15/CDK7 to form the CDK-activating kinase. *Cell*, **78**, 713.
46. Makela, T. P., Tassan, J.-P., Nigg, E. A., Frutiger, S., Hughes, G. J., and Weinberg, R. A. (1994) A cyclin associated with the CDK-activating kinase MO15. *Nature*, **371**, 254.
47. Molz, L. and Beach, D. (1993) Characterization of the fission yeast mcs2 cyclin and its associated protein kinase activity. *EMBO J.*, **12**, 1723.
48. Valay, J. G., Simon, M., and Faye, G. (1993) The Kin28 protein kinase is associated with a cyclin in *Saccharomyces cerevisiae*. *J. Mol. Biol.*, **234**, 307.
49. Hunt, T. (1991) Cyclins and their partners: from a single idea to complicated reality. *Semin. Cell Biol.*, **2**, 213.
50. Kobayashi, H., Stewart, E., Poon, R., Adamczewski, J. P., Gannon, J., and Hunt, T. (1992) Identification of the domains in cyclin A required for binding to, and activation of, cdc2 and p32cdk2 protein kinase subunits. *Mol. Biol. Cell*, **3**, 1279.
51. Lees, E. M. and Harlow, E. (1993) Sequences within the conserved cyclin box of human cyclin A are sufficient for binding to and activation of cdc2 kinase. *Mol. Cell. Biol.*, **13**, 1194.
52. Tamura, K., Kanaoka, Y., Jinno, S., Nagata, A., Ogiso, Y., Shimizu, K., Hayakawa, T., Nojima, H., and Okayama, H. (1993) Cyclin G: a new mammalian cyclin with homology to fission yeast Cig1. *Oncogene*, **8**, 2113.
53. Lew, D. and Reed, S. (1992) A proliferation of cyclins. *Trends Cell. Biol.*, **2**, 77.
54. Glotzer, M., Murray, A. W., and Kirschner, M. W. (1991) Cyclin is degraded by the ubiquitin pathway. *Nature*, **349**, 132.
55. Pines, J. and Hunter, T. (1989) Isolation of a human cyclin cDNA: evidence for cyclin mRNA and protein regulation in the cell cycle and for interaction with cdc2. *Cell*, **58**, 833.

56. Pines, J. and Hunter, T. (1990) Human cyclin A is adenovirus E1A-associated protein p60, and behaves differently from cyclin B. *Nature*, **346,** 760.
57. Lu, X. P., Koch, K. S., Lew, D. J., Dulic, V., Pines, J., Reed, S. I., Hunter, T., and Leffert, H. L. (1992) Induction of cyclin mRNA and cyclin-associated histone H1 kinase during liver regeneration. *J. Biol. Chem.*, **267,** 2841.
58. Evans, T., Rosenthal, E. T., Youngblom, J., Distel, D., and Hunt, T. (1983) Cyclin: a protein specified by maternal mRNA in sea urchin eggs that is destroyed at each cleavage division. *Cell*, **33,** 389.
59. Minshull, J., Golsteyn, R., Hill, C. S., and Hunt, T. (1990) The A- and B-type cyclin-associated cdc2 kinases in *Xenopus* turn on and off at different times in the cell cycle. *EMBO J.*, **9,** 2865.
60. Hunt, T., Luca, F. C., and Ruderman, J. V. (1992) The requirements for protein synthesis and degradation, and the control of destruction of cyclins A and B in the meiotic and mitotic cell cycles of the clam embryo. *J. Cell. Biol.*, **116,** 707.
61. Léopold, P. and O'Farrell, P. H. (1991) An evolutionarily conserved cyclin homolog from *Drosophila* rescues yeast deficient in G1 cyclins. *Cell*, **66,** 1207.
62. Brizuela, L., Draetta, G., and Beach, D. (1987) p13suc1 acts in the fission yeast cell division cycle as a component of the cdc2 protein kinase. *EMBO J.*, **6,** 3507.
63. Lew, D. J., Dulic, V., and Reed, S. I. (1991) Isolation of three novel human cyclins by rescue of G1 cyclin (Cln) function in yeast. *Cell*, **66,** 1197.
64. Bunnel, B. A., Heath, L. S., Adams, D. E., Lahti, J. M., and Kidd, V. J. (1990) Increased expression of a 58kDa protein kinase leads to changes in the CHO cell cycle. *Proc. Natl. Acad. Sci. USA*, **87,** 7467.
65. McGowan, C. H., Russell, P., and Reed, S. I. (1990) Periodic biosynthesis of the human M-phase promoting factor catalytic subunit p34 during the cell cycle. *Mol. Cell. Biol.*, **10,** 3847.
66. Dalton, S. (1992) Cell cycle regulation of the human cdc2 gene. *EMBO J.*, **11,** 1797.
67. Welch, P. J. and Wang, J. Y. (1992) Coordinated synthesis and degradation of cdc2 in the mammalian cell cycle. *Proc. Natl. Acad. Sci. USA*, **89,** 3093.
68. Broek, D., Bartlett, R., Crawford, K., and Nurse, P. (1991) Involvement of cdc2 in establishing the dependency of S phase on mitosis. *Nature*, **349,** 388.
69. Buendia, B., Draetta, G., and Karsenti, E. (1992) Regulation of the microtubule nucleating activity of centrosomes in *Xenopus* egg extracts: role of cyclin A-associated protein kinase. *J. Cell Biol.*, **116,** 1431.
70. Bailly, E., Pines, J., Hunter, T., and Bornens, M. (1992) Cytoplasmic accumulation of cyclin B1 in human cells: association with a detergent-resistant compartment and with the centrosome. *J. Cell Sci.*, **101,** 529.
71. Maldonado, C. G. and Glover, D. M. (1992) Cyclins A and B associate with chromatin and the polar regions of spindles, respectively, and do not undergo complete degradation at anaphase in syncytial *Drosophila* embryos. *J. Cell Biol.*, **116,** 967.
72. Verde, F., Dogterom, M., Stelzer, E., Karsenti, E., and Leibler, S. (1992) Control of microtubule dynamics and length by cyclin A- and cyclin B-dependent kinases in *Xenopus* egg extracts. *J. Cell Biol.*, **118,** 1097.
73. Pines, J. and Hunter, T. (1991) Human cyclins A and B are differentially located in the cell and undergo cell cycle-dependent nuclear transport. *J. Cell Biol.*, **115,** 1.
74. Gallant, P. and Nigg, E. A. (1992) Cyclin B2 undergoes cell cycle-dependent nuclear translocation and, when expressed as a non-destructible mutant, causes mitotic arrest in HeLa cells. *J. Cell Biol.*, **117,** 213.

75. Ookata, K., Hisanaga, S.-I., Okano, T., Tachnibana, K., and Kishimoto, T. (1992) Relocation and distinct subcellular localization of cdc2–cyclin B complex at meiosis reinitiation in starfish oocytes. *EMBO J.*, **5,** 1763.
76. Peter, M., Nakagawa, J., Dorée, M., Labbé, J. C., and Nigg, E. A. (1990) *In vitro* disassembly of the nuclear lamina and M-phase specific phosphorylation of lamins by cdc2 kinase. *Cell*, **61,** 591.
77. Karsenti, E. (1991) Mitotic spindle morphogenesis in animal cells. *Semin. Cell Biol.*, **2,** 251.
78. Pagano, M., Pepperkok, R., Verde, F., Ansorge, W., and Draetta, G. (1992) Cyclin A is required at two points in the human cell cycle. *EMBO J.*, **11,** 961.
79. Wang, J., Chenivesse, X., Henglein, B., and Bréchot, C. (1990) Hepatitis B virus integration in a cyclin A gene in a hepatocellular carcinoma. *Nature*, **343,** 555.
80. Wang, J., Zindy, F., Chenivesse, X., Lamas, E., Henglein, B., and Brechot, C. (1992) Modification of cyclin A expression by hepatitis B virus DNA integration in a hepatocellular carcinoma. *Oncogene*, **7,** 1653.
81. Girard, F., Strausfeld, U., Fernandez, A., and Lamb, N. J. C. (1991) Cyclin A is required for the onset of DNA replication in mammalian fibroblasts. *Cell*, **67,** 1169.
82. Zindy, F., Lamas, E., Chenivesse, X., Sobczak, J., Wang, J., Fesquet, D., Henglein, B., and Brechot, C. (1992) Cyclin A is required in S phase in normal epithelial cells. *Biochem. Biophys. Res. Commun.*, **182,** 1144.
83. Guadagno, T. M., Ohtsubo, M., Roberts, J. M., and Assoian, R. K. (1993) A link between cyclin A expression and adhesion-dependent cell cycle progression. *Science*, **262,** 1572.
84. Koff, A., Cross, F., Fisher, A., Schumacher, J., Phillipe, M., and Roberts, J. M. (1991) Cyclin E, a new class of human cyclin that can activate the cdc2/CDC28 kinase. *Cell*, **66,** 1217.
85. Dulic, V., Lees, E., and Reed, S. I. (1992) Association of human cyclin E with a periodic G1–S phase protein kinase. *Science*, **257,** 1958.
86. Knoblich, J. A., Sauer, K., Jones, L., Richardson, H., Saint, R., and Lehner, C. F. (1994) Cyclin E controls S phase progression and its down-regulation during *Drosophila* embryogenesis is required for the arrest of cell proliferation. *Cell*, **77,** 107.
87. Lehner, C. F. and O'Farrell, P. H. (1989) Expression and function of *Drosophila* cyclin A during embryonic cell cycle progression. *Cell*, **56,** 957.
88. Ohtsubo, M. and Roberts, J. M. (1993) Cyclin-dependent regulation G1 in mammalian fibroblasts. *Science*, **259,** 1908.
89. Keyomarsi, K. and Pardee, A. B. (1993) Redundant cyclin overexpression and gene amplification in breast cancer cells. *Proc. Natl. Acad. Sci. USA*, **90,** 1112.
90. Ewen, M. E., Sluss, H. K., Whitehouse, L. L., and Livingston, D. M. (1993) TGFb inhibition of Cdk4 synthesis is linked to cell cycle arrest. *Cell*, **74,** 1009.
91. Koff, A., Ohtsuki, M., Polyak, K., Roberts, J. M., and Massague, J. (1993) Negative regulation of G1 mammalian cells: inhibition of cyclin E-dependent kinase by TGF-beta. *Science*, **260,** 536.
92. Dulic, V., Kaufmann, W. K., Wilson, S. J., Tisty, T. D., Lees, E., Harper, W. J., Elledge, S. J., and Reed, S. I. (1994) p53-dependent inhibition of cyclin-dependent kinase activities in human fibroblasts during radiation-induced G1 arrest. *Cell*, **76,** 1013.
93. Mudryj, M., Devoto, S. H., Hiebert, S., Hunter, T., Pines, J., and Nevins, J. R. (1991) Cell cycle regulation of the E2F transcription factor involves an interaction with cyclin A. *Cell*, **65,** 1243.

94. Devoto, S. H., Mudryj, M., Pines, J., Hunter, T., and Nevins, J. R. (1992) A cyclin A–protein kinase complex possesses sequence-specific DNA binding activity: $p33^{cdk2}$ is a component of the E2F–cyclin A complex. *Cell*, **68,** 167.
95. Pagano, M., Draetta, G., and Jansen, D. P. (1992) Association of cdk2 kinase with the transcription factor E2F during S phase. *Science*, **255,** 1144.
96. Lees, E., Faha, B., Dulic, V., Reed, S. I., and Harlow, E. (1992) Cyclin E/cdk2 and cyclin A/cdk2 kinases associate with p107 and E2F in a temporally distinct manner. *Genes Dev.*, **6,** 1874.
97. Cross, F. R. and Tinkelenberg, A. H. (1991) A potential positive feedback loop controlling CLN1 and CLN2 gene expression at the start of the yeast cell cycle. *Cell*, **65,** 875.
98. Dirick, L. and Nasmyth, K. (1991) Positive feedback in the activation of G1 cyclins in yeast. *Nature (London)*, **351,** 754.
99. Nasmyth, K. and Dirick, L. (1991) The role of Swi4 and Swi6 in the activity of G1 cyclins in yeast. *Cell*, **66,** 995.
100. Primig, M., Sockanathan, S., Auer, H., and Nasmyth, K. (1992) Anatomy of a transcription factor important for the start of the cell cycle in *Saccharomyces cerevisiae*. *Nature*, **358,** 593.
101. Zhu, L., van den Heuvel, S., Helin, K., Fattaey, A., Ewen, M., Livingston, D., Dyson, N., and Harlow, E. (1993) Inhibition of cell proliferation by p107, a relative of the retinoblastoma protein. *Genes Dev.*, **7,** 1111.
102. Sekiguchi, T., Miyata, T., and Mishimoto, T. (1988) Molecular cloning of the cDNA of human X chromosomal gene (CCG1) which complements the temperature-sensitive G1 mutants, tsBN462 and ts13, of the BHK cell line. *EMBO J.*, **7,** 1683.
103. Hisatake, K., Hasegawa, S., Takada, R., Nakatani, Y., Horikoshi, M., and Roeder, R. (1993) The p250 subunit of native TATA box-binding factor TFIID is the cell cycle regulatory protein CCG1. *Nature*, **362,** 179.
104. Ruppert, S., Wang, E. H., and Tjian, R. (1993) Cloning and expression of human $TAF_{II}250$: a TBP-associated factor implicated in cell cycle regulation. *Nature*, **362,** 175.
105. Jackson, D. A., Hassan, A. B., Errington, R. J., and Cook, P. R. (1993) Visualization of focal sites of transcription within human nuclei. *EMBO J.*, **12,** 1059.
106. Hozák, P., Hassan, A. B., Jackson, D. A., and Cook, P. R. (1993) Visualization of replication factories attached to a nucleoskeleton. *Cell*, **73,** 361.
107. Ewen, M. E., Faha, B., Harlow, E., and Livingstone, D. M. (1992) Interaction of p107 with cyclin A independent of complex formation with viral oncoproteins. *Science*, **255,** 85.
108. Tommasino, M., Adamczewski, J. P., Carlotti, F., Barth, C. F., Manetti, R., Contorni, M., Cavalieri, F., Hunt, T., and Crawford, L. (1993) HPV16 E7 protein associates with the protein kinase p33cdk2 and cyclin A. *Oncogene*, **8,** 195.
109. Matsushime, H., Roussel, M. F., Ashmun, R. A., and Sherr, C. J. (1991) Colony-stimulating factor 1 regulates novel cyclins during the G1 phase of the cell cycle. *Cell*, **65,** 701.
110. Motokura, T., Keyomarsi, K., Kronenberg, H. M., and Arnold, A. (1992) Cloning and characterization of human cyclin D3, a cDNA closely related in sequence to the PRAD1/cyclin D1 proto-oncogene. *J. Biol. Chem.*, **267,** 20412.
111. Lukas, J., Pagano, M., Staskova, Z., Draetta, G., and Bartek, J. (1994) Cyclin D1 protein oscillates and is essential for cell cycle progression in human tumour cell lines. *Oncogene*, **9,** 707.

112. Won, K. A., Xiong, Y., Beach, D., and Gilman, M. Z. (1992) Growth-regulated expression of D-type cyclin genes in human diploid fibroblasts. *Proc. Natl. Acad. Sci. USA*, **89,** 9910.
113. Baldin, V., Lukas, J., Marcote, M. J., Pagano, M., and Draetta, G. (1993) Cyclin D1 is a nuclear protein required for cell cycle progression in G1. *Genes Dev.*, **7,** 812.
114. Quelle, D. E., Ashmun, R. A., Shurtleff, S. A., Kato, J., Bar-Sagi, D., Roussel, M. F., and Sherr, C. J. (1993) Overexpression of mouse D-type cyclins accelerates G_1 phase in rodent fibroblasts. *Genes Dev.*, **7,** 1559.
115. Motokura, T., Bloom, T., Kim, H. G., Jüppner, H., Ruderman, J. V., Kronenberg, H. M., and Arnold, A. (1991) A novel cyclin encoded by a bcl-linked candidate oncogene. *Nature (London)*, **350,** 512.
116. Withers, D. A., Harvey, R. C., Faust, J. B., Melnyk, O., Carey, K., and Meeker, T. C. (1991) Characterization of a candidate bcl-1 gene. *Mol. Cell Biol.*, **11, 4846.**
117. Lammie, G. A., Smith, R., Silver, J., Brookes, S., Dickson, C., and Peters, G. (1992) Proviral insertions near cyclin D1 in mouse lymphomas: a parallel for BCL1 translocations in human B-cell neoplasms. *Oncogene*, **7,** 2381.
118. Hanna, Z., Jankowski, M., Tremblay, P., Jiang, X., Milatovich, A., Francke, U., and Jolicoeur, P. (1993) The Vin-1 gene, identified by provirus insertional mutagenesis, is the cyclin D2. *Oncogene*, **8,** 1661.
119. Jiang, W., Kahn, S. M., Tomita, N., Zhang, Y. J., Lu, S. H., and Weinstein, I. B. (1992) Amplification and expression of the human cyclin D gene in esophageal cancer. *Cancer Res.*, **52,** 2980.
120. Leach, F. S., Elledge, S. J., Sherr, C. J., Willson, J. K., Markowitz, S., Kinzler, K. W., and Vogelstein, B. (1993) Amplification of cyclin genes in colorectal carcinomas. *Cancer Res.*, **53,** 1986.
121. Jiang, W., Kahn, S. M., Zhou, P., Zhang, Y.-J., Cacace, A. M., Infante, A. S., Doi, S., Santella, R. M., and Weinstein, I. B. (1993) Overexpression of cyclin D1 in rat fibroblasts causes abnormalities in growth control, cell cycle progression and gene expression. *Oncogene*, **8,** 3447.
122. Lukas, J., Müller, H., Bartkova, J., Spitkovsky, D., Kjerulff, A., Jansen-Dürr, P., Strauss, M., and Bartek, J. (1994) DNA tumor virus oncoproteins and retinoblastoma gene mutations share the ability to relieve the cell's requirement of cyclin D1 function in G1. *J. Cell. Biol.*, **125,** 625.
123. Kato, J., Matsushime, H., Hiebert, S. W., Ewen, M. E., and Sherr, C. J. (1993) Direct binding of cyclin D to the retinoblastoma gene product (pRb) and pRb phosphorylation by the cyclin D-dependent kinase cdk4. *Genes Dev.*, **7,** 331.
124. Hinds, P. W., Mittnacht, S., Dulic, V., Arnold, A., Reed, S. I., and Weinberg, R. A. (1992) Regulation of retinoblastoma protein functions by ectopic expression of human cyclins. *Cell*, **70,** 993.
125. Dowdy, S. F., Hinds, P. W., Louie, K., Reed, S. I., Arnold, A., and Weinberg, R. A. (1993) Physical interaction of the retinoblastoma protein with human D-type cyclins. *Cell*, **73,** 499.
126. Ewen, M. E., Sluss, H. K., Sherr, C. J., Matsushime, H., Kato, J.-Y., and Livingstone, D. M. (1993) Functional interactions of the retinoblastoma protein with the mammalian D-type cyclins. *Cell*, **73,** 487.
127. Jinno, S., Suto, K., Nagata, A., Igarashi, M., Kanaoka, Y., Nojima, H., and Okayama, H. (1994) Cdc25A is a novel phosphatase functioning early in the cell cycle. *EMBO J.*, **13,** 1549.

128. Hoffman, I., Draetta, G., and Karsenti, E. (1994) Activation of the phosphatase activity of human cdc25A by a cdk2-cyclin E dependent phosphorylation at the G_1/S transition. *EMBO J.*, **13**, 4302.
129. Gu, Y., Rosenblatt, J., and Morgan, D. O. (1992) Cell cycle regulation of cdk2 activity by phosphorylation of Thr160 and Tyr15. *EMBO J.*, **11**, 3995.
130. Gabrielli, B. G., Lee, M. S., Walker, D. H., Piwnica-Worms, H., and Maller, J. L. (1992) Cdc25 regulates the phosphorylation and activity of the *Xenopus* cdk2 protein kinase complex. *J. Biol. Chem.*, **267**, 18040.
131. Sebastian, B., Kakizuka, A., and Hunter, T. (1993) Cdc25M2 activation of cyclin-dependent kinases by dephosphorylation of threonine-14 and tyrosine-15. *Proc. Natl. Acad. Sci. USA*, **90**, 3521.
132. Kornbluth, S., Sebastian, B., Hunter, T., and Newport, J. (1994) Membrane localization of the kinase which phosphorylates cdc2 on threonine 14. *Mol. Biol. Cell*, **5**, 273.
133. Russell, P. and Nurse, P. (1987) The mitotic inducer *nim1*$^+$ functions in a regulatory network of protein kinase homologs controlling the initiation of mitosis. *Cell*, **49**, 569.
134. Coleman, T. R., Tang, Z., and Dunphy, W. G. (1993) Negative regulation of the Wee1 protein kinase by direct action of the nim1/cdr1 mitotic inducer. *Cell*, **73**, 919.
135. Tang, Z., Coleman, T. R., and Dunphy, W. G. (1993) Two distinct mechanisms for negative regulation of the Wee1 protein kinase. *EMBO J.*, **12**, 3427.
136. Chang, F. and Herskowitz, I. (1990) Identification of a gene necessary for cell cycle arrest by a negative growth factor of yeast: FAR1 is an inhibitor of a G1 cyclin, CLN2. *Cell*, **63**, 999.
137. Chang, F. and Herskowitz, I. (1992) Phosphorylation of FAR1 in response to alpha-factor: a possible requirement for cell-cycle arrest. *Mol. Biol. Cell*, **3**, 445.
138. Tyers, M. and Futcher, B. (1993) Far1 and Fus3 link the mating pheromone signal transduction pathway to three G1-phase Cdc28 kinase complexes. *Mol. Cell. Biol.*, **13**, 5659.
139. Hunter, T. (1993) Braking the cycle. *Cell*, **75**, 839.
140. Nasmyth, K. and Hunt, T. (1993) Dams and sluices. *Nature*, **366**, 634.
141. Pines, J. (1994) Arresting developments in cell cycle control. *Trends Biochem. Sci.*, **19**, 143.
142. Harper, J. W., Adami, G. R., Wei, N., Keyomarsi, K., and Elledge, S. J. (1993) The p21 Cdk-interacting protein Cip1 is a potent inhibitor of G1 cyclin-dependent kinases. *Cell*, **75**, 805.
143. Xiong, Y., Hannon, G. J., Zhang, H., Casso, D., Kobayashi, R., and Beach, D. (1993) p21 is a universal inhibitor of cyclin kinases. *Nature*, **366**, 701.
144. Noda, A., Ning, Y., Venable, S. F., Pereira-Smith, O. M., and Smith, J. R. (1994) Cloning of senescent cell-derived inhibitors of DNA synthesis using an expression screen. *Exp. Cell. Res.*, **211**, 90.
145. El-Deiry, W. S., Tokino, T., Velculesco, V. E., Levy, D. B., Parsons, R., Trent, J. M., Lin, D., Mercer, W. E., Kinzler, K. W., and Vogelstein, B. (1993) WAF1, a potential mediator of p53 tumor suppression. *Cell*, **75**, 817.
146. Gu, Y., Turck, C. W., and Morgan, D. O. (1993) Inhibition of cdk2 activity *in vivo* by an associated 20K regulatory subunit. *Nature*, **366**, 707.
147. Waga, S., Hannon, G. J., Beach, D., and Stillman, B. (1994) The p21 cyclin-dependent kinase inhibitor directly controls DNA replication via interaction with PCNA. *Nature*, **369**, 574.

148. Moreno, S. and Nurse, P. (1994) Regulation of progression through the G1 phase of the cell cycle by the *rum1*$^{+}$ gene. *Nature*, **367,** 236.
149. Nugroho, T. T. and Mendenhall, M. D. (1994) An inhibitor of yeast cyclin-dependent protein kinase plays an important role in ensuring the genomic integrity of daughter cells. *Mol. Cell. Biol.*, **14,** 3320.
150. Polyak, K., Lee, M.-H., Erdjument-Bromage, H., Koff, A., Roberts, J. M., Tempst, P., and Massagué, J. (1994) Cloning of p27Kip1, a cyclin-dependent kinase inhibitor and a potential mediator of extracellular antimitogenic signals. *Cell*, **78,** 59.
151. Toyoshima, H. and Hunter, T. (1994) p27, a novel inhibitor of G1 cyclin–Cdk protein kinase activity, is related to p21. *Cell*, **78,** 67.
152. Polyak, K., Kato, J.-V., Solomon, M., Sherr, C. J., Massague, J., Roberts, J. M., and Koff, A. (1944) p27^{Kip1} and cyclin D–cdk4, interacting regulators of cdk2, link TGF-B and contact inhibition to cell cycle arrest. *Genes Dev.*, **8,** 9.
153. Gyruis, J. E. G., Chertkov, H., and Brent, R. (1993) Cdi1, a human G1 and S phase protein phosphatase that associates with Cdk2. *Cell*, **75,** 791.
154. Xiong, Y., Zhang, H. and Beach, D. (1993) Subunit rearrangement of the cyclin-dependent kinases is associated with cellular transformation. *Genes and Development*, **7,** 1572.
155. Serrano, M., Hannon, G. J. and Beach, D. (1993) A new regulatory motif in cell cycle control causing specific inhibition of cyclin D/CDK4, *Nature*, **366,** 704.
156. Kamb, A., Gruis, N. A., Weaver-Feldhaus, J., Liu, Q., Harshman, K., Tavtigian, S. V., Stockert, E., Day, R. S., Johnson, B. E., and Skornik, M. H. (1994) A cell cycle regulator potentially involved in genesis of many tumor types. *Science*, **264,** 436.
157. Nobori, T., Miura, K., Wu, D. J., Lois, A., Takabayashi, K., and Carson, D. A. (1994) Deletions of the cyclin-dependent kinase-4 inhibitor gene in multiple human cancers. *Nature*, **368,** 753.
158. Bonetta, L. (1994) Open questions on p16. *Nature*, **370,** 180.
159. Hannon, G. and Beach, D. (1994) p15^{INK4B} is a potential effector of TGF-β-induced cell cycle arrest. *Nature*, **371,** 257.
160. Muschel, R. J., Zhang, H. B., Iliakis, G., and McKenna, W. G. (1991) Cyclin B expression in HeLa cells during the G2 block induced by ionizing radiation. *Cancer Res.*, **51,** 5113.
161. Muschel, R. J., Zhang, H. B., and McKenna, W. G. (1993) Differential effect of ionizing radiation on the expression of cyclin A and cyclin B in HeLa cells. *Cancer Res.*, **53,** 1128.
162. Kato, J. and Sherr, C. J. (1993) Inhibition of granulocyte differentiation by G1 cyclins D2 and D3 but not D1. *Proc. Natl. Acad. Sci. USA*, **90,** 11513.
163. Hellmich, M. R., Pant, H. C., Wada, E., and Battey, J. F. (1992) Neuronal cdc2-like kinase: a cdc2-related protein kinase with predominatly neuronal expression. *Proc. Natl. Acad. Sci. USA*, **89,** 10867.
164. Lew, J., Winkfein, R. J., Paudel, H. K., and Wang, J. H. (1992) Brain proline-directed protein kinase is a neurofilament kinase which displays high sequence homology to cdc2. *J. Biol. Chem.*, **267,** 25922.
165. Lew, J., Huang, Q.-Qi, Z., Winkfein, R. J., Aebersold, R., Hunt, T., and Wang, J. H. (1994) A brain-specific activator of cyclin-dependent kinase 5. *Nature*, **371,** 423.
166. Tsai, L.-H., Delalle, I., Caviness Jr, V. S., Chae, T., and Harlow, E. (1994) p35 is a neural-specific regulatory subunit of cyclin-dependent kinase 5. *Nature*, **371,** 419.
167. Hisanga, S., Kusubata, M., Okumura, E., and Kishimoto, T. (1991) Phosphorylation

of neurofilament H subunit at the tail domain by CDC2 kinase dissociates the association to microtubules. *J. Biol. Chem.*, **266,** 21798.

168. Lucibello, F. C., Sewing, A., Brüsselbach, S., Bürger, C., and Muller, R. (1993) Deregulation of cyclins D1 and E and suppression of cdk2 in senescent human fibroblasts. *J. Cell Sci.*, **105,** 123.
169. Kaffman, A., Herskowitz, I., Tjian, R., and O'Shea, E. K. (1994) Phosphorylation of the transcription factor PH04 by a cyclin–cdk complex, Pho80–Pho85. *Science*, **263,** 1153.
170. Schwob, E., Böhm, T., Mendenhall, M. D., and Nasmyth, K. (1994) The B-type cyclin kinase inhibitor $p40^{sic1}$ controls the G1/S transition in *Saccharomyces cerevisiae*. *Cell*, **79,** 233.

7 | S phase and its regulation

J. JULIAN BLOW

1. DNA replication and the cell cycle

During S phase of the eukaryotic cell cycle, chromosomal DNA is replicated precisely once as a prelude to its segregation to the daughter cells at mitosis. Replication is achieved by between 10^3 and 10^5 pairs of replication forks initiating at sites scattered throughout the genome, the number varying with the size of the genome and the length of S phase. These initiation events must be carefully regulated in at least three different ways. First, since most cells do not arrest in the G2 phase of the cycle, it is important that entry into S phase only occurs in cells that are legitimately committed to undergoing the complete cell division cycle. Second, it is important that initiation is coordinated with other cell cycle events so that the ploidy of the cell is maintained by the alternation of S and M phases. Third, it is important to regulate initiation events within each S phase, so that all the chromosomal DNA is completely replicated but no DNA is replicated more than once.

1.1 Basic events required for completion of S phase

Chromosome replication is a complex process requiring the coordinated activity of a large number of different proteins. This section briefly reviews some of these basic activities and mentions some of the proteins involved.

1.1.1 Initiation

The initiation of replication involves the assembly on to DNA of the complex of replication fork proteins required to replicate the chromosome. It is important that the cell correctly selects sites at which initiation occurs (replication origins) to ensure that they are spaced sufficiently close together so that there is enough time for all the intervening DNA to be replicated before entry into mitosis. It may also be advantageous to initiate at the 5′ end of genes so that replication forks travel in the same direction as transcription, instead of continually running head on into RNA polymerases (1). The exact requirement for establishing an origin of replication in higher eukaryotic cells is not understood. In budding yeast the situation is clearer, where replication origins require a number of defined DNA sequences including a highly conserved 12 bp ‘core origin’. This core origin is recognized and bound by a multiprotein Origin Recognition Complex (ORC) (2, 3). Since the rate and progression of replication forks is roughly constant (4), the crucial regulatory step in replication is the initiation of new forks.

1.1.2 Replication fork movement

As it moves along and replicates the template DNA, the replication fork has to perform a number of functions, as outlined in Fig. 1. Ahead of the fork, the two strands of duplex DNA are unwound, requiring a temporary disruption of the normal chromatin structure of the chromosome. Torsional stress produced by this unwinding must be relieved by topoisomerases. Since DNA polymerases can only add new dNTPs on to the 3′ end of a growing DNA molecule, only one nascent strand, the 'leading' strand, is synthesized in the same direction as the fork is travelling. The other, 'lagging', strand is synthesized in the opposite direction, requiring continual synthesis of RNA primers on to which the lagging strand Okazaki fragments can be synthesized. After synthesis, these RNA primers are removed, the gaps filled with DNA, and the fragments ligated together. Replicated DNA behind the fork is then reassembled into chromatin, which in addition to histones requires the reassembly of non-histone proteins such as transcription factors and the reconstitution of chromatin modifications and DNA methylation.

1.1.3 Termination

Replication forks progress along the DNA until they meet another fork coming from the opposite direction, thus ensuring complete replication of the DNA between adjacent origins. Sometimes replication forks appear to stall or pause at specific sites on the DNA. In ribosomal DNA, these pause sites appear to block forks from moving specifically into the 5′ end of the gene from the untranscribed spacer (5). When replication forks run into one another, they disassemble, probably leaving a short section of DNA between them unreplicated (6). The process

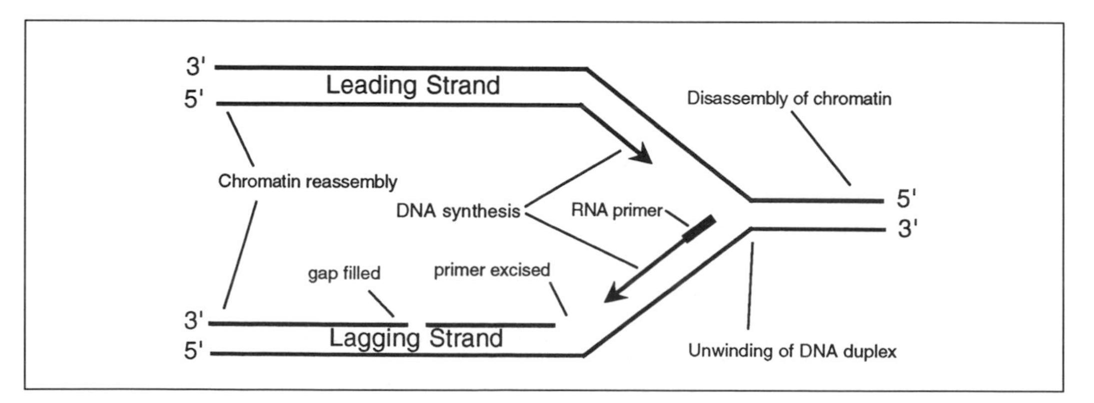

Fig. 1 General scheme for events occurring at the replication fork. DNA ahead of the fork is disassembled from its chromatin structure, and then unwound. The leading strand is synthesized continuously, but the lagging strand is synthesized in the form of Okazaki fragments 140–200 nt long. Behind the fork, DNA is reassembled into chromatin again.

of unwinding and replicating this remaining DNA is probably performed by a different mechanism than the replication of most of the DNA. Once this is complete, the nascent strands are ligated, and the duplex daughter strands are decatenated.

1.2 Timing of initiation events within S phase

In multicellular organisms, the total length of S phase (and the whole cell division cycle itself) often varies widely throughout development. Embryonic cells tend to have shorter S phases than differentiated cells. Since the rate of replication fork movement is roughly similar throughout development, this means that changes in S phase duration are largely due to changes in the location and timing of initiation events.

The shorter S phase of early embryonic cells can partly be explained by their use of a larger number of more closely spaced replication origins. The early *Drosophila* embryo, with an S phase duration of only 3.4 min, has a mean origin-to-origin spacing of about 7.9 kb as determined by electron microscopy, with initiation events occurring virtually synchronously at the start of S phase (7) (Fig. 2A). In

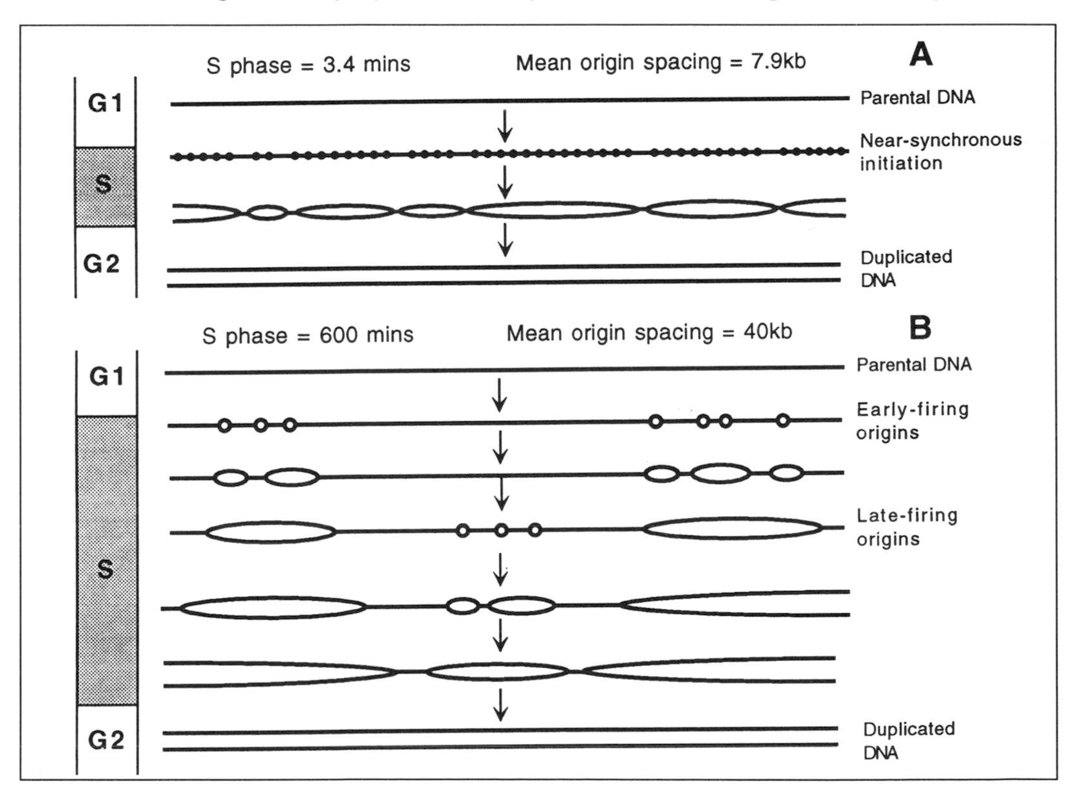

Fig. 2 Comparison of S phase events in embryos and somatic cells of *Drosophila* (7). Cartoon of replication structures seen on a single stretch of chromosomal DNA. (A) *Drosophila* embryo, with S phase duration of 3.4 min and mean origin spacing of 7.9 kb. (B) *Drosophila* somatic cell, with S phase duration of 600 min and mean origin spacing (of clustered origins by fibre autoradiography) of 40 kb.

contrast, adult *Drosophila* tissue culture cells, with an S phase duration of 600 min, have a mean origin-to-origin spacing of about 40 kb. Therefore, at least some origins active in the embryo are not used at later stages of development. Similar results have been obtained from a range of different organisms. Tissue culture cells typically show inter-origin spacing varying from 30 to 300 kb, whilst rapidly dividing embryos show tighter spacing. Similarly, early embryos also display a higher degree of promiscuity towards DNA sequences they can use for replication origins than do adult cells (8).

The five-fold increase in inter-origin spacing between embryonic and adult *Drosophila* cells is not sufficient to explain their nearly 200-fold difference in S phase duration. With the *Drosophila* fork rate of approximately 2.6 kb min^{-1}, most forks initiated within a cluster of origins spaced 40 kb apart would terminate in 10 min. The greater length of S phase in adult cells instead can be explained by these clusters of initiation events occurring at different times (4, 9) (Fig. 2B). This can be seen in labelling experiments where different chromosome segments have different but reproducible replication times (10). Since on the relatively short stretches of DNA used in fibre autoradiography, initiation events occur with approximate synchrony (4, 11), it appears that these different chromosome sections are each replicated by a cluster of near-synchronous initiation events, whilst different clusters are activated at different times.

The initiation timing of different chromosome sections appears to correlate at least partly with the transcriptional state of DNA. Transcriptionally inert heterochromatin and non-transcribed genes replicate late, whilst actively transcribed genes replicate early (12, 13). Replication timing seems to depend upon the chromosomal context of the origin, and can be changed by chromosome rearrangement (13). Similarly a late-firing telomeric origin in yeast became early-replicating when moved away from the telomere (14, 15).

2. Nucleus and cytoplasm: template and activator

In principle, replication might be controlled in the cell cycle by limiting the availability of some or all of the factors required at the replication fork. However, although there is some cell cycle control over their levels, most replication factors are present in the cell throughout the cell cycle. Part of the reason for this is that many are also involved in other cellular processes such as DNA repair. The continued presence of replication factors throughout the cell cycle means that DNA replication must be controlled by a relatively small number of 'S phase-inducers' whose function is to regulate when and where initiation takes place.

2.1 Cell fusion studies

How does the level of these S phase-inducing activities change during the cell cycle? In the classic cell fusion studies of Rao and Johnson (16), HeLa cells at different stages of the cell cycle were fused, and the replication timing of different

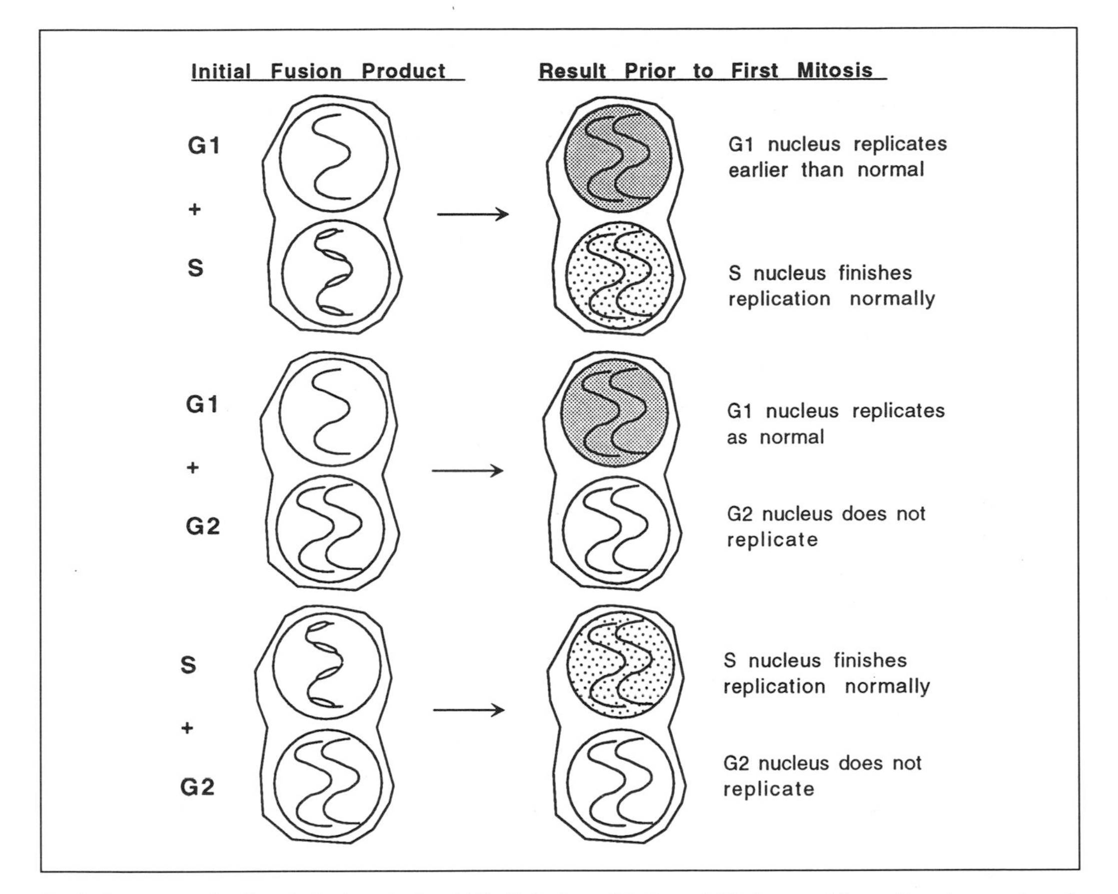

Fig. 3 Summary of cell cycle fusion studies (16). Cells from G1, S, and G2 stages of the cell cycle were fused, giving hybrids with two nuclei. G1 nuclei are indicated with a single chromosome, S nuclei with a chromosome containing replication bubbles, and G2 with a double chromosome. Prior to the first mitosis after fusion, all nuclei have fully replicated, and the extent of replication performed in the hybrid is indicated by the depth of shading in each nucleus.

nuclei measured. The results are summarized in Fig. 3. In fusions of G1 and S phase cells, the G1 nucleus was induced to enter S phase earlier than it would normally have done whilst the S phase nucleus continued replication as normal. The more S phase cells fused to a single G1 cell, the faster the G1 nucleus entered S phase. However, the S phase-inducers in S phase cells did not persist into G2, since in fusions between G1 and G2 cells, the G1 nucleus entered S phase approximately on schedule. However, cells that have very short G1 periods may contain S phase-inducers throughout the cell cycle. When metaphase V79–8 cells, whose G1 and G2 phases account for only about 5 per cent of interphase, were fused with G1 HeLa cells, the HeLa nuclei were rapidly induced to enter S phase before being drawn into premature chromatin condensation (17).

These cytoplasmic S phase-inducers can only act on G1 nuclei, and cannot induce G2 nuclei to undergo another round of DNA replication. In S phase/G2 and G1/G2 HeLa cell fusions the G2 nuclei did not replicate until after passage through mitosis (16) (Fig. 3). This suggests that the absence of DNA replication in normal G2 cells is not due to the presence of diffusive inhibitors in the G2 cytoplasm. Possible mechanisms preventing the re-replication of G2 nuclei are discussed in more detail below.

The apparently normal replication of G1 and S phase nuclei in G2 cytoplasm also suggests that ordered progression through S phase resulting in complete chromosome replication does not require a parallel sequence of signals or activators in the cytoplasm. This can be shown directly by monitoring the replication patterns of fusions between cells at defined points in S phase (18). Fusion of G1 and late S cells caused the G1 nuclei to enter S phase early, with replication patterns of nuclei in early S phase, rather than in late S phase; conversely, the late S nuclei maintained a late S replication pattern. This implies that the S phase-inducers present in the cytoplasm induce responsive nuclei to undergo the normal sequence of S phase events, but are not themselves responsible for dictating the stage of replication the nucleus should perform.

Figure 4 shows a model for the relationship between cell cycle progress and DNA replication that can be constructed from these experiments. Replication control can be divided into two distinct components: the DNA template in the nucleus, and activities present in the cytoplasm which act on this nuclear substrate. The nucleus can be in either of two states: either capable or incapable of responding to S phase-inducers by undergoing DNA replication. Similarly, the cytoplasm provides two types of activity, one that induces a responsive (G1) nucleus to undergo S phase (S phase-inducers), and another (related to passage through mitosis, as discussed below) that changes a refractory G2 nucleus into a responsive G1 nucleus. Once S phase has been induced, the progress of the nucleus through the different stages of DNA replication (including initiation at early- and late-firing origins) occurs without the need for specific signals from the

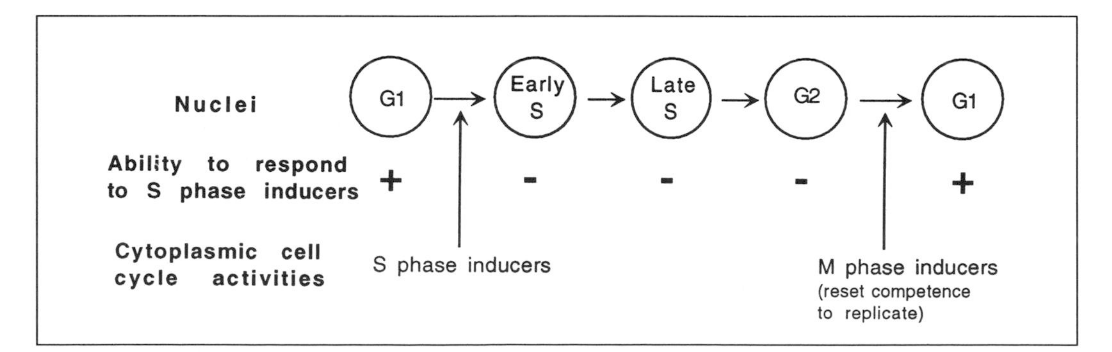

Fig. 4 Model to explain the results from cell fusion experiments. A single nucleus is shown as it passes through a complete cell cycle. In G1 it is capable of responding to S phase inducers by undergoing S phase, after which it becomes refractory. On passage through mitosis, the nucleus once again becomes competent to undergo S phase again.

cytoplasm. In the following sections, the role of the nuclear DNA template and the cytoplasmic signals acting upon it are considered separately.

3. The nuclear DNA template

Cell fusion experiments show that unreplicated G1 nuclei can respond to cytoplasmic S phase-inducers by undergoing DNA replication, whilst G2 nuclei cannot. What is the nature of the substrate for the S phase-inducers, and what constitutes the difference between G1 and G2 nuclei?

3.1 The nucleus as a fundamental unit of DNA replication

Cell-free extracts of *Xenopus* eggs can support the basic events of the cell cycle, including passage through S phase and mitosis (19, 20). When DNA is added to these cell-free extracts, it is replicated precisely once during each *in vitro* interphase (20). The cell-free system also assembles exogenously added DNA into structures resembling normal interphase nuclei (19). These *in vitro*-assembled nuclei are surrounded by a double unit envelope studded with nuclear pores, and are capable of selectively importing soluble nuclear proteins (21).

Possession of a complete nuclear envelope is required before the initiation of DNA replication can occur in *Xenopus* extracts. When extracts are centrifuged to remove membrane material, the supernatant ('high-speed supernatant') is unable to support the initiation of DNA replication (22–24). High-speed supernatants are fully active in complementary strand synthesis of single-stranded DNA and can assemble double-stranded DNA into large chromatin aggregates, though they cannot assemble nuclear envelope. Re-addition of pelleted membrane material permits high-speed supernatants to both assemble nuclei and to support the initiation of replication (22–24). A similar dependence of initiation on nuclear assembly is seen in extracts of *Drosophila* embryos (25). A plausible explanation for this nuclear envelope requirement is that it permits the selective nuclear accumulation of proteins involved in initiation. Consistent with this, wheat germ agglutinin, a lectin that binds to nuclear pores and prevents nuclear protein accumulation (26, 27), renders extracts unable to support the initiation of DNA replication though capable of elongating previously initiated replication forks (28). Nuclear assembly may also be required to provide structural components of the nucleus required for DNA replication (see Section 3.2 below). In *Xenopus* extracts, the nuclear lamina is assembled in the nuclear interior once nuclear envelope assembly is complete (22). If extracts are immunodepleted of lamin precursors, subsequent DNA replication does not occur, although nuclear envelopes are assembled (29, 30).

The dependence of initiation on the presence and function of the nuclear envelope means that the *Xenopus* system can be used to study the way that cytoplasmic signals induce nuclei to initiate DNA replication. Nascent DNA was labelled *in vitro* with biotinylated dUTP, and at different times nuclei were isolated and stained for DNA (Fig. 5A) and biotin (Fig. 5B) (31). The DNA and biotin content of individual nuclei was measured using flow cytometry and is presented as a contour plot in Fig. 5C. This shows a mixture of nuclei: completely unreplicated (1N DNA content, background biotin), fully replicated (2N DNA content, maximum biotin content), and replicating (intermediate DNA and biotin contents). At different times during the incubation *in vitro*, nuclei entered S phase and moved from the initial 1N peak to the fully replicated 2N peak (31). Once they had entered S phase, each nucleus replicated relatively fast, requiring a burst of approximately 50 000 near-synchronous initiation events. Even though nuclei continued to enter S phase over a significant period of time, no nuclei replicated more than once. This

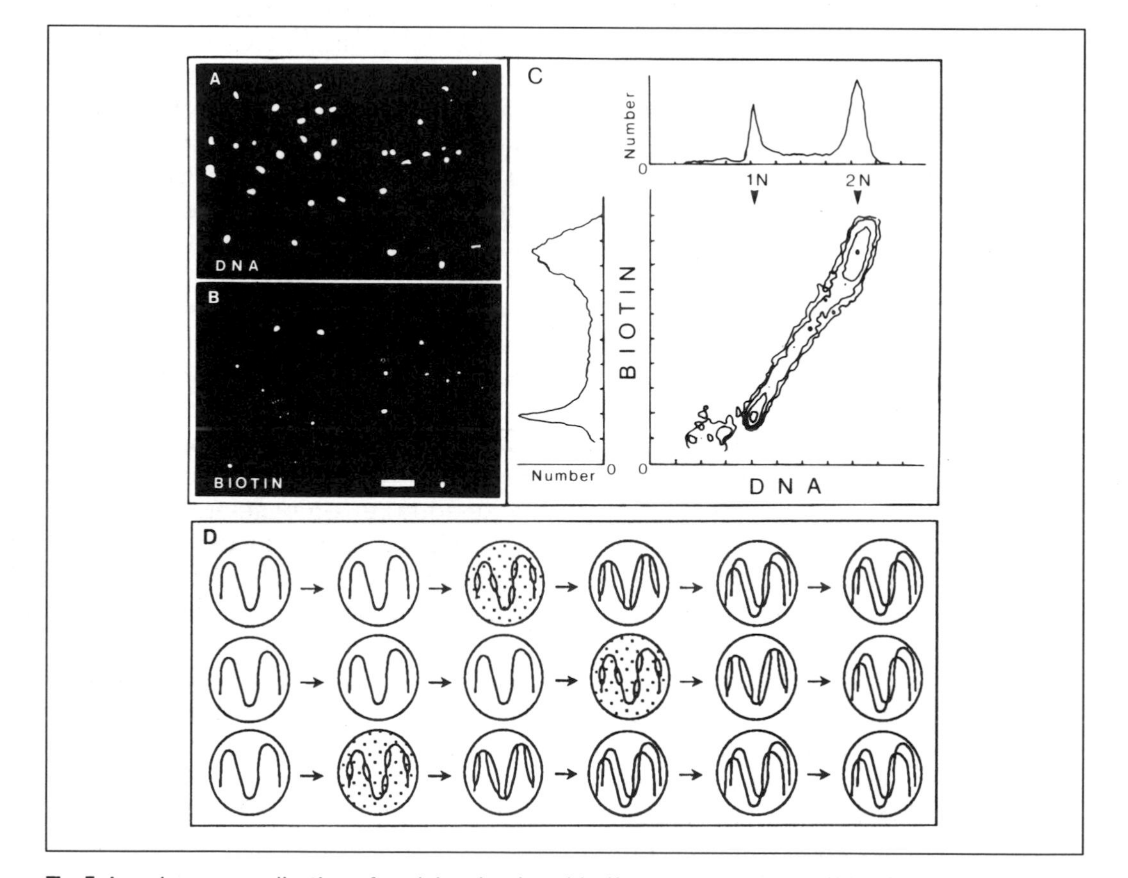

Fig. 5 Asynchronous replication of nuclei co-incubated in *Xenopus* egg extracts (31). Haploid sperm nuclei were incubated together in *Xenopus* egg extract, and nascent DNA was labelled with biotin-11-dUTP. Midway through the replication reaction, nuclei were isolated and stained for DNA (A) and biotin (B) (scale bar 50 μm). The DNA and biotin content of individual nuclei in the population were measured by flow cytometry, and the results expressed as a contour plot (C). Frequency profiles of DNA and biotin contents are shown flanking the contour plot. 1N and 2N show haploid and diploid DNA contents. Reproduced from ref. 31. Overall replication kinetics are shown in cartoon form (D). Three nuclei are shown replicating together in the cell-free system, each with a single chromosome. Different nuclei receive the signal to start replication at different times (shading). As each nucleus starts to replicate, it undergoes a burst of near synchronous initiation events.

pattern of replication is shown in cartoon form in Fig. 5D, and is consistent with data from cell fusion experiments and the model presented in Fig. 4. Nuclei act as individual 'units' whereby they receive a signal to replicate from the cytoplasm, which causes them to undergo once and only once the entire sequence of S phase events, without a requirement for a parallel sequence of events occurring in the cytoplasm.

Further experiments indicate that the feature that defines this 'unit' of replication is indeed the nuclear envelope (32). When demembranated chicken erythrocyte nuclei were incubated in *Xenopus* extract, they were replicated following nuclear envelope reassembly. The chicken nuclei often aggregated, so that the *Xenopus* extract surrounded a number of different nuclei with a continuous nuclear envelope. When each of these 'multinuclear aggregates' started to replicate, all its DNA started replication at about the same time despite originally being derived from a number of chicken nuclei. However, replication synchrony was only observed within the multinuclear aggregates, and different multinuclear aggregates started to replicate at different times. These results provide further evidence that the nucleus represents a fundamental unit of DNA replication, and that this unit is demarcated by the nuclear envelope.

3.2 Physical organization of DNA replication

At any one time, an S phase nucleus is likely to contain more than a thousand active replication forks, and early embryonic cells with short S phases may have many times this number. The subnuclear localization of these active replication forks can be monitored by following the incorporation of labelled dNTPs into nascent DNA. This shows that instead of being randomly distributed throughout the nuclear interior, active replication forks are clustered together into a hundred or so 'replication foci' (33–37) (Fig. 6). The distribution of these replication foci changes during S phase: early S phase nuclei have a relatively larger number of small foci scattered throughout the nuclear interior, whilst at later stages of S phase, foci become fewer and larger, often lining the nuclear envelope (38, 39). Part of this redistribution reflects the different replication timing of transcriptionally active and inactive chromatin. Transcriptionally active euchromatin, present throughout the nuclear interior, is replicated early in S phase, whilst transcriptionally inactive heterochromatin, distributed around the nuclear periphery, is replicated late. The correlation between the location of these replication foci and the type of chromatin being replicated suggests that each replication focus might represent a group of clustered origins within a chromosomal domain that fire coordinately at a particular time in S phase.

The physical clustering of replication forks within S phase nuclei is consistent with an association between replication forks and a nucleoskeleton (40). Newly replicated DNA is associated with a residual nuclear 'matrix', formed when nuclei are isolated and exposed to high salt. This association is maintained in actively replicating nuclei encapsulated in agarose beads, where nascent DNA behaves as

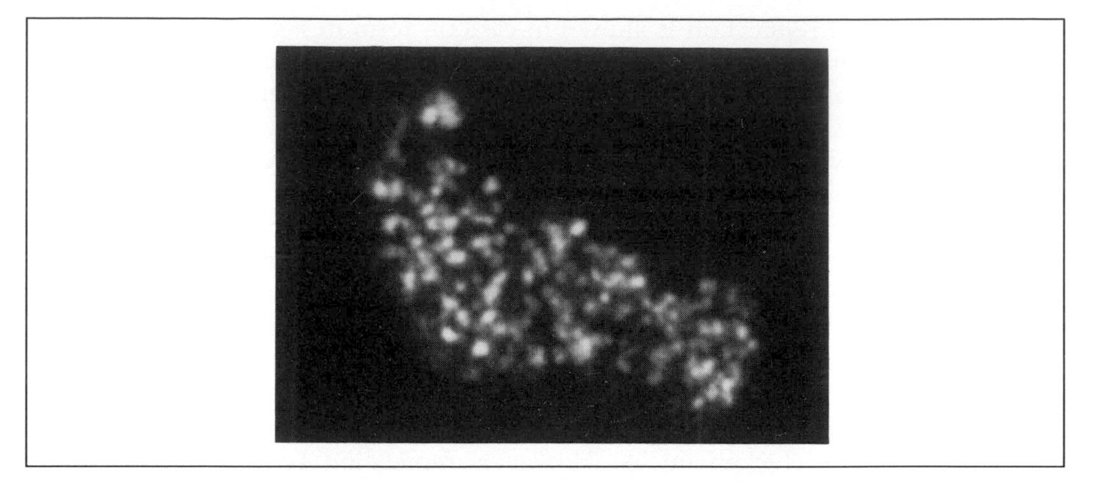

Fig. 6 Replication foci in a nucleus replicating in the *Xenopus* cell-free system. Sperm chromatin replicating in the *Xenopus* cell-free system was pulse labelled with biotinylated dUTP. Sites of biotin incorporation were labelled with fluorescent streptavidin, and then analysed by confocal microscopy. A single optical section is shown. Each bright spot contains several hundred replication forks that remain clustered together throughout S phase. Reproduced from ref. 36.

though attached to very high molecular weight material (41). One possible explanation for the clustering of replication forks into foci is that it depends on the sequential nature of chromosome replication, which requires not only the replication of DNA, but also the duplication of other features such as chromatin modifications, DNA methylation, and the presence of transcription factors (Fig. 1). The presence of clusters of replication forks may serve to localize these sequential processes into functional zones, so that, for example, DNA methylation occurs before the binding of proteins to nascent DNA (42).

3.3 Prevention of re-replication in a single S phase

3.3.1 The Licensing Factor model

The cell fusion experiments outlined in Fig. 3 and summarized in Fig. 4 show that only G1 nuclei can initiate DNA replication in response to cytoplasmic signals; G2 nuclei must pass through mitosis before they respond positively to these signals. The nature of this difference in responsiveness has been investigated in the *Xenopus* cell-free system. Although different nuclei replicated at different times in the same extract (Fig. 5), no DNA was replicated more than once (20, 31). Replicated 'G2' nuclei remained incapable of further replication even after isolation and transfer to fresh extract (43). However, if G2 nuclei were allowed to progress into mitosis and undergo nuclear envelope breakdown and chromosome condensation, they could then undergo a further round of DNA replication when added back to fresh extract. Some metaphase function therefore permitted G2 nuclei to revert to the responsive G1 state. G2 nuclei were subjected to various treatments to see whether

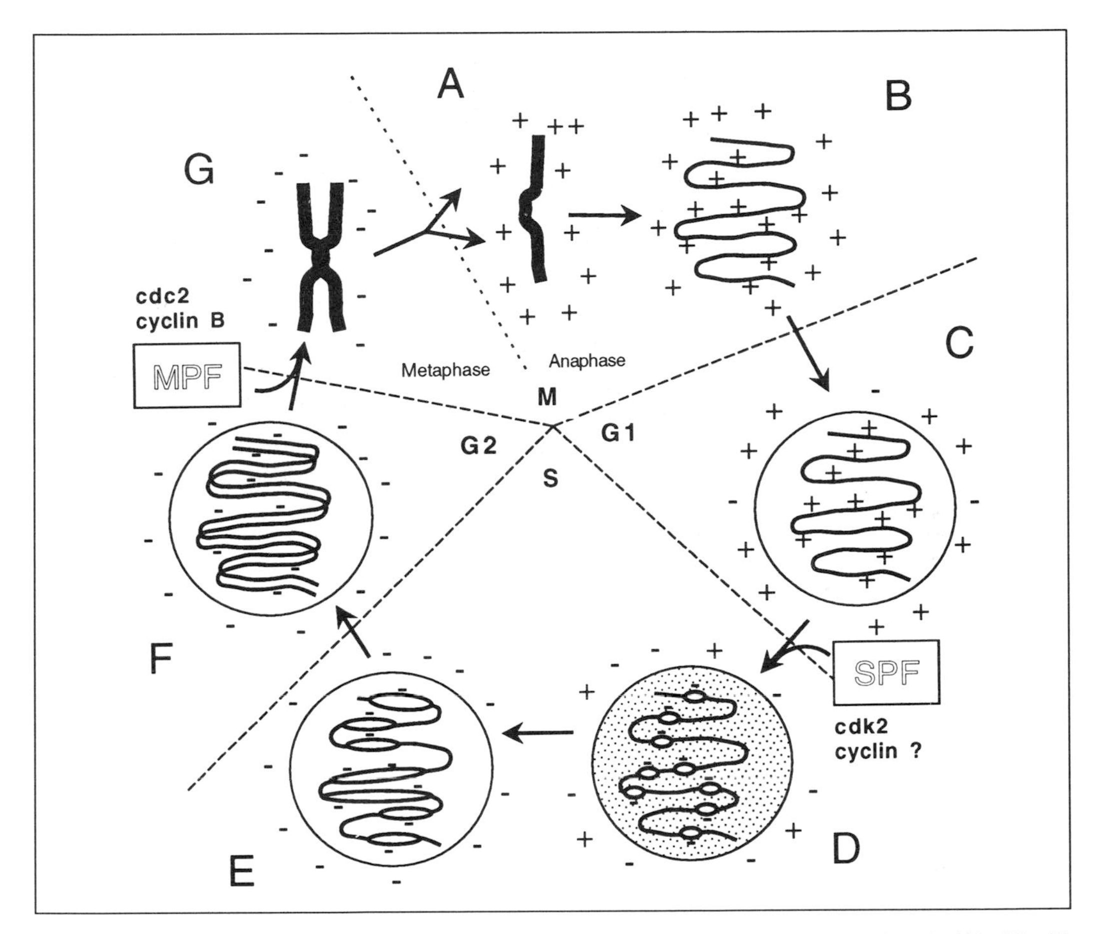

Fig. 7 The Licensing Factor Model to explain why DNA is replicated only once in each cell cycle (43, 46). (A) On exit from metaphase, inactive Licensing Factor (−) is rapidly activated (+) and the chromosomal DNA decondenses. (B) Licensing Factor (+) binds to decondensed chromosomal DNA at future sites of initiation. (C) DNA is assembled into a nucleus surrounded by a complete nuclear envelope. (D) A cytoplasmic activity ('SPF'), generates an intranuclear signal (stippled) that leads to initiation at sites where Licensing Factor is bound to DNA. Each molecule of Licensing Factor can support only a single initiation event, after which it is inactivated or destroyed (−). (E) The chromosome is replicated by forks initiated at Licensed sites. (F) Fully replicated G2 nuclei cannot re-replicate due to exclusion of Licensing Factor from the DNA by the nuclear envelope. (G) Entry into mitosis is accompanied by nuclear envelope breakdown and chromosome condensation. (H) Sister chromatids separate and are partitioned to the daughter cells. During the earlier parts of the cycle (A–E) cytoplasmic Licensing Factor activity has slowly decayed. Reproduced from ref. 46.

they could induce G2 nuclei to re-replicate in fresh extract without entering metaphase (43). Agents that caused nuclear envelope permeabilization, such as lysolecithin or phospholipase, left the G2 nuclei capable of re-replication in fresh extract. Similar results have been obtained in cell-free extracts of *Drosophila* eggs (25) and on introduction of mammalian tissue culture nuclei into *Xenopus* eggs (44) or egg extracts (45).

Figure 7 outlines a model proposed to explain these results (43, 46). An essential replication factor, called 'Licensing Factor' can bind to DNA during late mitosis or early interphase before nuclear assembly has occurred. Licensing Factor cannot cross the nuclear envelope, so that once nuclear assembly is complete, Licensing Factor is only present in the nucleus where bound to DNA. On entry into S phase, each molecule of Licensing Factor bound to DNA supports a single initiation event after which it is inactivated or destroyed. Thus in G2, no active Licensing Factor remains in the nucleus, and the nuclear envelope must be transiently permiabilized (as normally occurs during mitosis) to allow a further round of DNA replication to be licensed.

Further experiments have confirmed this general model (46, 47). When replication-competent permeabilized G2 nuclei were resealed using nuclear envelope precursors, they became once more incapable of re-replicating in *Xenopus* extract (47). However, when permeabilized nuclei were exposed to *Xenopus* egg extract prior to resealing, they retained their replication-competence. In a different approach, *Xenopus* extracts were treated with the protein kinase inhibitor 6-dimethylaminopurine (6-DMAP) prior to exit from metaphase, resulting in a block to the initiation of replication (46). 6-DMAP-treated extracts behaved as though they lacked Licensing Factor, since nuclei assembled in them were incapable of initiating replication in untreated extract without first undergoing nuclear envelope permeabilization. DNA exposed to normal extract rapidly became 'licensed' to undergo only a single round of replication in 6-DMAP-treated extract.

The ability of 6-DMAP to render extracts functionally devoid of Licensing Factor was used to measure its normal levels in the cytoplasm during the *Xenopus* cell cycle (46) (Fig. 7). Licensing Factor levels were low during the early stages of mitosis. However, shortly after the metaphase–anaphase transition, levels rose very abruptly so that the decondensing telophase chromosomes could become licensed prior to complete nuclear envelope assembly. The appearance of Licensing Factor at this stage of the *Xenopus* cell cycle took place in the absence of new protein synthesis and was presumably due to post-translational modification of a protein already present. This post-translational modification is likely to be the real target of inhibition by 6-DMAP (46, 48). Cytoplasmic Licensing Factor that failed to bind DNA was unstable, and had disappeared by mid-S phase. This provides a further mechanism to prevent the illegitimate re-replication of DNA, since even if the nuclear envelope became damaged in G2, no cytoplasmic Licensing Factor was present at this time so no further initiation could take place.

3.3.2 The *CDC46* gene family

The product of the yeast *CDC46* gene, which is required for a very early event in DNA replication, behaves as if it were a Licensing Factor homologue (49). Cdc46 protein is nuclear in G1, where it apparently associates with chromatin, but disappears from the nucleus at the start of S phase (Fig. 8). It is seen in the cytoplasm throughout G2 and mitosis, but at the end of mitosis it translocates back to the nucleus (yeasts do not undergo nuclear envelope breakdown during mitosis). Although Cdc46 protein levels are approximately constant throughout the cell

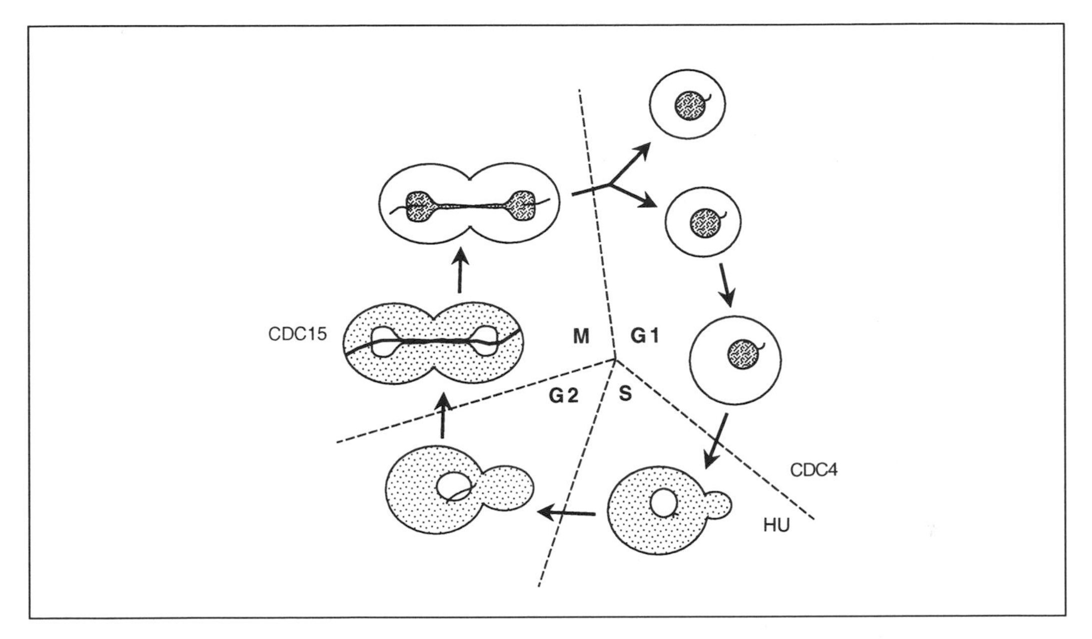

Fig. 8 Subcellular localization of CDC46 during the budding yeast cell cycle (49). During G1, CDC46 (shading) is found in the nucleus, probably associated with chromatin. At some point between the CDC4 and hydroxyurea (HU) block points, CDC46 disappears from the nucleus. During S, G2 and early mitosis CDC46 is cytoplasmic, but at some stage in late mitosis, after the CDC15 block point, it translocates back into the nucleus.

cycle, transcription peaks in early G1 (49). This is strikingly similar to the cell cycle behaviour of Licensing Factor in *Xenopus* eggs (compare Figs 7 and 8). *MCM5* ('minichromosome maintenance'), a gene identified by its role in the replicative stability of minichromosomes, is identical to the *CDC46* gene (50). Mutant *MCM* genes show increased loss of minichromosomes due to inefficient replication, and the severity of this defect varies with the replication origin used by the minichromosome. This suggests that the Cdc46 protein interacts with yeast origins of replication. *CDC46/MCM5* shows strong sequence homology to two other *MCM* genes, *MCM2* and *MCM3*, which also show the same subcellular translocation as *CDC46/MCM5* (51, 52). Members of this gene family are found throughout the eukaryotic kingdom, with homologues in fission yeast (51, 53), *Xenopus* (53), mouse, and humans (54). *CDC46*, and another gene, *CDC47*, were originally identified as suppressors of two other cell cycle genes involved in DNA replication, *CDC45* and *CDC54* (55, 56). These four genes show a complicated pattern of suppression and synthetic lethality, which suggest that they all interact in some common event or pathway required for DNA replication (56).

3.3.3 Late mitotic/early interphase events required for S phase

One prediction of the Licensing Factor model is that essential changes are made to replication origins at the end of mitosis which persist throughout G1 and which are required for their subsequent activity in S phase. Two additional lines of evidence support this idea:

1. Proteins thought to be involved in the initiation of DNA replication are associated with replication origins throughout G1. ORC is a protein complex that specifically binds to the core element of replication origins in budding yeast (2, 3). Genomic footprinting of asynchronous cells shows at least 80 per cent occupancy of the origin by ORC, suggesting that it is bound for most, if not all, the cell cycle. Potential initiation proteins are also chromatin bound for extensive periods in *Xenopus*. RP-A (replication protein A) is a cellular single-stranded DNA binding protein required for both initiation and elongation stages of SV40 viral DNA replication (57–59). Prior to nuclear assembly in the *Xenopus* cell free system, RP-A binds to chromatin at a small number of discrete foci (60). Once initiation occurs, these RP-A-bound foci colocalize with replication foci, and when DNA replication ceases, RP-A disappears from these foci. Thus RP-A appears to bind at 'pre-replication foci' where initiation will subsequently take place, and it seems likely that the presence of RP-A at these sites is an indication that the DNA is 'licensed' for subsequent initiation.

2. Genes whose major functions is in late mitosis can subsequently affect the initiation of DNA replication. An increased loss of minichromosomes was observed in several cdc mutants temporarily incubated at the non-permissive temperature (61). In two of these mutants, *cdc6* and *cdc14*, the plasmid loss rate could be significantly reduced by increasing the number of replication origins on each minichromosome, suggesting that the minichromosome loss was due to inefficient origin usage (61). This conclusion is easily understood for *CDC6*, which functions close to the initiation of replication (see below). However, the major function of *CDC14* in the cell cycle is at the end of mitosis, as mutants block in late telophase. Therefore it is possible that part of the function of *CDC14* is to permit licensing of replication origins to occur in late M or early G1. Similarly, the DBF2 gene, first identified as a mutant causing delayed DNA synthesis, performs its major cell cycle function in late mitosis (62). The fission yeast *cdc2* gene is required for both G1/S and G2/M transitions, as discussed in more detail below. When temperature-sensitive *cdc2* strains are heat-shocked during G2, mutant cdc2 protein is destroyed and cells perform another round of DNA replication on return to the permissive temperature (63). This suggests that cdc2 may be required for the cell to assess the stage of the cell cycle that it is in, and hence destruction of cdc2 allows inappropriate licensing of replicated DNA.

4. Inducers of S phase: activities required for initiation

A G1 cell, containing a properly assembled and responsive nucleus, must assess development, nutritional, and environmental information to make the correct decision about whether to undergo DNA replication and cell division. Among the activities generated on commitment to cell division are the S phase-inducers that cause G1 nuclei to undergo DNA replication. Genes that function during this process have been identified in yeast, some of which have close parallels in higher eukaryotes. However, there has been relatively little biochemical analysis of these activities, except in the SV40 DNA replication system, where initiation is dependent on the viral T antigen and a viral origin of replication. Further understanding of these events is likely to emerge from the analysis of initiation of chromosomal DNA replication.

4.1 Cell division cycle mutants in yeast

In both budding yeast (*Saccharomyces cerevisiae*) and fission yeast (*Schizosaccharomyces pombe*) a number of genes have been identified that are required to progress from the uncommitted state to the initiation of replication. The commitment event is called START, after which the cell becomes committed to DNA replication and cell division (64–67). Genes that act from START to early S phase in both yeasts are outlined in Fig. 9. Events occurring at START in budding yeast are dealt with in more detail in Chapter 3.

4.1.1 Periodic transcription at START

START corresponds to the execution point of the *CDC28* gene in budding yeast and the *cdc2* gene in fission yeast (reviewed in refs 68, 69), members of the widespread family of cyclin-dependent protein kinases (cdks), whose activity is dependent on association with proteins from the cyclin family. cdc2 and CDC28

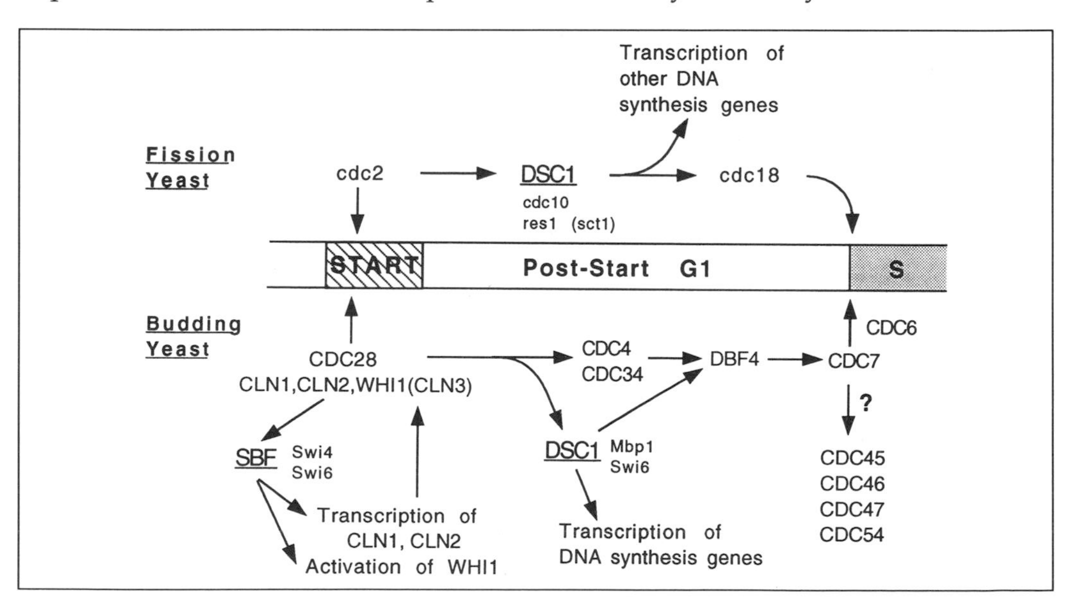

Fig. 9 Diagram of events occurring in budding and fission yeast between START and the initiation of S phase. Gene functions required at or after START for the initiation of DNA replication in budding yeast and fission yeast are shown. The beginning of S phase is marked by the first initiation events. See text for further details.

both have further functions at the G2/M transition. At START, Cdc28 interacts with at least three G1 cyclins, Cln1, Cln2, and Cln3. These cyclins are functionally redundant as any one of them can substitute for the others (70, 71). The Cln proteins appear to be rate-limiting for START, since if they are overexpressed cells traverse G1 more rapidly and undergo START at a reduced size (70, 72–74). This suggests that the Cdc28/Cln complex performs a function analogous to the S phase-inducers postulated from the cell fusion experiments described above in Section 2 (see also Chapter 3).

A major function of cdc2/Ccd28 at START is to induce transcription of genes required for S phase. In budding yeast two transcription complexes, called SBF and DSC1 (MBF), have been identified. SBF contains at least two gene products, Swi4 and Swi6. Since the function of SBF is dependent on *CDC28* at START (75) and the transcription of both *CLN1* and *CNL2* is SBF-dependent, a positive feedback loop appears to exist whereby low levels of active Cdc28 kinase lead to increased expression of the CLN genes and further activation of Cdc28 (76–78) (Fig. 9). This would in effect define a switch, which when thrown, induces entry into S phase.

The second transcription complex in budding yeast, DSC1, contains at least two gene products, Swi6 and Mbp1 (79–82). DSC1 induces transcription from a large group of genes required for S phase including DNA polymerases, RP-A, DNA ligase, and DNA primase (reviewed in refs 66, 67). The histone genes, which are also transcribed periodically during S phase, are controlled downstream of the *CDC4* gene function (83). The DSC1 expression system is conserved in fission yeast (84). *cdc10*, which has homology to both *SWI6* and *SWI4*, is required for events at START in fission yeast and is found in a DSC1-like complex (84), where it appears to act in partnership with another gene product, res1 (sct1) (85–87). The conservation of the DSC1 transcription system between two evolutionarily divergent yeasts suggests that it may be widespread throughout the eukaryotic kingdom.

4.1.2 Post-START functions required for DNA replication

A number of genes have been identified in yeast that act after START but before the initiation of DNA replication. These include *CDC4*, *CDC34*, *DBF4*, and *CDC7* in budding yeast, and cdc18 in fission yeast (Fig. 9).

CDC4 and *CDC34* are required after START for the initiation of DNA replication. They may primarily be involved in general cell cycle progression, since mutants show an unusual morphology. The *CDC34* gene encodes a ubiquitin conjugating enzyme (88). Since cyclin proteolysis may be mediated by a ubiquitin-dependent system (89), it is possible that *CDC34* is required for destruction of G1-specific cyclins.

Unlike the other post-START G1 genes, *CDC7* mutants arrested at the non-permissive temperature can initiate DNA synthesis in the absence of further protein synthesis. *CDC7* encodes a serine/threonine protein kinase (90, 91). It is possible that Cdc7 kinase acts at initiation itself by, for example, phosphorylation of origin-binding proteins, though there is currently no direct evidence to support this. Interaction occurs between *DBF4* and *CDC7*, since overexpressors of both can suppress temperature-sensitive mutations in the other (92). However, *dbf4* mutants, unlike *cdc7*, cannot initiate DNA synthesis in the absence of ongoing protein synthesis. *DBF4* transcription is cell cycle regulated, and appears to be under the control of the *DSC1* transcription system. This provides at least one route by which the *DBF4*/*CDC7* function is regulated by passage through START.

cdc18 mutants in fission yeast blocks in either late G1 or early S phase. *cdc18* appears to be the major target for the *DSC1* system since *cdc10* mutants at the non-permissive temperature can be induced to undergo S phase by constitutive expression of *cdc18* (93). However, deletion of the *cdc10* gene cannot be rescued in the same way, suggesting that other essential genes are transcribed by cdc10. The *cdc18* gene has homology with the budding yeast *CDC6* gene which is periodically expressed by the *DCS1* system and which functions near the beginning of S phase (94, 95).

In budding yeast, *CDC28* may also be required for a late S phase function. Certain mutants of B-type cyclins (Clbs, which have major functions in mitosis) also have difficulties in completing DNA replication, suggesting that a Cdc28/Clb complex may be required for a late stage in S phase (refs 96, 97, and see Chapter 3). Although neither *cdc28* or *cdc2* mutants show any obvious late S phase defect, this might be masked by a quantitatively greater requirement for these genes at START and at G2/M.

4.2 Cyclin-dependent kinases in metazoan chromosome replication

cdc2/*CDC28* homologues have been identified in a range of higher eukaryotic organisms, including humans, *Xenopus*, and *Drosophila*, forming a family of cyclin-dependent kinases, cdk1–5 (68, 69, 98). cdk1, the cdc2/CDC28 homologue, can functionally replace the yeast genes, and can support transit through START, whilst cdk2 and cdk3 can at least partially do so (99–103). Similarly, potential G1 cyclins have been identified in metazoans by their ability to rescue the triple *CLN* deletion in budding yeast. Three new classes of cyclins, C, D, and E, are all capable of substituting for the *CLN* function in this assay (104–108). However, the ability to substitute for the yeast genes may not be strong evidence for a G1/S function in the cells they come from, since mitotic B-type cyclins are also capable of rescuing the triple *CLN* defect (106).

Different cyclin–cdk complexes are assembled as cells traverse through G1, from commitment to replication to S phase itself (reviewed in refs 109, 110). D-type cyclins tend to appear early after growth factor stimulation, followed by cyclin E. At around the start of S phase cyclin A is detected, and finally cyclin B levels rise prior to entry into mitosis. In the early stages of the cycle, cyclins A and E tend to be found complexed with cdk2, whilst in G2 and mitosis, cyclins A and B are found complexed with cdc2 (cdk1). It is currently unclear what DNA replication function, if any, these cyclin–cdk complexes are performing. Their appearance in G1 is weakly suggestive of functions related to entry into S phase, though in certain cases, more direct evidence is available.

4.2.1 cdk involvement in S phase control

A number of observations suggest that cdk2 is required for DNA replication in higher eukaryotes. In the *Xenopus* cell-free system, depletion of cdks left extracts unable to replicate added DNA, but capable of assembling nuclei and supporting replication fork elongation (111). Replication could be restored by re-addition of fractions enriched for active cdks. A similar block to replication was obtained by immunodepleting extracts already in interphase with antibodies specifically directed against *Xenopus* cdk2 (112) or by addition of the cdk inhibitor Cip1 (141). The cdc2 protein may have a different effect on DNA replication, acting earlier in the cell cycle. When extracts arrested in metaphase were immunodepleted with antibodies directed against cdc2, they were incapable of initiating replication on progress into interphase (111). This is likely to result from a failure to activate the replication Licensing Factor that normally occurs during anaphase and telophase (Fig. 8) (46, 48). However, once *Xenopus* extracts have passed into interphase, DNA replication is apparently no longer dependent on cdc2 function (112).

Results consistent with these are obtained with mammalian cells. When antibodies directed against the cdk2 protein were microinjected into human fibroblasts, subsequent DNA replication was inhibited (113). Conversely, microinjection of antibodies directed against cdc2 blocked progress in mitosis but not DNA replication (114). The FT210 cell line, with a temperature-sensitive *cdc2* gene, failed to enter metaphase at the non-permissive temperature but was still capable of initiating DNA replication (115). However, a role for cdc2 in DNA replication is suggested by experiments where antisense oligonucleotides against cdc2 mRNA inhibited DNA synthesis in lymphocytes entering S phase from quiescence (116).

4.2.2 Cyclin involvement in S phase control

If cdk2 is involved in controlling S phase, it is likely to require activation by one or more cyclins. In mammalian cells cyclin A protein is observed within the nucleus at about the start of S phase (117–120). Microinjection of antibodies directed against cyclin A left cells incapable of replicating DNA (118, 119). Similarly, antisense constructs against cyclin A message blocked DNA replication, and the inhibition was reversible by coinjection of cyclin A protein (118, 121). During S phase, cyclin A appears to form part of a kinase complex with cdk2 (119, 122), whilst its activation of the cdc2 kinase does not appear until later in G2 (119, 123). In replicating myotubes, cyclin A and cdk2 can be seen to localize specifically to active replication foci, suggesting that the function of the cyclin A–cdk2 complex in DNA replication is fairly direct (124). Similarly, PCNA, the DNA polymerase δ auxiliary factor which also colocalizes to replication foci, can also be found complexed with a range of cyclin–cdk complexes (125).

However, cyclin A does not appear to be necessary for S phase in either *Drosophila* or *Xenopus* early embryos. Deletion of the *Drosophila* cyclin A gene caused arrest in G2, consistent with a role for cyclin A in mitotic progression, but no S phase defect could be detected (126, 127). Similarly, when protein synthesis is blocked in *Xenopus* embryos during metaphase, they are still capable of progressing through the subsequent S phase and arrest in G2, although most of the cyclin A is destroyed as the embryo passes out of mitosis (8, 123). Nevertheless, cyclin A is capable of inducing the initiation of DNA replication in *Xenopus* extracts whose cdk2 activity has been blocked by the Cip1 inhibitor (141).

Different lines of evidence suggest a role for cyclin E in G1 progression in higher eukaryotes. Mammalian fibroblasts overexpressing cyclin E had a shorter G1 and entered S phase at a reduced size (128). However, these changes were compensated by an increased length of S and G2 phases. By influencing the rate of progression from G1 to S phase, cyclin E is behaving in a comparable way to the CLNs of budding yeast and is performing a function analogous to the S phase-inducers postulated from the cell fusion experiments described above in Section 2. Cyclin E can also induce the initiation of DNA replication in *Xenopus* extracts whose cdk2 has been blocked by the Cip1 inhibitor (141). Similarly, in *Drosophila* embryos, ectopic expression of cyclin E induces cells that would normally have stopped dividing to undergo a further S phase (142). Mutant embryos lacking an active cyclin E gene also show an S phase defect. These results suggest that under different conditions both cyclins A and E may have a role in progression through S phase.

4.3 The SV40 model system

SV40 is a virus that infects monkey and human cells. Replication of viral DNA requires only a single virally-encoded origin-binding protein, the SV40 large T antigen. All other activities required for viral DNA replication are provided by the host cell, thus making SV40 a good model for chromosomal DNA replication. However, it should be noted that since SV40 is a virus, its aim is to circumvent many of the controls that normally restrict DNA replication within the cell cycle, and direct extrapolation from SV40 to chromosomal replication may not be valid. SV40 DNA replication can be reconstituted *in vitro*, and initiation and chain elongation have been reconstituted with purified proteins (129–131). Cell cycle control of SV40 DNA replication has been shown by a number of approaches.

SV40 replication extracts can be prepared from cells at different cell cycle stages. G1 extracts support SV40 DNA replication inefficiently compared to S or G2 extracts (109). One of the early events in the initiation of SV40 DNA replication is the unwinding of the DNA at the origin of replication, and this origin-unwinding activity is defective in G1 extracts. An activity from S phase cells capable of stimulating replication in G1 extracts could be inactivated by removal of cdc2-like proteins (132), and efficient origin-unwinding can be restored to G1 extracts by addition of purified cyclin A or cdc2–cyclin B complex (132, 133).

Origin-unwinding can be reconstituted *in vitro* with three proteins: T antigen, topoisomerase I, and RP-A (Fig. 10). T antigen binds to the SV40 origin as a double hexamer. In the presence of RP-A, a single-stranded DNA binding protein, extensive unwinding of the origin region can then occur (58, 134). SV40 DNA replication is potentially subject to cell cycle control by phosphorylation of T antigen. T antigen can be phosphorylated *in vitro* by cdc2–cyclin B kinase at threonine 124, a site which is also phosphorylated *in vivo* (135). Phosphorylation of this residue alters the binding of T antigen to the SV40 replication origin and enhances its ability to initiate replication by more than ten-fold (Fig. 10A–C). It is unlikely that the stimulation of replication in G1 extracts by the cdc2–cyclin B complex described above is directly due to phosphorylation of T antigen, since T antigen phosphorylated by cdc2 cannot complement the defect in G1 extracts (132). However, it is possible that T antigen is inactivated by components in the G1 extract which are themselves inactivated by the cdc2–cyclin B kinase.

T antigen activity is also stimulated by dephosphorylation at certain serine residues by protein phosphatase 2A (PP2A) (Fig. 10C, D). PP2A enhances the cooperativity of double hexamers in binding to the SV40 origin which is required for origin-unwinding (132, 136, 137). PP2A is also capable of enhancing SV40 replication in G1 extracts, though whether this occurs by dephosphorylating T antigen is not known.

RP-A undergoes complex phosphorylation during the cell cycle of both yeast and human cells (133, 138, 139). It is largely dephosphorylated in G1 but becomes phosphorylated in S phase. When human cells were fractionated for proteins able to phosphorylate RP-A, >50 per cent of the activity was associated with a cdc2–cyclin B complex which phosphorylated RP-A on sites normally phosphorylated on progress into S phase (Fig. 10E) (133). Similar RP-A-kinase activity was identified in a cyclin A immunoprecipitate. However, the purified origin-unwinding reaction consisting of RP-A, T antigen, and topoisomerase I was only slightly stimulated by cdk addition, suggesting that cdk phosphorylation of RP-A is not required for the initiation of SV40 replication. Instead the cdk stimulation of origin-unwinding in G1 extracts may act by counteracting an inhibitory activity which they contain. Another phosphorylation that RP-A undergoes follows the association of RP-A with single-stranded DNA that occurs at the initiation of replication (Fig. 10F, G) (139). The kinase responsible for this phosphorylation is present throughout the cell cycle, and does not appear to be a cdk. RP-A undergoes similar phosphorylation events in *Xenopus* egg extracts (140). At mitosis, RP-A is directly phosphorylated by cdc2; in interphase RP-A is phosphorylated when associated with single-stranded DNA by a kinase which is not a cdk.

5. Summary

During S phase of the cell cycle the genome is precisely duplicated by a mechanism that involves the initiation, progression, and termination of replication forks from replication origins which usually fire in a distinct sequence. Cell fusion studies and experiments using *Xenopus* egg extracts suggest that nuclei behave as autonomous

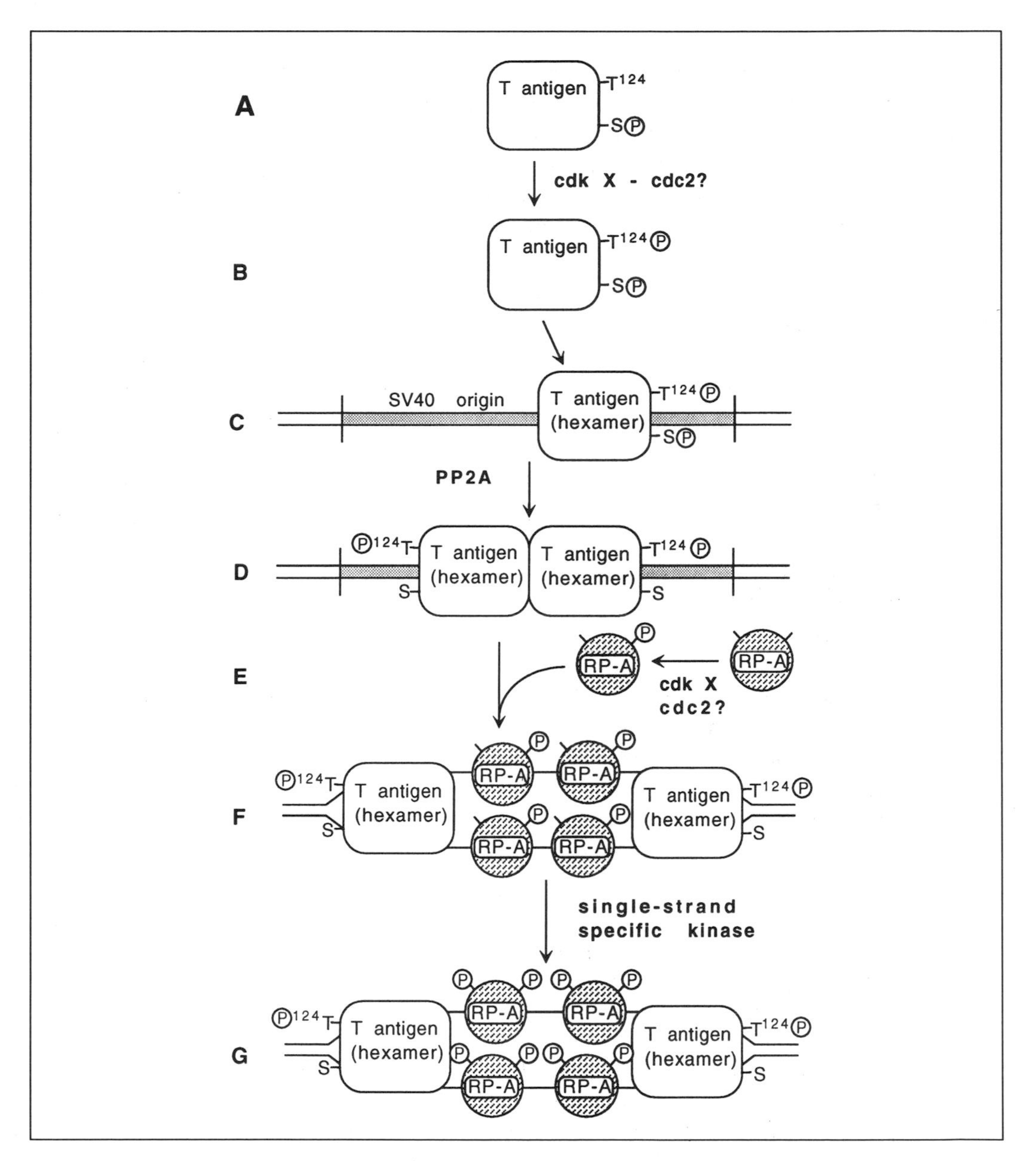

Fig. 10 Regulatory events occurring at the initiation of SV40 DNA replication. T antigen is phosphorylated by a cdk (possibly cdc2) at threonine 124 (A–B); this enhances its ability to bind to the SV40 origin (C). Protein phosphatase 2A dephosphorylates T antigen on certain serine residues, enhancing the cooperativity of double hexamer formation at the origin (C–D). RP-A is phosphorylated by a cdk (E); this has only a slight effect on initiation activity. In the presence of RP-A, T antigen causes a local unwinding of the origin region (F). RP-A associated with single-stranded DNA is then phosphorylated by a single-strand specific kinase (G).

'units' of DNA replication which receive two different types of signal from the cytoplasm: one signal causing a replication-competent G1 nucleus to undergo the entire sequence of S phase events, and a second type of signal returning a replication-incompetent G2 nucleus back to the competent G1 state. Replication-competence can be explained in terms of a Licensing Factor which modifies DNA in late mitosis but which is unable to gain access to DNA in the nucleus during interphase, thus ensuring that DNA is replicated only during each cell cycle. The signal to undergo the replication sequence seems likely to be a cdk. In yeast, functions that are performed after receipt of the signal are beginning to be understood. Although the ultimate targets of these functions are not clear, the SV40 system provides examples of the way that cdks and other cell cycle activities interact with the DNA replication machinery.

Acknowledgements

Thanks to James Chong and Margret Michalski-Blow for critical reading of the manuscript. Julian Blow is a Lister Institute Centenary Fellow.

References

1. Brewer, B. J. (1988) When polymerases collide: replication and the transcriptional organisation of the *E. coli* chromosome. *Cell*, **33**, 679.
2. Bell, S. P. and Stillman, B. (1992) ATP-dependent recognition of eukaryotic origins of DNA replication by a multiprotein complex. *Nature*, **357**, 128.
3. Diffley, J. F. X. and Cocker, J. H. (1992) Protein–DNA interactions at a yeast replication origin. *Nature*, **357**, 169.
4. Hand, R. (1978) Eukaryotic DNA: organization of the genome for replication. *Cell*, **15**, 317.
5. Brewer, B. J., Lockshon, D., and Fangman, W. L. (1992) The arrest of replication forks in the rDNA of yeast occurs independently of transcription. *Cell*, **71**, 267.
6. Sundin, O. and Varshavsky, A. (1980) Terminal stages of SV40 DNA replication proceed via multiple intertwined catenated dimers. *Cell*, **21**, 103.
7. Blumenthal, A. B., Kriegstein, H. J., and Hogness, D. S. (1973) The units of DNA replication in *Drosophila melanogaster* chromosomes. *Cold Spring Harbor Symp. Quant. Biol.*, **38**, 205.
8. Harland, R. M. and Laskey, R. A. (1980) Regulated DNA replication of DNA micro-injected into eggs of *Xenopus laevis*. *Cell*, **21**, 761.
9. Vassilev, L. and Russev, G. (1988) Kinetics of replicon initiation during S phase of Chinese hamster ovary cells. *Biochim. Biophys. Acta*, **949**, 138.
10. Stubblefield, E. (1975) Analysis of the replication pattern of Chinese hamster chromosomes using 5-bromodeoxyuridine suppression of 33258 Hoechst fluorescence. *Chromosoma*, **53**, 209.
11. Dubey, D. D. and Raman, R. (1987) Factors influencing replicon organization in tissues having different S-phase durations in the mole rat, *Bandicota bengalensis*. *Chromosoma*, **95**, 285.
12. Goldman, M. A., Holmquist, G. P., Gray, M. C., Caston, L. A., and Nag, A. (1984) Replication timing of genes and middle repetitive sequences. *Science*, **224**, 686.

13. Calza, R. E., Eckhardt, L. A., DeGiudice, T., and Schildkraut, C. L. (1984) Changes in gene position are accompanied by a change in time of replication. *Cell*, **36,** 689.
14. Ferguson, B. M. and Fangman, W. L. (1992) A position effect on the time of replication origin activation in yeast. *Cell*, **68,** 333.
15. Fangman, W. L. and Brewer, B. J. (1992) A question of time: replication origins of eukaryotic chromosomes. *Cell*, **71,** 363.
16. Rao, P. N. and Johnson, R. T. (1970) Mammalian cell fusion: studies on the regulation of DNA synthesis and mitosis. *Nature*, **225,** 159.
17. Rao, P. N., Wilson, B. A. and Sunkara, P. S. (1978) Inducers of DNA synthesis present during mitosis of mammalian cells lacking G1 and G2 phases. *Proc. Natl. Acad. Sci. USA*, **75,** 5043.
18. Yanishevsky. R. M. and Prescott, D. M. (1978) Late S phase cells (Chinese hamster ovary) induce early S phase DNA labelling patterns in G1 phase nuclei. *Proc. Natl. Acad. Sci. USA*, **75,** 3307.
19. Lohka, M. J. and Masui, Y. (1983) Formation *in vitro* of sperm pronuclei and mitotic chromosomes induced by amphibian ooplasmic contents. *Science*, **220,** 719.
20. Blow, J. J. and Laskey, R. A. (1986) Initiation of DNA replication in nuclei and purified DNA by a cell-free extract of *Xenopus* eggs. *Cell*, **47,** 577.
21. Newmeyer, D. D., Lucocq, J. J., Burglin, T. R., and De Robertis, E. M. (1986) Assembly *in vitro* of nuclei active in nuclear protein transport: ATP is required for nucleoplasmin accumulation. *EMBO J.*, **5,** 501.
22. Newport, J. (1987) Nuclear reconstitution *in vitro*: stages of assembly around protein-free DNA. *Cell*, **48,** 205.
23. Sheehan, M. A., Mills, A. D., Sleeman, A. M., Laskey, R. A., and Blow, J. J. (1988) Steps in the assembly of replication-competent nuclei in a cell-free system from *Xenopus* eggs. *J. Cell Biol.*, **106,** 1.
24. Blow, J. J. and Sleeman, A. M. (1990) Replication of purified DNA in *Xenopus* egg extracts is dependent on nuclear assembly. *J. Cell Sci.*, **95,** 383.
25. Crevel, G. and Cotterill, S. (1991) DNA replication in cell-free extracts from *Drosophila melanogaster*. *EMBO J.*, **10,** 4361.
26. Finlay, D. R., Newmeyer, D. D., Price, T. M. and Forbes, D. J. (1987) Inhibition of *in vitro* nuclear transport by a lectin that binds to nuclear pores. *J. Cell Biol.*, **104,** 189.
27. Finlay, D. R. and Forbes, D. J. (1990) Reconstitution of biochemically altered nuclear pores: transport can be eliminated and restored. *Cell*, **60,** 17.
28. Cox, L. S. (1992) DNA replication in cell-free extracts from *Xenopus* eggs is prevented by disrupting nuclear envelope function. *J. Cell Sci.*, **101,** 43.
29. Newport, J. W., Wilson, K. L., and Dunphy, W. G. (1990) A lamin-independent pathway for nuclear envelope assembly. *J. Cell Biol.*, **111,** 2247.
30 Meier, J., Campbell, K. H., Ford, C. C., Stick, R., and Hutchison, C. J. (1991) The role of lamin LIII in nuclear assembly and DNA replication, in cell-free extracts of *Xenopus* eggs. *J. Cell Sci.*, **98,** 271.
31. Blow, J. J. and Watson, J. V. (1987) Nuclei act as independent and integrated units of replication in a *Xenopus* cell-free system. *EMBO J.*, **6,** 1997.
32. Leno, G. H. and Laskey, R. A. (1991) The nuclear membrane determines the timing of DNA replication in *Xenopus* egg extracts. *J. Cell Biol.*, **112,** 557.
33. Nakamura, H., Morita, T., and Sato, C. (1986) Structural organisations of replicon domains during DNA synthetic phase in mammalian cells. *Exp. Cell Res.*, **165,** 291.
34. Bravo, R. and Macdonald, B. H. (1987) Existence of two populations of cyclin/

proliferating cell nuclear antigen during the cell cycle: association with DNA replication sites. *J. Cell Biol.*, **105,** 1549.

35. Nakayasu, H. and Berezney, R. (1989) Mapping replicational sites in the eucaryotic cell nucleus. *J. Cell Biol.*, **108,** 1.
36. Mills, A. D., Blow, J. J., White, J. G., Amos, W. B., Wilcock, D., and Laskey, R. A. (1989) Replication occurs at discrete foci spaced throughout nuclei replicating *in vitro*. *J. Cell Sci.*, **94,** 471.
37. Hozak, P., Hassan, A. B., Jackson, D. A., and Cook, P. R. (1993) Visualization of replication factories attached to a nucleoskeleton. *Cell*, **73,** 361.
38. Fox, M. H., Arndt-Jovin, D. J., Jovin, T. M., Baumann, P. H., and Robert-Nicoud, M. (1991) Spatial and temporal distribution of DNA replication sites localized by immunofluorescence and confocal microscopy in mouse fibroblasts. *J. Cell Sci.*, **99,** 247.
39. O'Keefe, R. T., Henderson, S. C., and Spector, D. L. (1992) Dynamic organization of DNA replication in mammalian cell nuclei: spatially and temporally defined replication of chromosome-specific alpha-satellite DNA sequences. *J. Cell Biol.*, **116,** 1095.
40. Cook, P. R. (1991) The nucleoskeleton and the topology of DNA replication. *Cell*, **66,** 625.
41. Jackson, D. A. and Cook, P. R. (1986) Replication occurs at a nucleoskeleton. *EMBO J.*, **5,** 1403.
42. Blow, J. J. (1993) DNA replication: methyltransferases in foci. *Nature*, **361,** 684.
43. Blow, J. J. and Laskey, R. A. (1988) A role for the nuclear envelope in controlling DNA replication within the cell cycle. *Nature*, **332,** 546.
44. De Roeper, A., Smith, J. A., Watt, R. A., and Barry, J. M. (1977) Chromatin dispersal and DNA synthesis in G1 and G2 HeLa cell nuclei injected into *Xenopus* eggs. *Nature*, **265,** 469.
45. Leno, G. H., Downes, C. S., and Laskey, R. A. (1992) The nuclear membrane prevents replication of human G2 nuclei but not G1 nuclei in *Xenopus* egg extract. *Cell*, **69,** 151.
46. Blow, J. J. (1993) Preventing re-replication of DNA in a single cell cycle: evidence for a Replication Licensing Factor. *J. Cell Biol.*, **122,** 993.
47. Coverley, D., Downes, C. S., Romanowski, P., and Laskey, R. A. (1993) Reversible effects of nuclear membrane permeabilisation on DNA replication: evidence for a positive Licensing Factor. *J. Cell Biol.*, **122,** 985.
48. Vesely, J., Havlicek, L., Letham, D. S., Strnad, M., Blow, J. J., Donnella-Deana, A., Pinna, L., Kato, J., Detivaud, L., Leclerc, S., and Meijer, L. (1994). Inhibition of cyclin-dependent kinases by purine analogues. *Eur. J. Biochem.*, **224,** 771.
49. Hennessy, K. M., Clark, C. D., and Botstein, D. (1990) Subcellular localization of yeast CDC46 varies with the cell cycle. *Genes Dev.*, **4,** 2252.
50. Chen, Y., Hennessy, K. M., Botstein, D., and Tye, B. (1992) CDC46/MCM5, a yeast protein whose subcellular localization is cell cycle-regulated, is involved in DNA replication at autonomously replicating sequences. *Proc. Natl. Acad. Sci. USA*, **89,** 10459.
51. Yan, H., Gibson, S., and Tye, B. K. (1991) Mcm2 and Mcm3, two proteins important for ARS activity, are related in structure and function. *Genes Dev.*, **5,** 944.
52. Yan, H., Merchant, A. M., and Tye, B. K. (1993) Cell cycle-regulated nuclear localization of MCM2 and MCM3, which are required for the initiation of DNA synthesis at chromosomal origins in yeast. *Genes Dev.*, **7,** 2149.
53. Coxon, A., Maundrell, K., and Kearsey, S. E. (1992) Fission yeast cdc21^{+} belongs to a family of proteins involved in an early step of chromosome replication. *Nucleic Acids Res.*, **20,** 5571.

54. Thömmes, P., Fett, R., Schray, B., Burkhart, R., Barnes, M., Kennedy, C., Brown, N. C., and Knippers, R. (1992) Properties of the nuclear P1 protein, a mammalian homologue of the yeast Mcm3 replication protein. *Nucleic Acids Res.*, **20,** 1069.
55 Moir, D., Stewart, S. E., Osmond, B. C., and Botstein, D. (1982) Cold-sensitive cell-division-cycle mutants of yeast: properties and pseudo-reversion studies. *Genetics*, **100,** 547.
56. Hennessy, K. M., Lee, A., Chen, E., and Botstein, D. (1991) A group of interacting yeast DNA replication genes. *Genes Dev.*, **5,** 958.
57. Wobbe, C. R., Weissbach, L., Borowiec, J. A., Dean, F. B., Murakami, Y., Bullock, P., and Hurwitz, J. (1987) Replication of simian virus 40 origin-containing DNA *in vitro* with purified proteins. *Proc. Natl. Acad. Sci. USA*, **84,** 1834.
58. Wold, M. S. and Kelly, T. (1988) Purification and characterization of replication protein A, a cellular protein required for *in vitro* replication of simian virus 40 DNA. *Proc. Natl. Acad. Sci. USA*, **85,** 2523.
59. Fairman, M. P. and Stillman, B. (1988) Cellular factors required for multiple stages of SV40 replication *in vitro*. *EMBO J.*, **7,** 1211.
60. Adachi, Y. and Laemmli, U. K. (1992) Identification of nuclear pre-replication centers poised for DNA synthesis in *Xenopus* egg extracts: immunolocalization study of replication protein A. *J. Cell Biol.*, **119,** 1.
61. Hogan, E. and Koshland, D. (1992) Addition of extra origins of replication to a minichromosome suppresses its mitotic loss in cdc6 and cdc14 mutants of *Saccharomyces cerevisiae*. *Proc. Natl. Acad. Sci. USA*, **89,** 3098.
62. Johnston, L. H., Eberly, S. L., Chapman, J. W., Araki, H., and Sugino, A. (1990) The product of the *Saccharomyces cerevisiae* cell cycle gene *DBF2* has homology with protein kinases and is periodically expressed in the cell cycle. *Mol. Cell. Biol.*, **10,** 1358.
63. Broek, D., Bartlett, R., Crawford, K., and Nurse, P. (1991) Involvement of p34cdc2 in establishing the dependency of S phase on mitosis. *Nature*, **349,** 388.
64. Bartlett, R. and Nurse, P. (1990) Yeast as a model system for understanding the control of DNA replication in eukaryotes. *Bioessays*, **12,** 457.
65. Forsburg, S. L. and Nurse, P. (1991) Cell cycle regulation in the yeasts *Saccharomyces cerevisiae* and *Schizosaccharomyces pombe*. *Annu. Rev. Cell Biol.*, **7,** 227.
66. Johnston, L. H. and Lowndes, N. F. (1992) Cell cycle control of DNA synthesis in budding yeast. *Nucleic Acids Res.*, **20,** 2403.
67. Merrill, G. F., Morgan, B. A., Lowndes, N. F., and Johnston, L. H. (1992) DNA synthesis control in yeast: an evolutionarily conserved mechanism for regulating DNA synthesis genes? *Bioessays*, **14,** 823.
68. Nurse, P. (1990) Universal control mechanism regulating onset of M-phase. *Nature*, **344,** 503.
69. Reed, S. I. (1992) The role of p34 kinases in the G1 to S-phase transition. *Annu. Rev. Cell Biol.*, **8,** 529.
70. Richardson, H. E., Wittenberg, C., Cross, F., and Reed, S. I. (1989) An essential G1 function for cyclin-like proteins in yeast. *Cell*, **59,** 1127.
71. Cross, F. R. (1990) Cell cycle arrest caused by *CLN* gene deficiency in *Saccharomyces cerevisiae* resembles START-I arrest and is independent of the mating-pheromone signalling pathway. *Mol. Cell. Biol.*, **10,** 6482.
72. Sudbery, P. E., Goodey, A. R., and Carter, B. L. A. (1980) Genes which control cell proliferation in the yeast *Saccharomyces cerevisiae*. *Nature*, **288,** 401.

73. Cross, F. R. (1988) *DAF1*, a mutant gene affecting size control, pheromone arrest and cell cycle kinetics of *Saccharomyces cerevisiae*. *Mol. Cell. Biol.*, **8**, 4675.
74. Nash, R., Tokiwa, G., Anand, S., Erickson, K., and Futcher, A. B. (1988) The $WHI1^+$ gene of *Saccharomyces cerevisiae* tethers cell division to cell size and is a cyclin homolog. *EMBO J.*, **7**, 4335.
75. Breeden, L. and Nasmyth, K. (1987) Cell cycle control of the yeast *HO* gene: *cis*- and *trans*-acting regulators. *Cell*, **48**, 389.
76. Nasmyth, K. and Dirick, L. (1991) The role of SWI4 and SWI6 in the activity of G1 cyclins in yeast. *Cell*, **66**, 995.
77. Ogas, J., Andrews, B. J., and Herskowitz, I. (1991) Transcriptional activation of CLN1, CLN2, and a putative new G1 cyclin (HCS26) by SW14, a positive regulator of G1-specific transcription. *Cell*, **66**, 1015.
78. Cross, F. R. and Tinkelenberg, A. H. (1991) A potential positive feedback loop controlling *CLN1* and *CLN2* gene expression at the start of the yeast cell cycle. *Cell*, **65**, 875.
79. Lowndes, N. F., Johnson, A. L., and Johnston, L. H. (1991) Coordination of expression of DNA synthesis genes in budding yeast by a cell-cycle regulated *trans* factor. *Nature*, **350**, 247.
80. Lowndes, N. F., Johnson, A. L., Breeden, L., and Johnston, L. H. (1992) SWI6 protein is required for transcription of the periodically expressed DNA synthesis genes in budding yeast. *Nature*, **357**, 505.
81. Dirick, L., Moll, T., Auer, H., and Nasmyth, K. (1992) A central role for SWI6 in modulating cell cycle Start-specific transcription in yeast. *Nature*, **357**, 508.
82. Koch, C., Moll, T., Neuberg, M., Ahorn, H., and Nasmyth, K. (1993) A role for the transcription factors Mbp1 and Swi4 in progression from G1 to S phase. *Science*, **261**, 1551.
83. White, J. H. M., Green, S. R., Barker, D. G., Dumas, L. B., and Johnston, L. H. (1987) The CDC8 transcript is cell cycle regulated in yeast and is expressed coordinately with CDC9 and CDC21 at a point preceding histone transcription. *Exp. Cell Res.*, **171**, 223.
84. Lowndes, N. F., McInery, C. J., Johnson, A. L., Fantes, P. A. and Johnston, L. H. (1992) Control of DNA synthesis genes in fission yeast by the cell-cycle gene $cdc10^+$. *Nature*, **355**, 449.
85. Marks, J., Fankhauser, C., Reymond, A., and Simanis, V. (1992) Cytoskeletal and DNA structure abnormalities result from bypass of requirement for the *cdc10* start gene in the fission yeast *Schizosaccharomyces pombe*. *J. Cell Sci.*, **101**, 517.
86. Tanaka, K., Okazaki, K., Okazaki, N., Ueda, T., Sugiyama, A., Nojima, H., and Okayama, H. (1992) A new *cdc* gene required for S phase entry of *Schizosaccharomyces pombe* encodes a protein similar to the $cdc10^+$ and *SWI4* gene products. *EMBO J.*, **11**, 4923.
87. Caligiuri, M. and Beach, D. (1993) Sct1 functions in partnership with cdc10 in a transcription complex that activates cell cycle START and inhibits differentiation. *Cell*, **72**, 607.
88. Goebl, M. G., Yochem, J., Jentsch, S., McGrath, J. P., Varshavsky, A., and Byers, B. (1988) The yeast cell cycle gene *CDC34* encodes a ubiquitin-conjugating enzyme. *Science*, **241**, 1331.
89. Glotzer, M., Murray, A. W., and Kirschner, M. W. (1991) Cyclin is degraded by the ubiquitin pathway. *Nature*, **349**, 132.
90. Patterson, M., Sclafani, R. A., Fangman, W. L., and Rosamond, J. (1986) Molecular

characterization of cell cycle gene *CDC7* from *Saccharomyces cerevisiae*. *Mol. Cell. Biol.*, **6,** 1590.

91. Hollingsworth, R. J. and Sclafani, R. A. (1990) DNA metabolism gene *CDC7* from yeast encodes a serine (threonine) protein kinase. *Proc. Natl. Acad. Sci. USA*, **87,** 6272.
92. Kitada, K., Johnston, L. H., Sugino, T., and Sugino, A. (1992) Temperature-sensitive cdc7 mutations of *S. cerevisiae* are suppressed by the *DBF4* gene, which is required for the G1/S transition. *Genetics*, **131,** 21.
93. Kelly, T. J., Martin, G. S., Forsburg, S. L., Stephen, R. J., Russo, A., and Nurse, P. (1993) The fission yeast *cdc18*$^{+}$ gene product couples S phase to START and mitosis. *Cell*, **74,** 371.
94. Zhou, C. and Jong, A. (1990) CDC6 mRNA fluctuates periodically in the yeast cell cycle. *J. Biol. Chem.*, **265,** 19904.
95. Bueno, A. and Russell, P. (1992) Dual functions of CDC6: a yeast protein required for DNA replication also inhibits nuclear division. *EMBO J.*, **11,** 2167.
96. Epstein, C. B. and Cross, F. R. (1992) CLB5: a novel B cyclin from budding yeast with a role in S phase. *Genes Dev.*, **6,** 1695.
97. Richardson, H., Lew, D. J., Henze, M., Sugimoto, K., and Reed, S. I. (1992) Cyclin-B homologs in *Saccharomyces cerevisiae* function in S phase and in G2. *Genes Dev.*, **6,** 2021.
98. Norbury, C. and Nurse, P. (1992) Animal cell cycles and their control. *Annu. Rev. Biochem.*, **61,** 441.
99. Paris, J., Le, G. R., Couturier, A., Le, G. K., Omilli, F., Camonis, J., MacNeill, S., and Philippe, M. (1991) Cloning by differential screening of a *Xenopus* cDNA coding for a protein highly homologous to cdc2. *Proc. Natl. Acad. Sci. USA*, **88,** 1039.
100. Tsai, L. H., Harlow, E., and Meyerson, M. (1991) Isolation of the human *cdk2* gene that encodes the cyclin A- and adenovirus E1A-associated p33 kinase. *Nature*, **353,** 174.
101. Ninomiya, T. J., Nomoto, S., Yasuda, H., Reed, S. I., and Matsumoto, K. (1991) Cloning of a human cDNA encoding a CDC2-related kinase by complementation of a budding yeast cdc28 mutation. *Proc. Natl. Acad. Sci. USA*, **88,** 9006.
102. Elledge, S. J. and Spottiswood, M. R. (1991) A new human p34 protein kinase, CDK2, identified by complementation of a cdc28 mutation in *Saccharomyces cerevisiae*, is a homolog of *Xenopus* Eg1. *EMBO J.*, **10,** 2653.
103. Meyerson, M., Enders, G. H., Wu, C. L., Su, L. K., Gorka, C., Nelson, C., Harlow, E., and Tsai, L. H. (1992) A family of human cdc2-related protein kinases. *EMBO J.*, **11,** 2909.
104. Xiong, Y., Connolly, T., Futcher, B., and Beach, D. (1991) Human D-type cyclin. *Cell*, **65,** 691.
105. Koff, A., Cross, F., Fisher, A., Schumacher, J., Leguellec, K., Philippe, M., and Roberts, J. M. (1991) Human cyclin E, a new cyclin that interacts with two members of the *CDC2* gene family. *Cell*, **66,** 1217.
106. Lew, D. J., Dulic, V., and Reed, S. I. (1991) Isolation of three novel human cyclins by rescue of G1 cyclin (Cln) function in yeast. *Cell*, **66,** 1197.
107. Leopold, P. and O'Farrell, P. H. (1991) An evolutionarily conserved cyclin homolog from *Drosophila* rescues yeast deficient in G1 cyclins. *Cell*, **66,** 1207.
108. Lahue, E. E., Smith, A. V., and Orr, W. T. (1991) A novel cyclin gene from *Drosophila* complements CLN function in yeast. *Genes Dev.*, **5,** 2166.
109. Roberts, J. M. and D'Urso, G. (1988) An origin unwinding activity regulates initiation of DNA replication during mammalian cell cycle. *Science*, **241,** 1486.
110. Sherr, C. J. (1993) Mammalian G1 cyclins. *Cell*, **73,** 1059.

111. Blow, J. J. and Nurse, P. (1990) A cdc2-like protein is involved in the initiation of DNA replication in *Xenopus* egg extracts. *Cell*, **62,** 855.
112. Fang, F. and Newport, J. W. (1991) Evidence that the G1–S and G2–M transitions are controlled by different cdc2 proteins in higher eukaryotes. *Cell*, **66,** 731.
113. Pagano, M., Pepperkok, R., Lukas, J., Baldin, V., Ansorge, W., Bartek, J., and Draetta, G. (1993) Regulation of the cell cycle by the cdk2 protein kinase in cultured human fibroblasts. *J. Cell Biol.*, **121,** 101.
114. Riabowol, K., Draetta, G., Brizuela, L., Vandre, D., and Beach, D. (1989) The cdc2 kinase is a nuclear protein that is essential for mitosis in mammalian cells. *Cell*, **571,** 393.
115. Th'ng, J. P., Wright, P. S., Hamaguchi, J., Lee, M. G., Norbury, C. J., Nurse, P., and Bradbury, E. M. (1990) The FT210 cell line is a mouse G2 phase mutant with a temperature-sensitive *CDC2* gene product. *Cell*, **63,** 313.
116. Furukawa, Y., Piwnica, W. H., Ernst, T. J., Kanakura, Y., and Griffin, J. D. (1990) *cdc2* gene expression at the G1 to S transition in human T lymphocytes. *Science*, **250,** 805.
117. Pines, J. and Hunter, T. (1991) Human cyclins A and B1 are differentially located in the cell and undergo cell cycle-dependent nuclear transport. *J. Cell Biol.*, **115,** 1.
118. Girard, F., Strausfeld, U., Fernandez, A., and Lamb, N. J. (1991) Cyclin A is required for the onset of DNA replication in mammalian fibroblasts. *Cell*, **67,** 1169.
119. Pagano, M., Pepperkok, R., Verde, F., Ansorge, W., and Draetta, G. (1992) Cyclin A is required at two points in the human cell cycle. *EMBO J.*, **11,** 961.
120. Marraccino, R. L., Firpo, E. J., and Roberts, J. M. (1992) Activation of the p24 CDC2 protein kinase at the start of S phase in the human cell cycle. *Mol. Cell. Biol.*, **3,** 389.
121. Zindy, F., Lamas, E., Chenivesse, X., Sobczak, J., Wang, J., Fesquet, D., Henglein, B., and Brechot, C. (1992) Cyclin A is required in S phase in normal epithelial cells. *Biochem. Biophys. Res. Commun.*, **182,** 1144.
122. Pines, J. and Hunter, T. (1990) Human cyclin A is adenovirus E1A-associated protein p60 and behaves differently from cyclin B. *Nature*, **346,** 760.
123. Minshull, J., Golsteyn, R., Hill, C. S., and Hunt, T. (1990) The A- and B-type cyclin associated cdc2 kinases in *Xenopus* turn on and off at different times in the cell cycle. *EMBO J.*, **9,** 2865.
124. Cardoso, M. C., Leonhardt, H., and Nadal-Ginard, B. (1993) Reversal of terminal differentiation and control of DNA replication: cyclin A and cdk2 specifically localize at subnuclear sites of DNA replication. *Cell*, **74,** 979.
125. Xiong, Y., Zhang, H., and Beach, D. (1992) D-type cyclins associate with multiple protein kinases and the DNA replication and repair factor PCNA. *Cell*, **71,** 504.
126. Lehner, C. F. and O'Farrell, P. H. (1990) The roles of *Drosophila* cyclins A and B in mitotic control. *Cell*, **61,** 535.
127. Knoblich, J. A. and Lehner, C. F. (1993) Synergistic action of *Drosophila* cyclins A and B during the G2–M transition. *EMBO J.*, **12,** 65.
128. Ohtsubo, M. and Roberts, J. M. (1993) Cyclin-dependent regulation of G1 in mammalian fibroblasts. *Science*, **259,** 1908.
129. Stillman, B. (1989) Initiation of eukaryotic DNA replication *in vitro*. *Annu. Rev. Cell Biol.*, **5,** 197.
130. Tsurimoto, T., Melendy, T., and Stillman, B. (1990) Sequential initiation of lagging and leading strand synthesis by two different polymerase complexes at the SV40 DNA replication origin. *Nature*, **346,** 534.

131. Weinberg, D. H., Collins, K. L., Simancek, P., Russo, A., Wold, M. S., Virshup, D. M., and Kelly, T. J. (1990) Reconstitution of simian virus 40 DNA replication with purified proteins. *Proc. Natl. Acad. Sci. USA*, **87,** 8692.
132. D'Urso, G., Marraccino, R. L., Marshak, D. R., and Roberts, J. M. (1990) Cell cycle control of DNA replication by a homologue from human cells of the p34cdc2 protein kinase. *Science*, **250,** 786.
133. Dutta, A. and Stillman, B. (1992) cdc2 family kinases phosphorylate a human cell DNA replication factor, RPA, and activate DNA replication. *EMBO J.*, **11,** 2189.
134. Borowiec, J. A., Dean, F. B., Bullock, P. A., and Hurwitz, J. H. (1990) Binding and unwinding—how T antigen engages the SV40 origin of DNA replication. *Cell*, **60,** 181.
135. McVey, D., Brizuela, L., Mohr, I., Marshak, D. R., Gluzman, Y., and Beach, D. (1989) Phosphorylation of large tumour antigen by cdc2 stimulates SV40 DNA replication. *Nature*, **341,** 503.
136. Virshup, D. M., Kauffman, M. G., and Kelly, T. J. (1990) Activation of SV40 DNA replication *in vitro* by cellular protein phosphatase 2A. *EMBO J.*, **8,** 3891.
137. Virshup, D. M., Russo, A. A., and Kelly, T. J. (1992) Mechanism of activation of simian virus 40 DNA replication by protein phosphatase 2A. *Mol. Cell. Biol.*, **12,** 4883.
138. Din, S., Brill, S. J., Fairman, M. P., and Stillman, B. (1990) Cell-cycle-regulated phosphorylation of DNA replication factor A from human and yeast cell. *Genes Dev.*, **4,** 968.
139. Fotedar, R. and Roberts, J. M. (1992) Cell cycle regulated phosphorylation of RPA-32 occurs within the replication initiation complex. *EMBO J.*, **11,** 2177.
140. Fang, F. and Newport, J. (1993) Distinct roles of cdk2 and cdc2 in RP-A phosphorylation during the cell cycle. *J. Cell Sci.*, **106,** 983.
141. Strausfeld, U. P., Howell, M., Rempel, R., Maller, J. L., Hunt, T., and Blow, J. J. (1994) Cip1 blocks the initiation of DNA replication in *Xenopus* extracts by inhibition of cyclin-dependent kinases. *Current Biol.*, **4,** 876.
142. Knoblich, J. A., Sauer, K., Jones, L., Richardson, H., Saint, R., and Lehner, C. F. (1994) Cyclin E controls S phase progression and its down-regulation during *Drosophila* embryogenesis is required for the arrest of cell proliferation. *Cell*, **77,** 107.

8 Cell cycle progression and cell growth in mammalian cells: kinetic aspects of transition events

ANDERS ZETTERBERG and OLLE LARSSON

1. Introduction

Most studies of the control of animal cell proliferation have been performed in various model systems *in vitro*, in which cell proliferation can be modulated in a controlled fashion. Although each *in vitro* system has its own particular features and limitations, and although it is unclear to what extent *in vitro* data can be extrapolated in the *in vivo* situation, some general features of proliferation control of animal cells have emerged from the *in vitro* studies.

Normal cells usually cease to proliferate in a cell-cycle-specific way. They arrest in G1 or enter a state of quiescence (G0) from G1 after depletion of serum or growth factors (1–3) or nutrients (4) or after cell crowding (5, 6). This is also consistent with the general opinion that arrested cells *in vivo*, e.g. terminally differentiated cells, contain a G1 amount of DNA. It has, however, been reported that cells can occasionally be arrested in the G2 phase under physiological conditions (7–9) and under certain experimental conditions (10).

In contrast to normal cells, cells transformed to tumourigenicity or cells of tumour origin often respond differently to suboptimal culture conditions, e.g. growth factor starvation. Instead of being arrested in G1 or entering G0, they continue slowly through the cell cycle until they eventually die as a consequence of the environmental restraints (11–15). Consequently, the ability of normal cells, as opposed to tumour cells, to arrest in G0, as a response to changes in environmental conditions, reflects a fundamental growth regulatory mechanism that operates stringently in untransformed cells but is defective in transformed cells.

Research focused on the processes that lie behind G1 arrest is therefore of particular interest in tumour biology research. Studies of the molecular basis of these growth control events in G1 would be facilitated if such studies could focus on a defined and very limited stage within G1 that is of particular importance for the specific G0 arrest. To search for such a stage and to map its precise location within G1 are therefore important.

In this chapter, we discuss certain aspects of commitment to DNA replication and mitosis and exit from the cell cycle, as well as the coordination between cell growth (in size) and transit through the cell cycle.

2. Kinetic analysis of the cell cycle by time-lapse video recording

Time-lapse video recording of cells in culture enables detailed kinetic analysis of various aspects of transition events in the cell cycle. In contrast to alternative methods such as thymidine labelling and flow cytometry, which only describe the behaviour of the average cell in the population, time-lapse video recording enables detailed measurements of individual cells in an unperturbed, asynchronously growing cell population. In particular, this method makes it possible to map the cell cycle in detail with regard to response to *brief* environmental manipulations such as growth factor depletion, inhibition of protein synthesis, and inhibition of mevalonic acid synthesis (16–21). As is evident from these studies, time-lapse video recording is a powerful method in the analysis of cell cycle kinetics. Figure 1 illustrates schematically the basic principle of the analysis and the expected type of information that can be obtained. Cell age (time elapsed from last cell division)

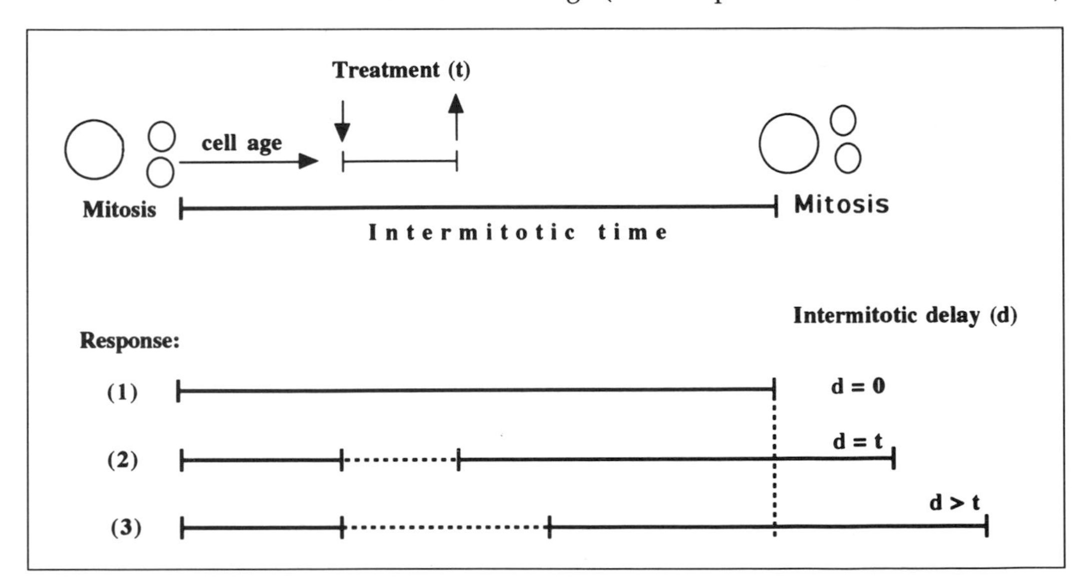

Fig. 1 Schematic diagrams demonstrating the utility of time-lapse analysis in the study of cell cycle progression in proliferating cells. In the top part of the figure, onset and withdrawal of a brief exposure to environmental change (e.g. serum depletion) as related to cell age (i.e. time spent after last mitosis) is indicated. The bottom section of the figure shows three possible consequences of the brief treatment on cell cycle progression: no response at all (1); a prolongation equal to the length of the treatment period (2); a prolongation that exceeds the time of treatment (3).

and intermitotic time (time between two subsequent cell divisions) are recorded for each individual cell in the microscopic field of vision in response to treatment of the cells (type and duration of treatment). Three types of responses to treatment can be expected: (1) no response, or response in terms of a delayed cell division (intermitotic delay) following the treatment either (2) equal in time to the duration of the treatment or (3) longer than the time of treatment. As will be discussed below, all three types of responses may occur depending on cell cycle position, cell type, and type of treatment.

The method allows the following aspects of the cell cycle to be studied:

(a) Response in relation to precise cell cycle position in unperturbed, asynchronously growing cell populations. This permits exact timing of point of commitment to go through the cell cycle (restriction point) and its relation to initiation of DNA replication.

(b) Response as a consequence of treatment for a brief time period (< 1 h). This reflects readiness of response.

(c) Duration of response with respect to duration of treatment. This allows a distinction to be made between temporary arrest in the cell cycle or set back in the cell cycle (exit to G0).

(d) Response of each individual cell. This reveals intercellular variability in responsiveness.

3. Transition events in G1

3.1 Two distinct subpopulations of G1 cells

In Fig. 2 a typical result from time-lapse cinematographic analysis is illustrated. Cell cycle progression is rapidly interrupted in postmitotic, early G1 cells by a short period of growth factor starvation. This response is detected as a postponed mitosis (delayed intermitotic time). Only cells younger than 3 h (time after mitosis) responded, whereas cells older than 4 h were not arrested by growth factor starvation, but advanced through the remaining part of the cell cycle with the same speed as untreated control cells. The subpopulation of postmitotic G1 cells arrested by growth factor starvation was denoted G1-pm cells, and the remaining G1 cells, which are able to initiate DNA replication in the absence of growth factors, were denoted G1-ps cells (pre-S phase) (16). The transition from growth factor dependence in G1-pm cells to growth factor independence in G1-ps cells is equivalent to commitment (1) to the chromosone cycle (DNA replication and mitosis) (22) or the restriction point (2) and probably corresponds to 'START' (23, 24) in yeast. The time-lapse analysis reveals that virtually all G1-pm cells in the population undergo this transition within the narrow time period of 1 h (between the third and the fourth hour after mitosis), i.e. a small intercellular variability with respect to the length of the G1-pm-period as opposed to a large intercellular variability seen with respect to initiation of DNA replication (see below).

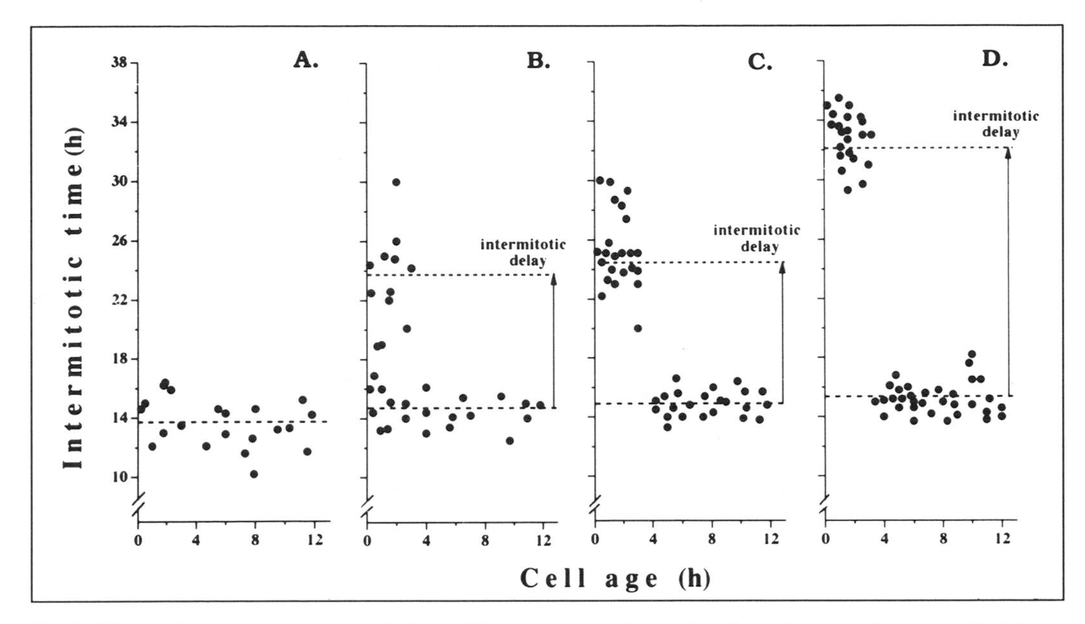

Fig. 2 Effects of transient serum depletion, with respect to cell age (i.e. time elapsed after last mitosis), on intermitotic times. Exponentially growing 3T3 cells cultured in medium containing serum were exposed to serum-free medium for 0 h (control) (A), 0.25 h (B), 1 h (C), or 8 h (D), whereupon they were shifted back to medium containing serum. The cell ages at the time of onset of serum-free exposures and intermitotic time for individual cells were determined by time-lapse video recording.

Time-lapse cinematography analysis in combination with very brief exposures to growth factor-free medium further reveals that G1-pm cells respond quickly. Some of these cells are in fact arrested by such a short growth factor starvation period as 15 min (Fig. 2B). A 1 h starvation period is required to arrest all G1-pm cells (Fig. 2C).

A situation similar to that seen after very short growth factor starvation (15 min) is also observed after a partial growth factor starvation performed in 0.5 per cent serum (data not shown). Of principal interest is the finding that the cells are arrested in all parts of G1-pm and not only at the restriction point. The synthetic programme operating in G1-pm and leading to commitment of the chromosome cycle is thus equally sensitive to inhibition by growth factor starvation or metabolic inhibitors (see Section 3.3) throughout the entire G1-pm period from mitosis to the restriction point.

To investigate whether the existence of a growth-factor-dependent G1-pm subphase and a growth-factor-independent G1-ps subphase is a general property of postembryonic animal cells, we have also carried out time-lapse cinematographic experiments on two different types of normal human cells, namely, human diploid fibroblasts (HDFs) and human mammary epithelial cells (HMECs). Growing populations of Swiss-3T3 cells, HDF, and HMEC were exposed to medium lacking growth factors, and the effect on cell cycle progression was studied. Cells younger than approximately 3–4 h (measured as time elapsed after last mitosis) at onset of growth factor depletion were not capable of undergoing a new mitosis in any of these three cell types (16, 25–27). This implies that these two human diploid cell types exhibit G1-pm properties similar to those of 3T3 cells, i.e. their cell cycle includes a 3–4 h postmitotic phase before commitment (restriction point).

Another finding of principal importance detected by time-lapse analysis is the fact that the mitotic delay seen in the G1-pm cells exceeds the actual starvation or treatment time by approximately 8 h. This 8 h set-back suggests exit from the cell cycle to G0. This will be discussed in greater depth below (see Sections 4 and 5).

3.2 Position of the restriction point and its relation to initiation of DNA replication

Time-lapse cinematographic analysis permits exact timing of the transition between G1-pm and G1-ps (commitment or restriction point) and transition between G1-ps and S (initiation of DNA replication) in individual cells in the population. Thus, detailed information about the temporary relationship between these two transitional events in the cell cycle can be obtained. Both in 3T3 cells and in HDF (Fig. 3 upper), the restriction point (G1-pm/G1-ps transition) is located between the third and the fourth hour after mitosis. DNA replication, on the other hand, is initiated from the third to the thirteenth hour after mitosis in most cells in the two cell populations (Fig. 3 upper). Thus, G1-pm is remarkably constant in length, whereas the length of G1-ps varies considerably (Fig. 3 lower). In fact, the G1-ps variability accounts for almost all variability of the whole cell cycle (25, 26). Thus, it seems as if the cells, which make the 'yes or no' decision in G1-pm about whether to continue through the cell cycle or not, have the capacity to decide, in G1-ps, 'when' they will enter the S phase. The differences in the kinetics between these two transitions (G1-pm/G1-ps vs G1-ps/S) suggest the involvement of different mechanisms in their control. In addition to the probable involvement of labile proteins in the 'G1-pm programme' leading up to commitment (see also Section 3.3), its constant length in time after mitosis suggests that other processes initiated at or immediately after mitosis may also be involved. Such processes might concern reorganization of the cytoskeleton or decondensation of the chromatin. In contrast, it is conceivable that more variable events may underlie the control of the G1-ps/S transition. Such a variable event might comprise overall accumulation of cellular proteins. As support for this hypothesis, preliminary results in our laboratory have shown that cells fail to grow as long as they are maintained in the G1-pm subphase. However, as soon as they have completed the G1-pm/G1-ps transition, they start to increase in size (data not shown). Therefore, it is tempting to speculate that the cells adjust their cell size in G1-ps before initiating DNA synthesis. A small G1-ps cell would thus need a relatively long period to accumulate sufficient protein content in order to traverse into S phase, whereas a large cell would require a short G1-ps period for this purpose. This would be in line with previous data on L cells (28, 29).

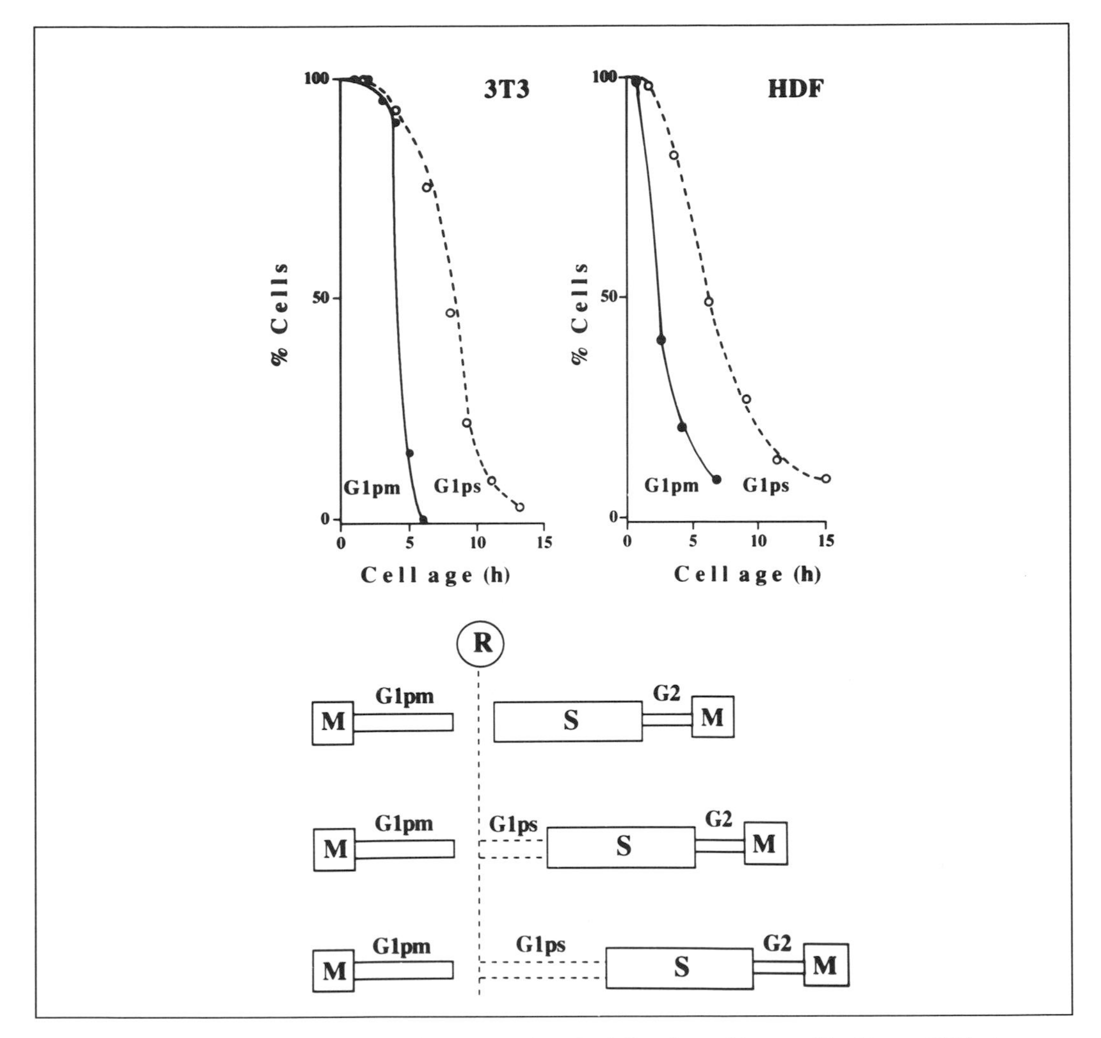

Fig. 3 Upper: cell age distribution of G1-pm and G1-ps in 3T3 cells and human fibroblasts (HDF). Lower: a model demonstrating the variable length of G1-ps.

3.3 Properties of early G1 (G1-pm)

Time-lapse analysis also reveals the way in which cells re-enter G1-pm upon replenishment of growth factors after termination of a brief depletion period. The cells, which first have to pass an 8 h entry phase, would theoretically re-enter G1-pm in three principally different ways. Firstly, the cells may enter G1-pm at the beginning of G1-pm (time-point 0), irrespective of G1-pm stage (cell age 0–4 h) at exit. Secondly, the cells may enter the cell cycle at the end of G1-pm (i.e. between the third and fourth hour after mitosis or 'restriction point' or 'START'). Thirdly, the cells may enter G1-pm at exactly the same stage as they exited. As is shown

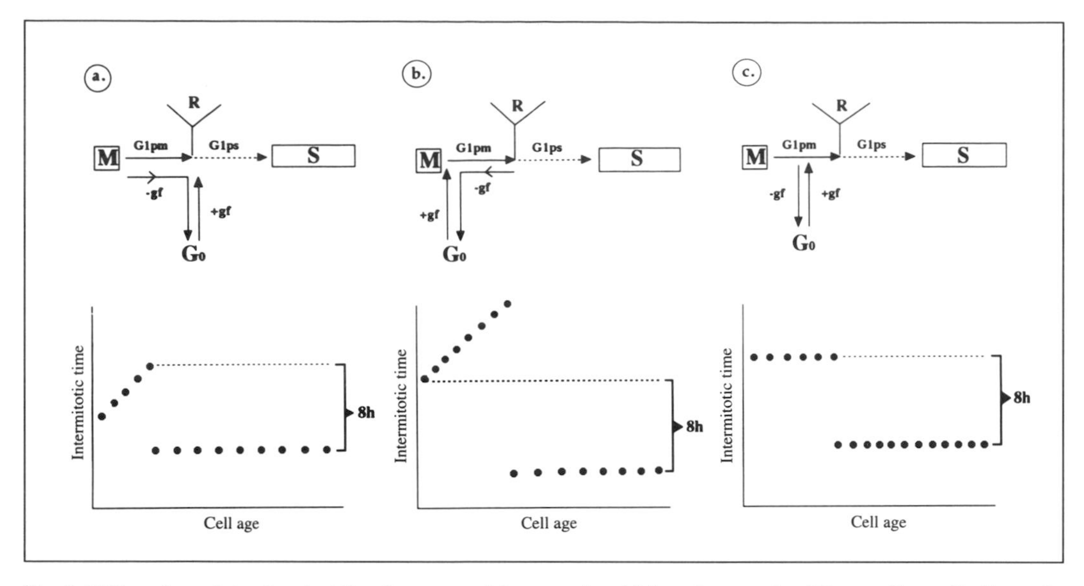

Fig. 4 Different models showing the three possible ways in which cells re-enter G1-pm after exit: the cells re-enter at the end of G1-pm (i.e. time-point 3–4 h) independent of cell position at exit (a); the cells re-enter at the beginning (time-point 0) of G1-pm independent of position at exit (b); the cells re-enter at the same position as they exited (c).

in Fig. 4, these three different modes of behaviour would give rise to three different distributions of intermitotic times following a transient growth factor depletion. After comparison with the real time-lapse data from a large number of cells (Fig. 5), the distribution pattern strongly suggests that the cells enter G1-pm at the same position as they left it (compare with Fig. 4). These data raise the possibility that cells 'recall' the G1-pm position from which they enter G0. The mechanism lying behind this kind of memory remains to be elucidated.

To study the role of purified serum growth factors on G1-pm/G1-ps transition of 3T3 cells, each of platelet-derived growth factor (PDGF) and insulin or insulin-like growth factor-1 (IGF-1) was added to the serum-free medium during the serum-starvation period and was removed when the serum-containing medium was added. The results are illustrated in Fig. 6. As can be seen, the intermitotic delay resulting from serum starvation for 8 h was efficiently prevented by PDGF whereas insulin, or IGF-1 (data not shown), had no effect (Fig. 6). This means that insulin or IFG-1 is insufficient for the final completion of the commitment process in G1-pm, PDGF, on the other hand, is alone sufficient for successful completion of the commitment process during G1-pm and the subsequent traverse of the rest of the cell cycle in Swiss 3T3 cells. Similar results have been obtained in human fibroblasts (25, 26, 30).

It has been demonstrated that treatment with an inhibitor of *de novo* protein synthesis (i.e. cycloheximide) and of 3-hydroxy-3-methylglutaryl Coenzyme A (HMG-CoA) reductase (i.e. 25-hydroxycholesterol and lovastatin) can induce an

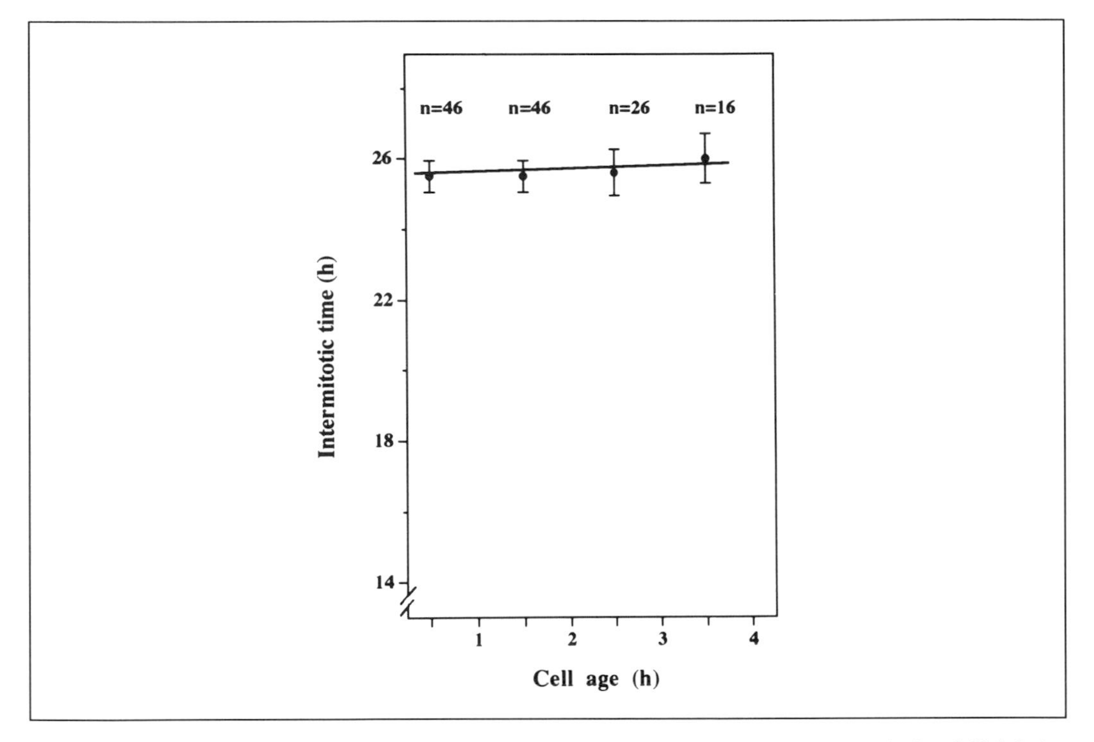

Fig. 5 The real intermitotic time distribution of G1-pm cells, based on a time-lapse analysis of 134 Swiss 3T3 cells which had been exposed to a 4 h exposure to serum-free medium. Mean values±SEM are shown.

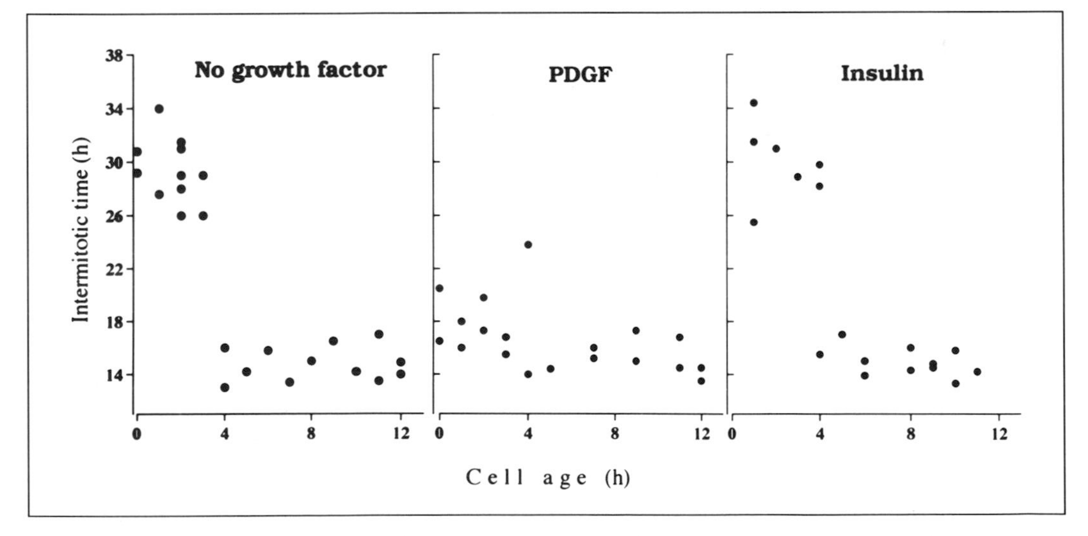

Fig. 6 Effects of growth factors on G1-pm—G1-ps transition. Exponentially growing 3T3 cells cultured in medium containing serum were exposed to serum-free medium (left) or serum-free medium containing either platelet-derived growth factor (25 ng ml^{-1}) or insulin (100 mg ml^{-1}). After 8 h the cells were transferred back to medium containing serum. The cell ages at the time of onset of serum-free exposures and intermitotic time for individual cells were determined by time-lapse video recording.

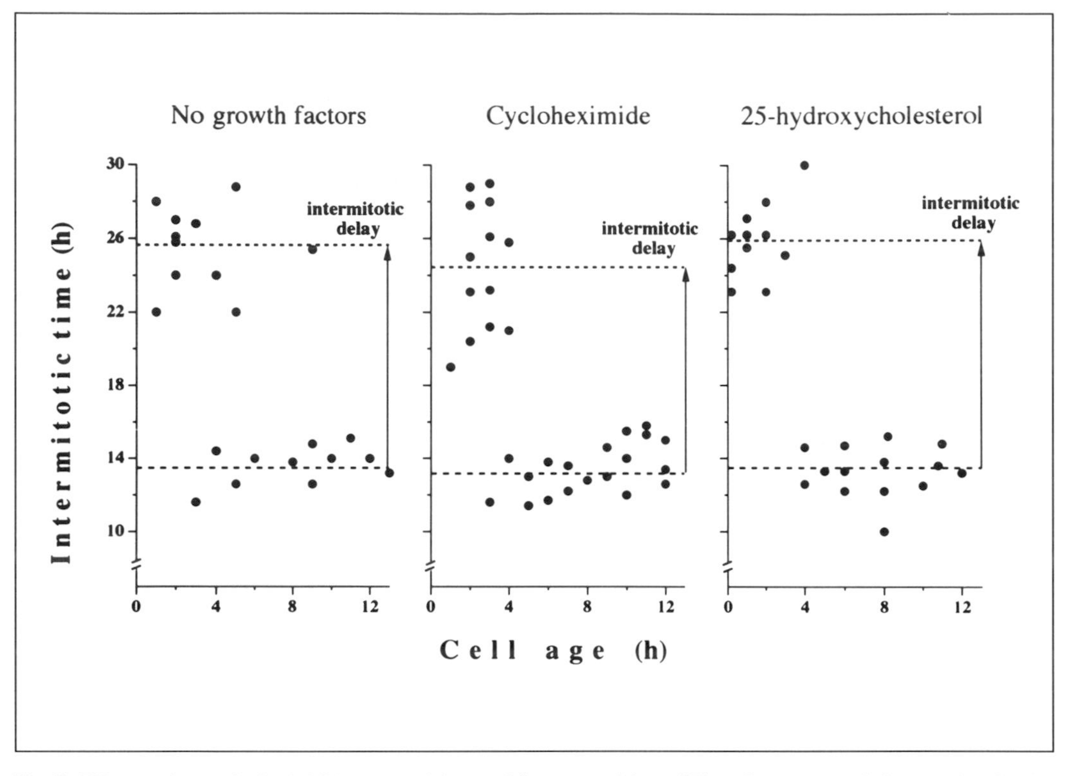

Fig. 7 Effects of metabolic inhibitors on G1-pm–G1-ps transition. 3T3 cells exponentially growing in the presence of serum were, as indicated, shifted to either serum-free medium (no growth factors), or serum-containing medium together with cycloheximide (100 ng ml^{-1}), or 25-hydroxycholesterol (1.5 mg ml^{-1}) (an inhibitor of HMG-CoA reductase); 4 h later, the cells were reshifted to serum-containing medium without supplements. The cell ages and intermitotic times were determined by time-lapse video recording.

intermitotic delay of G1-pm cells similar to that obtained by growth factor depletion (16, 17, 19, 20, 25, 26) (see also Fig. 7). A cycloheximide dose causing a 50 per cent inhibition of the rate of protein synthesis was sufficient to induce exit (16, 17, 25, 26). In fact, an inhibition of protein synthesis as low as approximately 20 per cent is sufficient to induce an intermitotic delay in a limited portion of G1-pm cells (16). These data are consistent with data obtained by other investigators (31–33), and suggest the existence of labile proteins e.g. cyclins in the control of G1-pm transition. In addition to cyclins one such labile cell cycle-regulatory protein might by HMG-CoA reductase. This enzyme regulates the formation of mevalonate, which constitutes the key metabolite in the biosynthesis of cholesterol and isoprenoid derivatives (34, 35). It is also well established that mevalonate is required for proliferation of mammalian cells (34–36). Several proteins have been shown to be post-translationally modified by isoprenoid residues (36–38). Such lipid residues are farnesyl and geranylgeranyl, which are covalently linked to the proteins through a thioether bond to cystein residues at the C-terminus of the

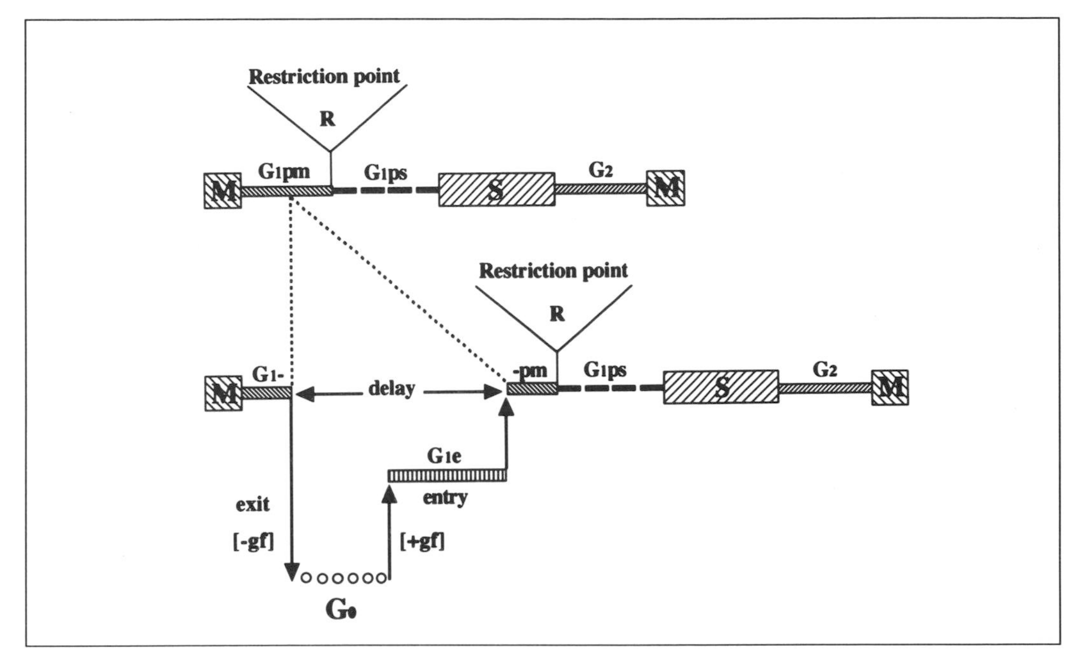

Fig. 8 A cell cycle model based on data obtained by time-lapse analysis of Swiss 3T3 cells.

proteins (36–38). Even if the entire role of protein prenylation is still unknown, it seems to be important for anchoring of certain membrane-associated protein (36–38). $p21^{ras}$ is processed in this manner, and there is evidence that inhibition of HMG-CoA reductase leads to inactivation of the ras protein (39). Since ras is involved in growth control of mammalian cells, it is possible that the cell cycle block induced by inhibition of HMG-CoA reductase may be mediated through this mechanism.

Another possible role of HMG-CoA reductase in regulating G1-pm/G1-ps transition is its regulatory influence on N-linked glucosylation of proteins. Hence, dolichyl phosphate (also a mevalonate product), which is a phosphoester of long-chain isoprenoid alcohols, acts as a carrier of oligosaccharides to proteins (40). It has been demonstrated that inhibition of HMG-CoA reductase using lovastatin or 25-hydroxycholesterol decreases the level of dolichyl phosphate and protein glycosylation (41). Among proteins that are glycosylated in this manner, growth factor receptors are of particular interest. It has, for instance, been shown that inhibition of HMG-CoA reductase decreases the expression of c-fos after stimulation with insulin-like growth factor-1 (42). This effect might be a consequence of a reduced number of functional IGF-1 receptors at the cell surface due to underglycosylation of receptor proteins.

Figure 8 shows our cell cycle model, which is based on data obtained by time-lapse analysis of 3T3 cells.

4. Exit from the cell cycle

It is well known that the time from G0 to mitosis is longer than intermitotic time in exponentially growing cells (3). Time-lapse analysis performed in our laboratory of quiescent 3T3 cells stimulated with serum growth factors shows that the average time from G0 to mitosis is about 23 h (data not shown). This is approximately 8 h longer than average intermitotic time (about 15 h) in exponentially proliferating 3T3 cells (see Fig. 2). As was also evident from Figs 2 and 7, the recorded intermitotic delay is approximately 8 h longer than the time of exposure to growth-factor-free medium or to metabolic inhibitors. Since an intermitotic delay of 8 h in addition to the actual exposure time occurs after both brief exposures (15 min to 1 h) and longer exposures, these data suggest that the cells rapidly (within less than 1 h) exit to G0 even after a brief treatment. They remain in G0 during the period of treatment, and after re-addition of growth factors or removal of metabolic inhibitors, the cells return to the cell cycle, which takes about 8 h. Although G1-pm cells respond immediately with an intermitotic delay (Fig. 9, left), time-lapse analysis of the second cell cycle reveals that committed cells beyond the restriction point (i.e. G1-ps, S, and G2 cells) also respond to a temporary exposure to growth-factor-free medium by a intermitotic delay observed in the second cell cycle (Fig. 9, right). The indication that in fact all cells in the population respond to growth factor starvation, irrespective of cell cycle position, is consistent with the finding that the

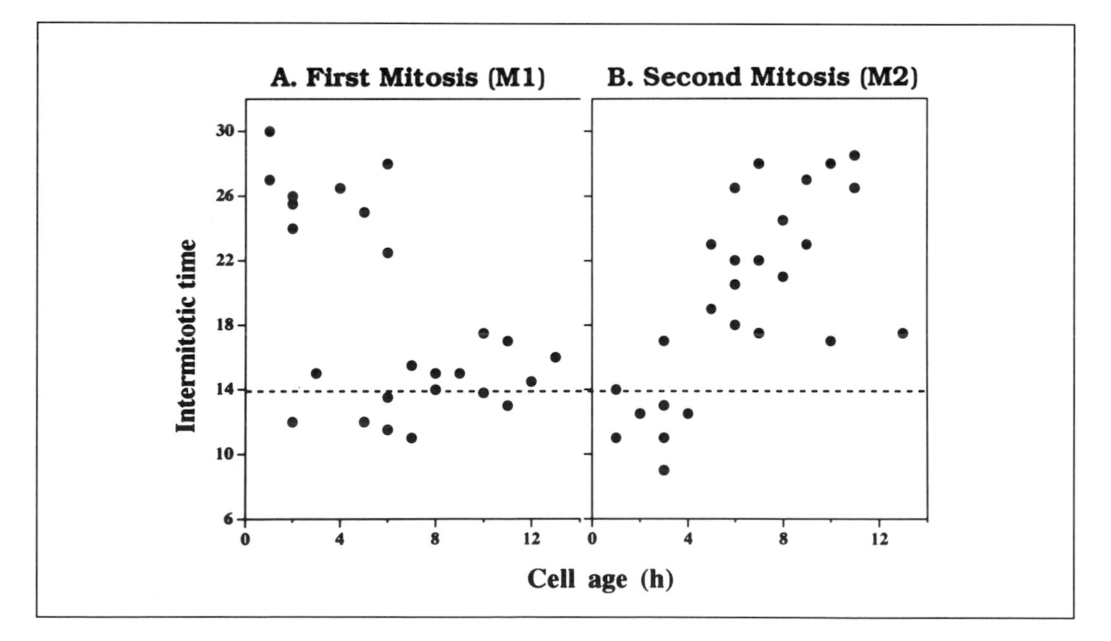

Fig. 9 Relationship between cell age at the onset of serum depletion, and intermitotic times. Exponentially growing 3T3 cells were exposed to a serum-free treatment for 4 h, after which they were again exposed to medium containing serum. Cell ages at the onset of serum depletion and intermitotic times during the first (left) and second (right) generation for individual cells were determined by time-lapse analysis.

rate of protein synthesis is suppressed rapidly after growth factor starvation in all cell cycle stages (16, 18, 21, 25).

If the ability to remain in the cell cycle depends on a high rate of protein synthesis to maintain a critical concentration of labile proteins of importance for the proliferative state (e.g. c-myc, HMG-CoA reductase, or cyclins), one would expect these proteins to be depleted rapidly in all cells in which protein synthesis is suppressed. A model taking all of these observations into consideration is presented in Fig. 10. Cells treated (growth-factor-starved or inhibited by metabolic inhibition) while in G1-pm exit immediately from the cell cycle. Cells treated after G1-pm, i.e. in G2-ps, S, or G2, also leave the cell cycle. However, the chromosome cycle (DNA replication and mitosis) is irreversibly initiated and runs on independently of the influence of growth factors on the cell, and the exit from the cycle is not observed until the cell enter the second cell cycle. In this case the second mitosis is delayed (Fig. 10). The time taken to proceed from G0 to mitosis in cells treated before commitment in G1-pm is equal (23 h) to the time from G0 to the second mitosis (23 h) in cells treated after commitment in G1-ps, S, or G2. In these latter cells, time to first mitosis must be ignored since the chromosome cycle is already irreversibly initiated in these cells at the time of treatment and runs independently of the growth factor situation in the cellular environment. A G1-ps, S, and G2 cell that has been exposed to growth-factor-free medium or metabolic inhibitors can thus be considered as a G0 cell with respect to proliferation

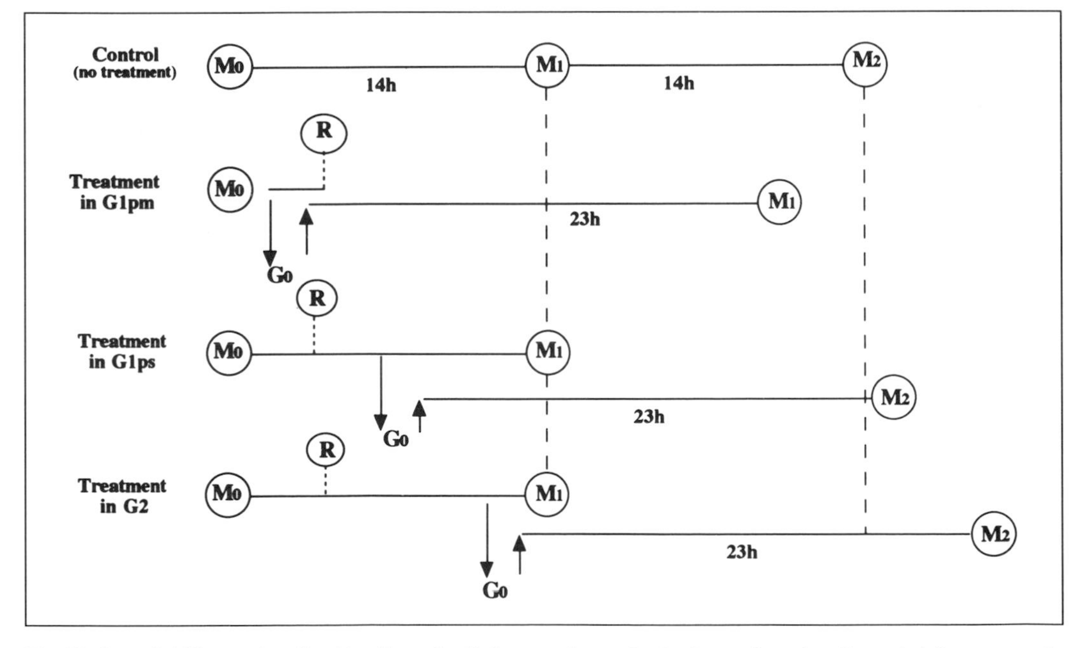

Fig. 10 A model illustrating the kinetics of exit from and re-entry to the cell cycle after a brief exposure to serum depletion or metabolic inhibitors (e.g. inhibition of *de novo* protein synthesis or depression of 3-hydroxy-3-methylglutaryl Coenzyme A reductase). M0 represents mitosis before treatment and M1 and M2 represent first and second mitosis after treatment, respectively. For further details, see text.

and growth control but is still in the cell cycle with respect to the chromosome cycle.

5. Exit to G0 versus G1 arrest

Unlike normal 3T3 cells, the SV40-transformed derivative (SV-3T3) does not respond by an intermitotic delay upon treatment with serum-free medium or cycloheximide (19, 20, 25). However, the transformed cells are arrested in a G1-pm-like phase when treated with the HMG-CoA reductase inhibitor 25-hydroxycholesterol (19, 20, 25). Figure 11, which is based on time-lapse data from several experiments, clearly shows that duration of intermitotic delay in SV-3T3 cells is identical to duration of treatment. Thus, in contrast to untransformed 3T3 cells, the transformed SV-3T3 cells are not set back in the cell cycle by the treatment, i.e. they do not exit from the cell cycle to G0 but are instead arrested in G1-pm as long as they are exposed to the inhibitor. The loss of the ability to exit from the cycle and become G0-arrested most likely reflects some fundamental defect in the cell cycle or growth-regulatory mechanisms of tumour-transformed cells. Thus, SV-3T3 cells also posses a 'G1-pm-programme', which must be completed before commitment to DNA synthesis and mitosis.

Several attempts to provide a biochemical characterization of the G0 phase have been performed during the last few years. In a study by Tay *et al.* (43) it has been shown that the allosteric M1 subunit of ribonucleotide reductase (M1-RR) is con-

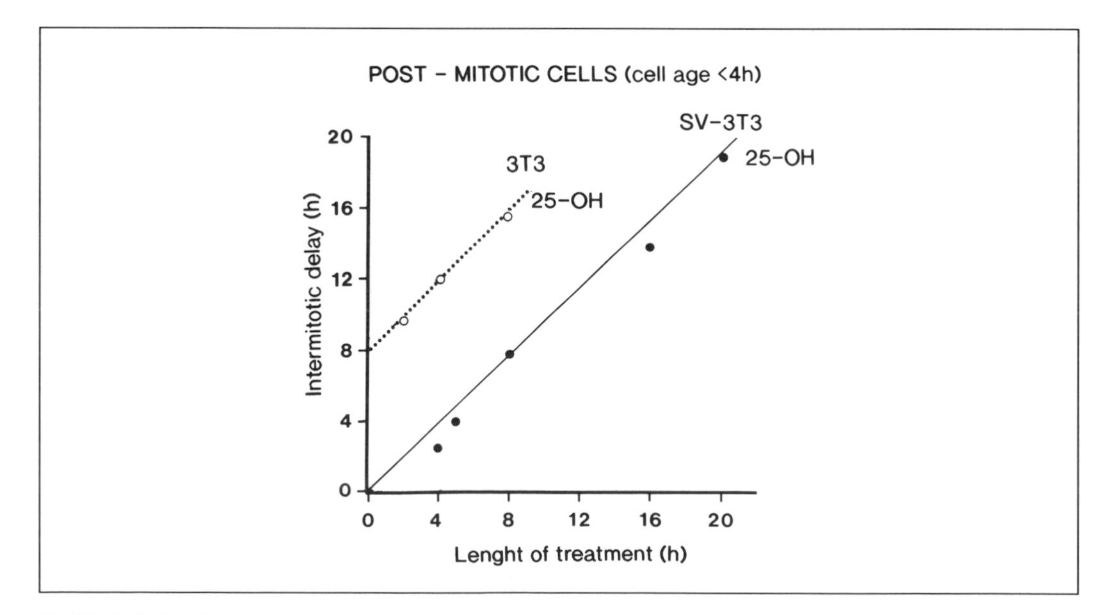

Fig. 11 Relationship between treatment time with 25-hydroxycholesterol and intermitotic delay for 3T3 and SV-3T3 cells. The mean intermitotic delay of G1-pm cells (i.e. cells younger than 4 h) following treatment with serum-free medium or 25-hydroxycholesterol for different periods was determined from several experiments.

stitutively expressed by cycling cells, but is lost during exit to G0. In studies by Schneider *et al.* (44), genes expressed during G0 but not during growing state in NIH 3T3 cells have been identified. Since then, a number of growth arrest-specific (gas) genes expressed in resting animal cells have been cloned, and the encoded proteins characterized (45–47). The products from the gas genes in WI-38 cells have been shown to be associated with extracellular matrix components (45). These proteins, called 'quiescins', seem to be expressed before cells reach confluency and become contact-inhibited (48). Another protein that may be of importance for transition into G0 is the intracellular 30 kDa protein prohibitin (49). In studies by Nuell *et al.* (50) it was shown that prohibitin could switch off proliferation in normal cells and HeLa cells. In yeast, several gene products (e.g. FAR-1) induce arrest through stimulating the alpha factor-induced inhibition of cyclins (51). It is possible that negative regulatory factors, e.g. transforming growth factor-beta (TGF-b), might act in a similar manner in mammalian cells.

However, despite all new observations of events associated with induction of G0, the accurate mechanism regulating this important transition still remains to be clarified.

6. Cellular growth

The importance of cell size in control of cell division has been discussed for several decades but still remains unclear. As early as 1956 Prescott (52) demonstrated that division in *Amoeba proteus* was postponed for several days following periodic amputation of the cytoplasm. The main conclusion from these experiments was that cells cannot undergo division unless they are allowed to reach a critical size (52). Killander and Zetterberg (28, 29) presented data suggesting that cellular enlargement in G1 was somehow involved in the regulation of entry into S phase of L cells. Further evidence for a size-controlled initiation of DNA synthesis was given by Donachie (53), who demonstrated that DNA synthesis in *Escherichia coli* is started at a fixed size independent of the growth rate. Similarly, a cell size control over initiation of DNA synthesis has been suggested in other systems such as the fission yeast *Schizosaccharomyces pombe* (54), the budding yeast *Saccharomyces cerevisiae* (55), the slime mould *Physarum polycephalum* (56), and the protist *Paramecium tetraurelia* (57, 58). More recent studies on yeast have concerned molecular aspects of cell size (59–61). Data from these studies have suggested that G1 cyclins may be involved in coordination between cell cycle commitment and 'START' and cell size in *S. cerevisiae*. It has been demonstrated that $p107^{Wee1}$ or mik1 protein kinases, which sense nutritional status, can regulate cell size in yeast by interfering with the cell division cycle through phosphorylation of cdc2 kinase (62).

It is reasonable that there is an interrelationship between the progression through the cell cycle and the growth in cell size, in the sense that cells approximately double in size prior to mitosis under physiological conditions, producing 'balanced cell growth'. Nevertheless, it has been shown that it is possible to separate cellular growth from cell cycle progression (21, 63–67). It has, for instance, been demonstrated that arrested 3T3 cells can be stimulated to undergo DNA synthesis and cell division in the absence of growth in cell size ('unbalanced growth') (64, 68, 69). In our laboratory we have been able to analyse the interrelationship between growth in size and cell cycle progression in exponentially growing cells (21). Whereas exposures to growth factor-free medium result in rapid G0-arrest of G1-pm cells (i.e. cells younger than 3–4 h), cells located in later cell cycle stages (i.e. cells in G1-ps, S, and G2) were fully capable of completing their cell cycle and thereby reaching mitosis (see Fig. 2). Since the rate of protein synthesis is decreased rapidly and drastically in all cell cycle stages following growth factor depletion (16, 21, 25), it is conceivable that the increase in cell size (= protein content) during G1-ps, S, and G2 would be reduced. To verify this possibility, the protein content of mitotic cells which before fixation had been subjected to short exposures (2–8 h) to growth factor-free medium was measured by microspectrometry (21). Growth factor depletion led to an immediate cessation of the increase in cellular enlargement, and the cells divide with cell sizes reduced by 30–40 per cent as compared to untreated cells (21, 25). In Fig. 12 it is shown that supplementation of the serum-free medium with superphysiological doses of insulin or physiological doses of insulin-like growth factor-1 (IFG-1) substitutes for serum and the cells divide by normal cell sizes. In contrast, addition of platelet-derived growth factor (PDGF) or fibroblast growth factor (FGF) only exerts a small increase in cell size (70).

In Fig. 13 (upper) it is shown that G0-blocked 3T3 cells, which had been serum-

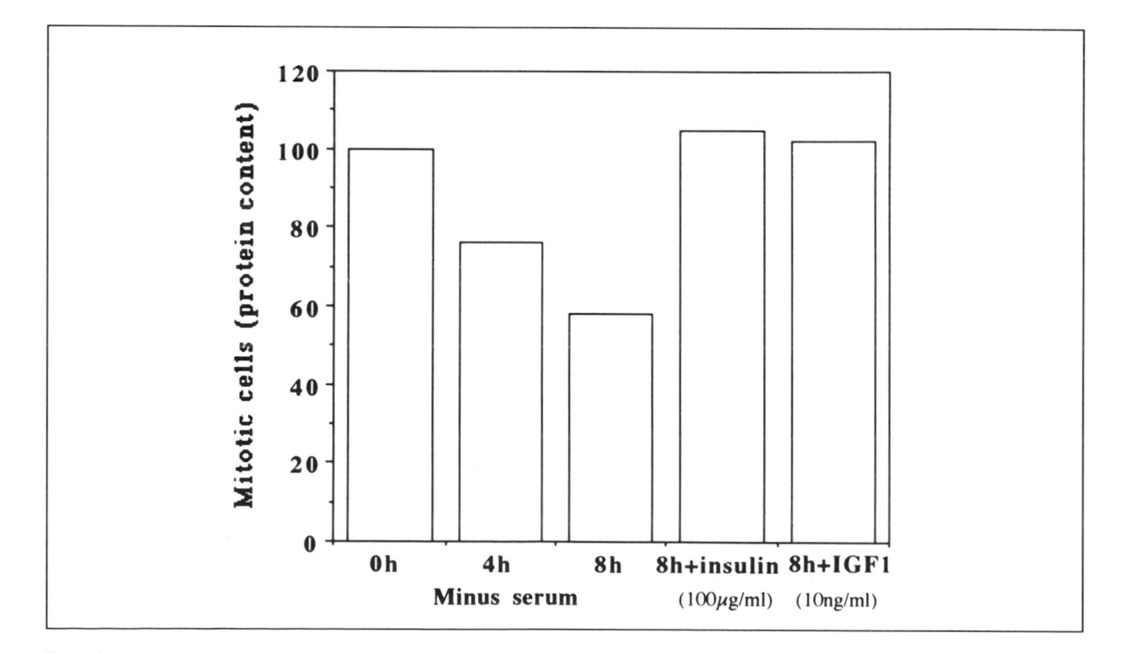

Fig. 12 The effects of different growth factors on cell size of mitotic cells. Proliferating 3T3 cells were shifted to new serum-free medium for 4 or 8 h, or to serum-free medium containing either insulin (100 mg ml^{-1}) or IGF-1 (10 ng ml^{-1}) for 8 h. Thereafter the cells were fixed and stained by Feulgen/Naphthol Yellow S. Mitotic cells were identified microscopically and DNA and protein content was determined by microspectrometry.

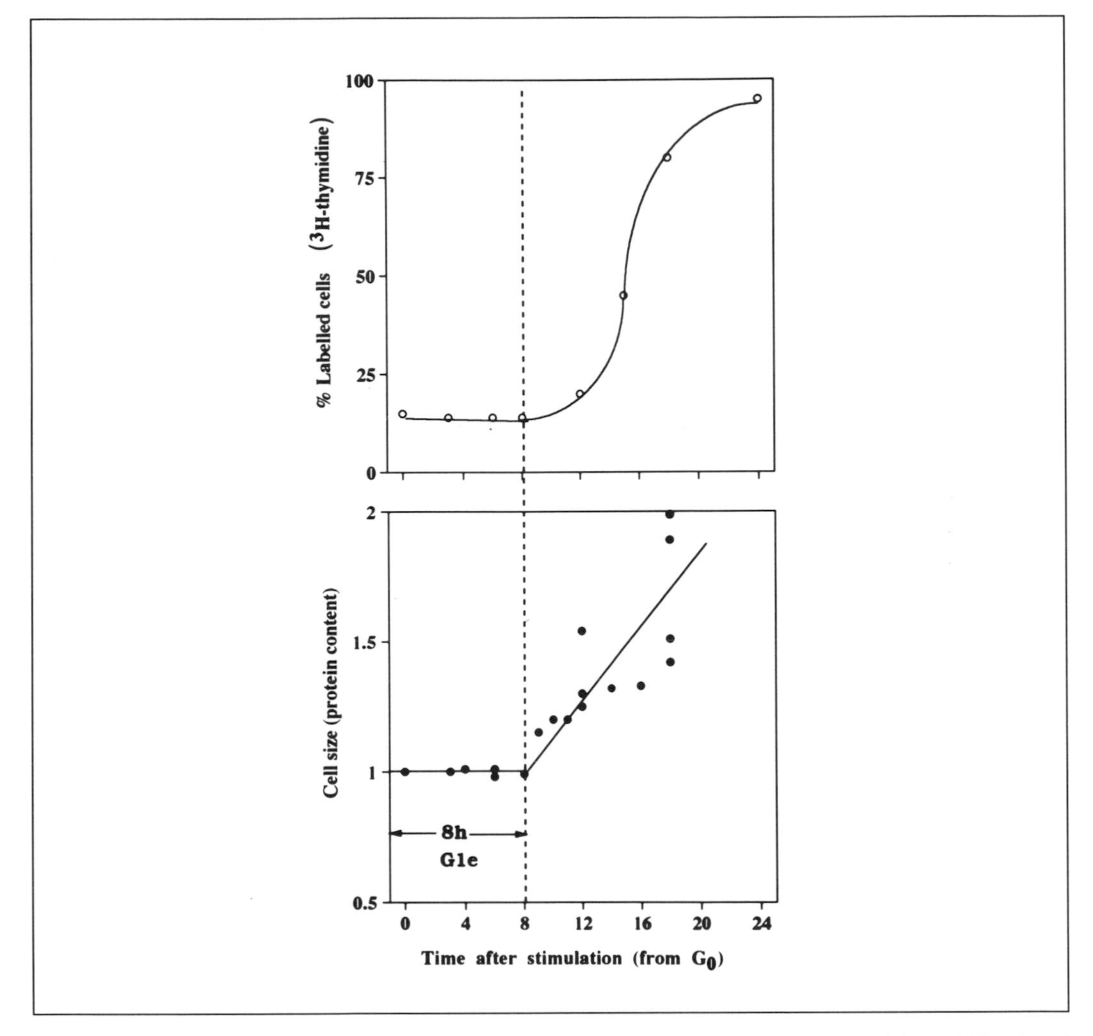

Fig. 13 Re-entry (G1e) to the cell cycle. The top section shows the relationship between G1e and kinetics of initiation of DNA synthesis in quiescent cells stimulated by serum, while the bottom section demonstrates the relationship between G1e and increase in cell size. For details, see text.

depleted for a 48 h period, started initiating DNA synthesis after approximately 12 h upon stimulation by serum growth factors. According to our terminology the first 8 h of this lag period represents the entry phase (G1e). Interestingly, on this re-entering to the cell cycle from G0, the cells do not start increasing in size (i.e. cellular protein content) until 8 h after the onset of serum stimulation (Fig. 13, lower). However, after this period all cells begin to increase in protein content and had approximately doubled in size prior to mitosis. These data indicate that G1e constitutes a functionally distinctive cell cycle stage, in which preparations critical for *entering* both the cell division cycle and growth cycle are made. However,

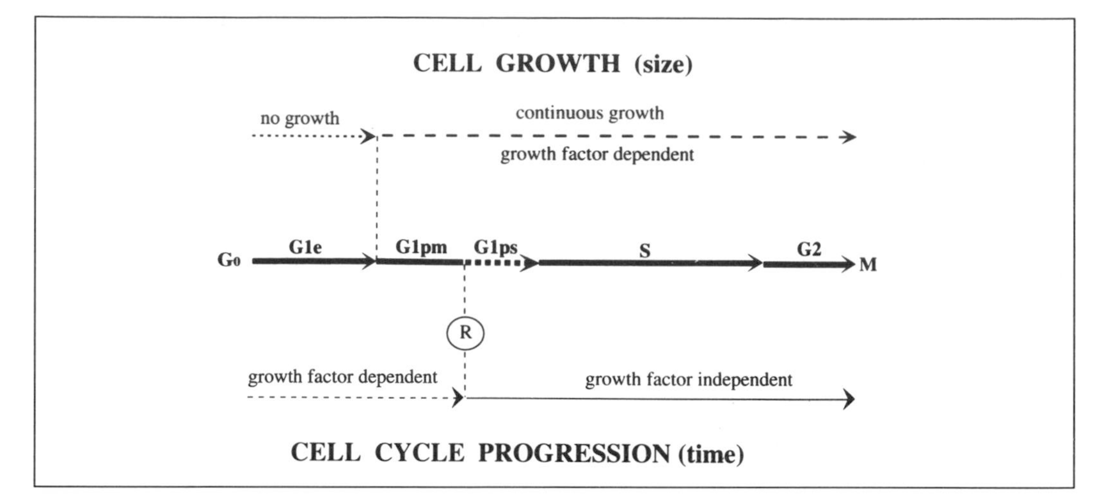

Fig. 14 A model describing the chromosome and growth cycle in mammalian cells.

increase in cellular protein content is *per se* not necessary for entry to the cell cycle. This finding is also in concurrence with the observation of Stiles *et al.* (71) that there is no requirement for amino acids during the first 6 h re-entry into the cell cycle in Balb-3T3 cells.

A comprehensive cell cycle model taking both the chromosome and growth cycle into consideration is presented in Fig. 14.

7. Different growth factor requirements for cell cycle progression and for growth in cell size

As was shown in Fig. 6, PDGF alone could substitute for the whole serum complement in driving 3T3 cells through the whole of G1-pm, including the restriction point and commitment to the chromosome cycle (16, 25). In contrast, insulin (or IGF-1) failed to do so. On the other hand, insulin or IGF-1 counteracted the depressive effects of treatment with growth-factor-free medium on protein synthesis, whereas PDGF only had a partial effect in this respect (see Fig. 12) (16, 17). These data suggest that a general increase in overall protein synthesis, as induced by insulin, is not sufficient to counteract exit from the cell cycle. In contrast, the stimulatory effect of insulin and IGF-1 on *de novo* protein synthesis seems to be a necessary event for cellular growth (see Section 6). Since PDGF did not increase the overall rate of protein synthesis much, the question may be raised as to whether PDGF instead induces the synthesis of specific cell cycle-regulatory proteins and thereby would overcome the intermitotic delay. This would be in line with the notion that the cellular decision to proceed through the cell cycle instead of becoming quiescent is dependent on the accumulation of critical cell-cycle-specific or growth-promoting proteins (32, 33, 72–76). To investigate this, the effects of insulin and PDGF on 3T3 cells treated with cycloheximide for 8 h were studied (17). It was shown that neither of PDGF nor insulin could counteract the cycloheximide-induced inhibition of protein synthesis (17). However, PDGF was capable of preventing the intermitotic delay, whereas insulin had no such effects (17). Similar results were obtained in human fibroblasts (25, 26). On the basis of these results, it is reasonable to assume that PDGF does not prevent intermitotic delay primarily by restoring the overall rate of protein synthesis to normal levels. Rather, it is likely that PDGF exerts its effect by altering the expression of cell-cycle-specific or growth-promoting genes encoding for proteins required for progression through G1-pm and commitment to the chromosome cycle. This opinion is in line with previous reports showing a preferential effect of PDGF on expression of *c-myc* when added to quiescent cells (77, 78).

References

1. Temin, H. (1971) Stimulation by serum of multiplication on stationary chicken cells. *J. Cell Physiol.*, **78,** 161.
2. Pardee, A. B. (1974) A restriction point for control of normal animal proliferation. *Proc. Natl. Acad. Sci. USA*, **71,** 1286.
3. Baserga, R. (1976) *Multiplication and division in mammalian cells*. Marcel Dekker, New York.
4. Prescott, D. M. (1976) *Reproduction of eukaryote cells*. Academic Press, New York.
5. Nielhausen, K. and Green, H. (1965) Reversible arrest of growth in G1 of an established fibroblast line (3T3). *Exp. Cell Res.*, **40,** 166.
6. Zetterberg, A. and Auer, G. (1970) Proliferative activity and cytochemical properties of nuclear chromatin related to local cell density of epithelial cells. *Exp. Cell Res.*, **62,** 262.
7. Gelfant, S. (1981) Cycling/noncycling cell transitions in tissue ageing, immunological surveillance, transformation, and tumor growth. *Inv. Rev. Cytol.*, **70,** 1.
8. Melchers, F. and Lernhardt, W. (1985) Three restriction points in the cell cycle of activated murine B lymphocytes. *Proc. Natl. Acad. Sci. USA*, **82,** 7681.
9. Gomez-Lechon, L. and Castell, J. V. (1987) Evidence for arrested G2 cell subpopulation rat liver inducible to mitosis. *Cell Tissue Kinet.* **20,** 583.
10. Yoshida, M. and Beppu, T. (1988) Reversible arrest of proliferation of rat 3Y1 fibroblast in both the G1 and G2 phases by trichostatin. *Exp. Cell Res.*, **177,** 122.
11. Zetterberg, A. and Sköld, O. (1969) The effect of serum starvation on DNA, RNA and protein synthesis during interphase in L-cells. *Exp. Cell Res.*, **57,** 114.
12. Paul, D. (1973) Quiescent SV-40 virus transformed 3T3-cells in culture. *Biochem. Biophys. Res. Commun.*, **53,** 745.
13. Pardee, A. B. and James, L. J. (1975) Selective killing of transformed baby hamster kidney (BHK) cells. *Proc. Natl. Acad. Sci. USA*, **72,** 4494.
14. Vogel, A. and Pollack, R. J. (1975) Isolation and characterization of revertant cell lines. *J. Cell Physiol.*, **85,** 151.
15. Medrano, E. E. and Pardee, A. B. (1980) Prevalent deficiency in tumor cells of cycloheximide in the cell cycle arrest. *Proc. Natl. Acad. Sci. USA*, **77,** 4123.
16. Zetterberg, A. and Larsson, O. (1985) Kinetic analysis of regulatory events in G1 leading to proliferation or quiescence of Swiss 3T3 cells. *Proc. Natl. Acad. Sci. USA*, **82,** 5365.

17. Larsson, O., Zetterberg, A., and Engström, W. (1985) Cell-cycle-specific induction of quiescence achieved by limited inhibition of protein synthesis: Counteractive effect of addition of purified growth factors. *J. Cell Sci.*, **75,** 375.
18. Larsson, O., Zetterberg, A., and Engström, (1985) Consequences of parental exposure to serum-free medium for progeny cell division. *J. Cell Sci.*, **75,** 259.
19. Larsson, O. and Zetterberg, A. (1986) Kinetics of G1-progression in 3T3 and SV-3T3 cells following treatment by 25-hydroxycholesterol. *Cancer Res.*, **46,** 1223.
20. Larsson, O. and Zetterberg, A. (1986) Effects of 25-hydroxycholesterol, cholesterol and isoprenoid derivatives on the G1-progression in Swiss 3T3-cells. *J. Cell Physiol.*, **129,** 94.
21. Larsson, O., Dafgård, E., Engström, W., and Zetterberg, A. (1987) Immediate effects of serum depletion on dissociation between growth in size and cell division in proliferating 3T3-cells. *J. Cell Physiol.*, **127,** 267.
22. Mitchison, J. M. (1971) *The biology of the cell cycle*. Cambridge University Press.
23. Hartwell, L. H., Culotti, J., Pringle, J. R., and Reid, B. J. (1974) Genetic control of the cell division cycle in yeast. *Science*, **183,** 46.
24. Nurse, P. (1981) A re-appraisal of 'Start' in the fungal nucleus. In *Mutants of fission yeast*. (K. Gull and S. Oliver, (ed.) Cambridge University Press, London, p. 331.
25. Zetterberg, A. and Larsson, O. (1991) Coordination between cell growth and cell cycle transit in animal cells. *Cold Spring Harbor Symp. Quant. Biol.*, **56,** 137.
26. Larsson, O., Latham, C., Zickert, P., and Zetterberg, A. (1989) Cell cycle regulation of human diploid fibriblasts: Possible mechanisms of platelet-derived growth factor. *J. Cell Physiol.*, **139,** 477.
27. Larsson, O., Blegen, H., Wejde, J., and Zetterberg, A. (1993) A cell cycle study of human mammary epithelial cells. *Cell Biol. Int.*, **17,** 565.
28. Killander, D. and Zetterberg, A. (1965) Quantitative cytochemical studies on interphase growth. Determination of DNA, RNA and mass content of age determined mouse fibroblasts *in vitro* and of intercellular variation in generation time. *Exp. Cell. Res.*, **38,** 272.
29. Killander, D. and Zetterberg, A. (1965) A quantitative cytochemical investigation of the relationship between cell mass and initiation of DNA synthesis in mouse fibroblasts *in vitro*. *Exp. Cell Res.*, **40,** 12.
30. Larsson, O., Barrios, C., Latham, C., Ruiz, J., Zetterberg, A., Zickert, P., and Wejde, J. (1989) Abolition of mevinolin-induced growth inhibition in human fibroblasts following transformation by simian virus-40. *Cancer Res.*, **49,** 5605.
31. Highfield, B. P. and Dewey, W. C. (1972) Inhibition of DNA synthesis in synchronized Chinese hamster cells treated in G1 or early S-phase with cycloheximide or puromycin. *Exp. Cell Res.*, **75,** 314.
32. Rossow, P. W., Riddle, V. G., and Pardee, A. B. (1979) Synthesis of labile serum-dependent protein in early G1 controls animal cell growth. *Proc. Natl. Acad. Sci. USA*, **76,** 4446.
33. Pardee, A. B., Medrano, E. E., and Rossow, P. W. (1981) A labile protein model for growth control of mammalian cells. In *The biology of human normal growth*. M. Ritzén, A. Aperia, K. Hall, A. Larsson, A. Zetterberg, and A. Zetterström (ed.). Raven Press, New York, p. 59.
34. Brown, M. S. and Goldstein, J. L. (1980) Multivalent feedback regulation of HMG-CoA reductase: A control mechanism coordinating isoprenoid synthesis and cell growth. *J. Lipid Res.*, **21,** 505.
35. Siperstein, M. D. (1984) Role of cholesterologenesis and isoprenoid synthesis in DNA-replication and cell growth. *J. Lipid Res.*, **25,** 1462.

36. Goldstein, J. L. and Brown, M. S. (1990) Regulation of the mevalonate pathway. *Nature*, **343,** 425.
37. Glomset, J. A., Gelb, M. H., and Farnsworth, C. C. (1990) Prenylated proteins in eukaryotic cells: a new type of membrane anchor. *Trends Biochem. Sci.*, **15,** 139.
38. Sinensky, M. and Lutz, R. J. (1992) The prenylation of proteins. *BioEssays*, **14,** 25.
39. Schafer, W. R., Kim, R., Sterne, R., Thorner, J., Kim, S.-H., and Rine, J. (1989) Genetic and pharmacological suppression of oncogenic mutations in *RAS* genes of yeast and humans. *Science*, **245,** 379.
40. Struck, D. K. and Lennarz, W. J. (1980) The function of saccharide-lipids in synthesis of glycoproteins. In *The biochemistry of glycoproteins and proteoglycans*. Lennarz, W. J. (ed.). Plenum Press, New York, p. 35.
41. Kabakoff, B. D., Doyle, J. W., and Kandutsch, A. A. (1990) Relationships among dolichyl phosphate, glycoprotein synthesis, and cell culture growth. *Arch. Biochem. Biophys.*, **276,** 382.
42. Vincent, T. S., Wulpert, E., and Merler, E. (1991) Inhibition of growth factor signaling pathways by lovastatin. *Biochem. Biophys. Res. Commun.*, **180,** 1284.
43. Tay, D. L. M., Bhatal, P. S., and Fox, R. M. (1991) Quantitation of GO and G1 phase cells in primary carcinomas. *J. Clin. Invest.*, **87,** 519.
44. Schneider, C., King, R. M., and Philipson, L. (1988) Genes specifically expressed at growth arrest of mammalian cells. *Cell*, **54,** 787.
45. Coppock, D. L. and Scandalis, S. (1990) Isolation and characterization of human cDNA clones preferentially expressed in quiescent W138 fibroblasts. *J. Cell Biochem.*, Suppl., **14C,** 280.
46. Kallin, B., de Martin, R., Etzold, T., Sorrentino, V., and Philipson, L. (1991) Cloning of a growth arrest-specific and transforming growth factor B-regulated gene, T11, from an eipthelial cell line. *Mol. Cell. Biol.*, **11,** 5338.
47. Del Sal, G., Ruaro, M. E., Philipson, L., and Schneider, C. (1992) The growth arrest-specific gene, gas 1, is involved in growth suppression. *Cell*, **70,** 595.
48. Coppock, D. L. and Kopman, C. (1991) Regulation of gene expression in human diploid fibroblasts (W138) during exit from the proliferative cell cycle. *J. Cell Biol.*, **115,** 275a.
49. McClung, J. K., Danner, D. B., Steward, D. A., Smith, J. R., Schneider, E. L., Lumpcin, C. K., Del'Orco, R. T., and Nuell, M. J. (1989). Isolation of a cDNA that hybrid selects antiproliferative mRNA from rat liver. *Biochem. Biophys. Res. Commun.*, **164,** 1316.
50. Nuell, M. J., Stuart, J. A., Walker, L., Friedman, V., Wood, C. M., Owens, G. A., Smith, J. R., Schneider, E. L., Del'Orco, R., Lumpkin, C. K., Danner, D. B., and McClung, J. K. (1991) Prohibitin, an evolutionarily conserved intracellular protein that blocks DNA synthesis in normal fibroblasts and HeLa cells. *Mol. Cell. Biol.*, **22,** 1372.
51. Herskowit, I., Ogas, J., Andrews, B. J., and Chang, F. (1991) Regulations of synthesis and activity of the G1 cyclins of budding yeast. *Cold Spring Harbor Symp. Quant. Biol.*, **56,** 33.
52. Prescott, D. M. (1956) Changes in nuclear volume and growth rate and prevention of cell division in *Amoeba proteus* resulting from cytoplasmic amputations. *Exp. Cell Res.*, **11,** 94.
53. Donachie, W. D. (1968) Relationship between cell size and time of initiation of DNA replication. *Nature*, **219,** 1077.
54. Fantes, P. and Nurse, P. (1977) Control of cell size at division in fission yeast by a growth modulated size control over nuclear division. *Exp. Cell Res.*, **107,** 377.
55. Johnston, G. C., Pringle, J. R., and Hartwell, L. H., (1977) Coordination of growth with cell division in the yeast *S. cerevisiae*. *Exp. Cell Res.*, **105,** 79.

56. Sachsenmaier, W. (1981) The mitotic cycle in physarum. In *The cell cycle*. P. C. C. John (ed.). Cambridge University Press, p. 139.
57. Berger, J. D. (1982) Effects of gene dosage on protein synthesis rate in *Paramecium tetraurelia. Exp. Cell Res.*, **141,** 261.
58. Rasmussen, C. D. and Berger, J. D. (1982) Downward regulation of cell size in *Paramecium tetraurelia*. Effects of increased cell size with or without increased DNA content on the cell cycle. *J. Cell Sci.*, **57,** 315.
59. Reed, S. I., Hadwiger, J. A., and Lorincz, A. T. (1985) Protein kinase activity associated with the product of the yeast cell division cycle gene CDC28. *Proc. Natl. Acad. Sci. USA*, **82,** 4055.
60. Cross, F. R. (1988) A mutant gene affecting size control, pheromone arrest and cell cycle kinetics of *Saccharomyces cerevisiae. Mol. Cell. Biol.*, **8,** 4675.
61. Nash, R., Tokawa, G., Anad, S., Erickson, K., and Futcher, A. B. (1988) The *WHII*$^{+}$ gene of *Saccharomyces cerevisiae* tethers cell division to cell size and is a cyclin homolog. *EMBO J.*, **13,** 4335.
62. Fantes, P. A., Warbrick, E., Hughes, D. A., and MacNeil, S. A. (1991) New elements in the mitotic control of the fission yeast *Schizosaccharomyces pombe. Cold Spring Harbor Symp. Quant. Biol.*, **56,** 605.
63. Auer, G., Zetterberg, A., and Foley, G. E. (1970) The relationship of DNA synthesis to protein accumulation in the cell nucleus. *J. Cell Physiol.*, **76,** 357.
64. Zetterberg, A., Engström, W., and Larsson, O. (1982) Growth activation of resting cells. *Ann. NY Acad. Sci.*, **397,** 130.
65. Das, H. R., Lavin, M., Sicuso, A., and Young, D. V. (1983) The uncoupling of macromolecular synthesis from cell division in SV-3T3 cells by glylcocorticoids. *J. Cell Physiol.*, **117,** 241.
66. Baserga, R. (1976) Growth in cell size and cell DNA-replication. *Exp. Cell Res.*, **151,** 1.
67. Mercer, H. E., Avignolo, C., Galanti, N., Ruse, K. M., Hyland, J. K., Jacob, S. T., and Baserga, A. (1984) Cellular DNA-replication is dependent of the synthesis and the accumulation of ribosomal RNA. *Exp. Cell Res.*, **150,** 118.
68. Zetterberg, A. and Engström, W. (1983) Indication of DNA synthesis and mitosis in the absence of cellular enlargement. *Exp. Cell Res.*, **144,** 199.
69. Rönning, B. and Petterson, E. (1984) Doubling in cell mass is not necessary to achieve cell division in cultured human cells. *Exp. Cell Res.*, **155,** 267.
70. Zetterberg, A., Engström, W., and Dafgård, E. (1984) The relative effects of different types of growth factors on DNA-replication, mitosis and cellular enlargement. *Cytometry*, **5,** 368.
71. Stiles, C. D., Pledger, W. J., Antoniades, H. N., and Scher, C. D. (1979) Control of the Balb/c-3T3 cells by nutrients and serum factors. *J. Cell. Physiol.*, **99,** 395.
72. Croy, R. and Pardee, A. B. (1983) Enhanced synthesis and stabilization of MW 68 000 protein in normal and virus transformed 3T3 cells. *Biochem. J.*, **214,** 695.
73. Hadwiger, J. A., Wittenberg, C., Richardson, H. E., De-Barras Lopes, M., and Reed, S. J. (1989) A family of cyclin homologs that control the G1 phase in yeast. *Proc. Natl. Acad. Sci. USA*, **86,** 6255.
74. Furakawa, Y., Piwnica-Worms, H., Ernst, T. J., Kanakura, Y., and Griffin, J. D. (1990) cdc2 gene expression at the G1 to S transition in human T lymphocytes. *Science*, **250,** 805.
75. Sutton, A., Lin, F., Sarabia, M. J. F., and Arndt, K. T. (1991) The SIT4 protein

phosphatase is required in late G1 for progression into S phase. *Cold Spring Harbor Symp. Quant. Biol.*, **56,** 75.

76. Lew, D. J. and Reed, S. I. (1992) A proliferation of cyclins. *Trends Cell. Biol.*, **2,** 77.
77. Kelly, K., Cochran, B., Stiles, C. B., and Leder, P. (1983) Cell cycle specific regulation of the *c-myc* gene by lymphocyte mitogens and platelet derived growth factor. *Cell*, **35,** 603.
78. Campisi, J., Grey, H. E., Pardee, A. B., Dean, M., and Sonenshein, G. E. (1984) Cell cycle control of *c-myc* but not *c-ras* expression is lost following chemical translocation. *Cell*, **36,** 241.

9 Cancer and the cell cycle

EMMA M. LEES and ED HARLOW

1. Introduction

The decision for a cell to divide is a tightly regulated process that integrates signals from many sources. These signals indicate when division is needed and appropriate. Two sources of these signals are now known, and these arise either from the extracellular environment or from intracellular checkpoint controls. Environmental signals help cells determine when there is a need to divide, that the required nutrients are available, and that there are no overriding reasons not to begin a replicative cycle. Signals from internal checkpoints ensure the preceding steps of the cell cycle have been completed correctly before the next stage of division takes place. At the centre of these signalling pathways lies a mechanism for integrating these signals, the proteins for basic cell cycle control. All processes that contribute to decisions about proliferation or that are regulated in a cell-cycle-dependent manner must ultimately interface with the basic controls of the cell cycle.

Normally this process runs smoothly. The steps of cell division occur uneventfully, and even when problems do arise, they are corrected efficiently. However, when major perturbations are introduced by mutation of key regulatory genes, the control of division can be disrupted permanently. Depending on the mutation, different manifestations of the loss of control are seen. In some circumstances mutations promote a selective growth advantage for affected cells, leading to excessive and inappropriate division. When this occurs in multicellular organisms, uncontrolled division can be the initiating event of a number of diseases, the most devastating of which is cancer. While other systems must also be deregulated for oncogenesis, losing cell cycle control is clearly one of the most important. The last few years have seen the first examples of direct connections between cell cycle control and mutations that lead to cancer. In this review we will look at how the loss of cell cycle control is involved in tumourigenesis.

2. Cell cycle control

Genetic alterations which promote cell proliferation could occur in any of the different systems that contribute to cell cycle control, including components of its basic machinery, extracellular signal transduction, or checkpoint control. Mutations in the extracellular signalling pathways have been well documented and include changes in oncogenes and tumour suppressor genes. This work is vast and will not be the focus of this review, except when these proteins are directly linked to the cell cycle machinery. Recently, the first examples of mutations in the components of basic cell cycle machinery and of checkpoint regulation have been detected in human cancers. Much of the work reviewed in this volume concerns the control of the cell cycle and how these controls can be disrupted by mutation.

Most of the regulatory mechanisms of cell cycle control impinge on the gap or G phases of the division cycle. Thus, the G1 phase between mitosis (M phase) and DNA synthesis (S phase) and G2 phase between DNA synthesis and mitosis are the times in which regulatory decisions of the cell cycle are made. To understand how cell cycle control might be lost, it is essential to understand how the transitions between G1/S and G2/M are regulated. In this section we will outline the components of the cell cycle machinery and the several mechanisms which regulate their activity. As will become apparent, alteration of either could promote proliferation.

2.1 Cell cycle transitions

The basic machinery that regulates cell cycle progression has been studied in many systems, but it was the pioneering work in yeast that identified the key regulatory proteins. Genetic studies in yeast identified a single protein kinase, CDC28 in *Saccharomyces cerevisiae* or its homologue cdc2 in *Schizosaccharomyces pombe*, that plays a central role in progression through the cell cycle (1–3). In yeast, CDC28 (or cdc2) is required for both the G1/S and the G2/M transitions. At each of these transition points, the activity of the kinase is regulated by association with regulatory subunits known as cyclins. Cyclins are synthesized and degraded in each cell cycle, and their temporal association with the catalytic subunit provides a major element of the timing of kinase activation (4–9). The activity of cdc2 in the G1/S and G2/M transitions is distinguished by the cyclin that interacts with the kinase subunit. In *S. cerevisiae* the cyclins that regulate the activity of CDC28 during G1/S are known as CLNs. To date three CLNs, CLN1, CLN2, and CLN3, have been identified (7, 8, 10–12). The cyclins that regulate CDC28 during G2 and M are collectively termed CLBs, and seven have been isolated to date (13–17, and see Chapter 3).

Another regulatory mechanism involved in the control of kinase activation is phosphorylation of the kinase subunit at distinct sites that serve both activating and inhibitory roles. Phosphorylation at threonine 167 in the yeast cdc2 kinase (18) is absolutely required for kinase activity. A second phosphorylation on tyrosine 15 inhibits kinase activity (9, 19–22), and at least for the G2/M transition, the dephosphorylation of this residue is the timing step that allows the coordinated activation of kinase activity. Mutations in these phosphorylation sites have the expected phenotype; inability to phosphorylate threonine 167 blocks kinase activity, while tyrosine 15 mutations lead to early activation of the kinase (see Chapter 5). In yeast then there are several points of regulation. These include the timing of cyclin synthesis and degradation, as well as the phosphorylation and dephosphorylation of the cdc2 kinase itself.

In multicellular eukaryotes, homologous proteins control the cell cycle. The human cdc2 gene was identified through functional complementation of a mutation in the *S. pombe* cdc2 gene (23), indicating that they not only encode for structurally related proteins used in these diverse organisms but that they are functionally interchangeable. For historical reasons, the 'cdc2' name rather than 'CDC28' has been used to identify homologues of this kinase in other species. Surprisingly, in the last three years it has become apparent that rather than using a single catalytic subunit, mammalian cells and probably all metazoans use a number of related kinases to serve different roles in the regulation of the cell cycle (24–31). This family of proteins has been called the cyclin-dependent kinases or cdks based on their requirement for association with a cyclin regulatory subunit for activity. At present there are six cyclin-dependent kinases known. These include the original cdc2, and cdk2 through cdk6 (see Chapter 7).

To complement this large set of catalytic subunits there is an even larger and more divergent family of cyclins in mammalian cells (32–39). Cyclins in mammalian cells are designated by letter for closely related family members, i.e. cyclins A–G, and then by number to distinguish the order of discovery within each group, e.g. cyclin D1, D2, and D3. In a similar fashion to yeast, there are cyclins that function at different points in the cell cycle. The precise mechanism by which cyclins exert their control over the kinase catalytic subunit is unclear, but physical association with a cyclin moiety is essential for kinase activity. As in yeast, phosphorylation state also plays a major role in controlling the timing of activation of the kinase (9, 20–22, 40). Table 1 lists the known mammalian cyclins and cyclin-dependent kinases and their likely roles in cell division.

In mammalian cells another type of regulation has recently come to light (41). A group of inhibitory proteins that bind to some of the G1/S cyclin/cdk complexes to inhibit their kinase activity have been described (42–45). At least one of these inhibitors appears to be activated by treatment of cells with the growth factor TGFβ (46, 47). One potential role of these inhibitors may be to block the activation of the G1/S cyclin/cdk complexes when extracellular conditions are not appropriate for division. However, how the function of these inhibitors might be regulated is still unclear, and therefore their potential link to carcinogenesis remains unknown.

Table 1 Cyclin/cdk complexes

Complex	Timing	Proposed function
Cyclin D/cdk4	G1	Progression through early or mid-G1
Cyclin D/cdk6	G1	Progression through early or mid-G1
Cyclin E/cdk2	G1/S	Progression through late G1 to S phase
Cyclin A/cdk2	S phase to M	Initiation and completion of DNA synthesis
Cyclin A/cdc2	G2	Link to M phase?
Cyclin B/cdc2	G2/M	Initiation and maintenance of M phase

2.2 The cdk/cyclin complexes as signal integrators

How do the signalling cues from the environment and from checkpoint controls connect to the activation of the cdk/cyclin kinase? As discussed above, several mechanisms regulating the activity of the cdk/cyclin complexes are now known. However, how these regulatory events are linked to signals from the environment or from checkpoint controls is not well understood. This is particularly true in mammalian cells where environmental cues such as growth factors are known to stimulate cell division. Yet how these signals are translated to the activation of the cdk/cyclin complexes is only just beginning to be elucidated.

One obvious point of regulation is the control of expression of the cyclins, since this in turn contributes to the timing of activation of the cdks. Particularly with the G1 cyclins, the timing of their synthesis must be an important element of G1 control, and as discussed below this regulation is lost in certain human tumours. Overexpression of D- and E-type cyclins in mammalian cells advances the cell cycle by shortening the length of G1 itself, clearly indicating the acute sensitivity of cells to cyclin levels (48–50). Extracellular signals are known to lead to the activation of a number of promoters. One set of genes that must be activated to allow transition through the G1/S boundary is the G1 cyclins.

Another point of integration is the control of the activating and inhibiting phosphorylation events. For yeast, many of the enzymes that carry out these events are known and well characterized (51–56). In addition, a link between extracellular signals and these modifications has been made, with mating factors ultimately impinging on CLN activity (57–59). For mammalian cells, a number of analogues of these kinases and phosphatases have been identified, but there are no examples of how an extracellular signalling event is linked to modifications of the cell cycle machinery.

Considerably more is known about how checkpoint controls pass signals to the cdk/cyclin complexes. The best understood example is the initiation of mitosis in yeast (reviewed in ref. 60). Here, signalling mechanisms monitor for the completion of DNA synthesis and help scan chromosomal DNA for damage. Independent pathways pass signals that block the dephosphorylation of tyrosine 15 on cdc2 until the monitoring is completed. The inhibition of tyrosine 15 dephosphorylation keeps the kinase inactive and thereby ensures that mitosis does not occur prior to the completion of DNA synthesis or until DNA repairs are complete. In mammalian cells the G1/S checkpoint appears to be regulated by p53. In the presence of damaged DNA elevated p53 activity leads to the production of p21, a universal inhibitor of the cyclin/cdk complexes. This will be discussed in more detail in a later section.

2.3 G1 decisions are key for tumourigenesis

In mammalian cells it appears that most of the decisions that affect cell cycle progression are taken in G1. Developmental, differentiation, and apoptotic decisions depend primarily on signals that are illicited during G1. Likewise, signals that are known to stimulate division, such as growth factor treatment, induce G1 changes. Hence this is the period where most of the mutations that are linked to tumourigenesis exert their effects.

Genetic evidence for the importance of G1 has been established for some time. In both yeast and animal cells a critical decision point in G1 was described as early as the mid-1920s (1, 61, 62). When cells proceed beyond this point, they are committed to complete the cycle and divide. In yeast this control point is referred to as 'start' and in mammalian cells as 'restriction point'. While it is not clear that start and restriction point are identical biochemical events, they have similar consequences for cell cycle progression. Not surprisingly, start in yeast corresponds to the activation of the G1 cyclin/cdk kinase. In mammalian cells, it seems likely that a similar event occurs, but it is not clear whether the decision is handled by one or many kinases. Nor is it known which cyclin/cdk complexes carry out this role. None the less, most of the mutations associated with tumourigenesis have been shown to affect events that are specific for G1 decision making, and this is the point at which most research on the links between cell cycle control and oncogenesis has been focused.

3. Errors in cell cycle control that contribute to cancer

Links between oncogenesis and cell cycle regulation have been identified in three areas. These include:

- changes in the regulation of the cyclin/cdk complexes
- the demonstration that cyclin/cdk complexes help regulate proteins that are important in tumourigenesis
- tumours often show severe chromosome defects that normally should have been corrected under checkpoint regulation.

In this section we will describe pre-existing examples of genetic alterations to the cell cycle machinery that have been identified in tumours and discuss the evidence that cell cycle proteins may be able to act as oncogenes.

3.1 Can cyclins and cyclin-dependent kinases act as oncogenes?

Most disruptions of cell cycle machinery would result in inappropriate activity of one or more of the cyclin-dependent kinase complexes. Compelling evidence is accumulating to implicate cyclins as potential oncogenes, with overexpression of such regulatory proteins likely to be a contributing factor to tumourigenesis.

3.1.1 G1 cyclins as targets for tumourigenesis

The different cyclins accumulate with variable kinetics, magnitude, and timing across the cell cycle (reviewed in ref. 63). Figure 1 shows the relative timing of

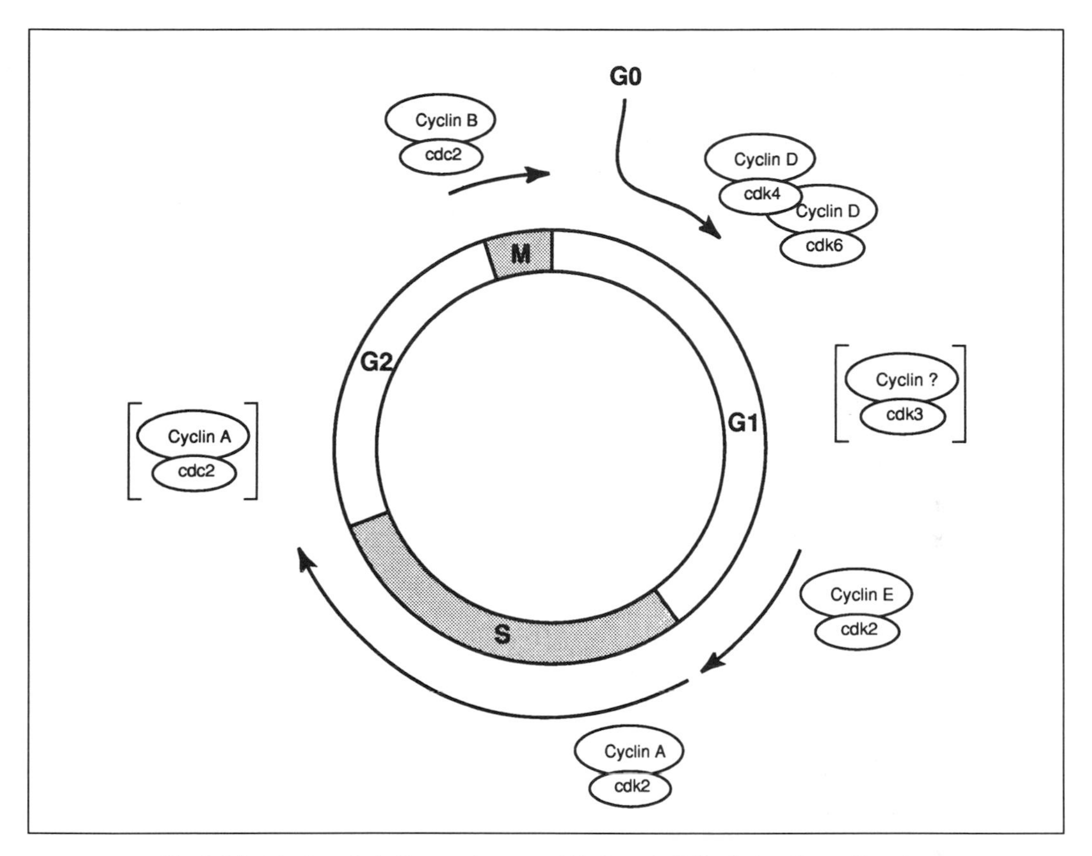

Fig. 1 Model of cyclin/cdk complexes and their periodicity during the cell cycle.

expression of the various mammalian cyclins during the cell cycle. The timing of cyclin protein appearance gives the first clue about the stage of the cell cycle that is regulated by that particular cyclin. For example, the level of cyclin B in both *S. pombe* and mammalian cells peaks at the G2/M boundary (4–6), and the need for cyclin B in G2/M decisions is well established (9, 64, 65). Since most decisions about cell cycle progression in mammalian cells are made in G1, the G1 cyclins are the best candidates for alterations giving rise to uncontrolled cell growth. This would include cyclins A, E, and the D-type cyclins.

As described below, overexpression of certain cyclins appears to be a common event in tumours. Specific translocation events in the vicinity of the cyclin D1 gene are the most persuasive evidence that cyclin amplification is causative to the onset of oncogenesis.

3.1.2 D-type cyclins

3.1.2.1 Isolation of the cyclin D1 cDNA

Cyclin D1 was originally isolated using several different genetic approaches. Functional complementation of the phenotype caused by disruption of all three

S. cerevisiae G1 cyclin genes, CLN1, CLN2, and CLN3, using a human glioblastoma cDNA library yielded sequences encoding a number of the cyclins, including cyclin D1 (34, 66). Cyclin D1 was independently isolated from mouse macrophages as a gene whose expression was induced upon treatment of cells with colony stimulating factor (CSF) during the G1 phase of the cell cycle (38). More pertinent to this discussion of the links between cancer and cell cycle control was the identification of human cyclin D1 (originally called PRAD-1) as a gene found rearranged with the parathyroid hormone locus in a subset of parathyroid tumours (36, 67). These three independent parathyroid adenomas each had a similar chromosomal inversion, with the 5′ regulatory region of the parathyroid hormone gene (PTH; on 11p15) transposed to the cyclin D1 coding region, normally located on 11q13. This translocation is associated with aberrant overexpression of cyclin D1, implicating cyclin D1 dysregulation as an important step in the formation of these lesions. Several other genes in this region of 11q13 have been examined intensely, including those encoding fibroblast growth factors INT2 and HST1 (68–70). However, these FGF oncogenes are not usually transcribed in the parathyroid adenomas.

3.1.2.2 Cyclin D1 overexpression is common in many cancers

Cyclin D1 overexpression is also found in a number of non-parathyroid neoplasms. A region of 11q13 is amplified in 15–20 per cent of human breast and squamous cell carcinomas of the head and neck (71). In a study of 20 human squamous oesophageal tumours, 25 per cent of these showed amplification of cyclin D1(72). In addition there is growing evidence to suggest that cyclin D1 is likely to be the BCL1 translocation involved in B cell lymphomas (73). This t(11;14) translocation involves the immunoglobulin heavy chain locus on chromosome 14 (74–76). The first transcriptional unit telomeric to the breakpoint is cyclin D1, located approximately 110 kb downstream. In centrocytic lymphoma, a tumour frequently associated with BCL1 rearrangements, steady-state levels of cyclin D1 mRNA and protein are highly elevated (67, 77). Recent studies have shown that in these tumours the 11q23 breakpoints can occur throughout this region, some of them being within 1 to 2 kb of the cyclin D1 gene (78). This argues strongly that cyclin D1 is the 'bcl-1 oncogene'. Cyclin D1 transcripts are truncated in such tumours (70), and lack 3′ non-translated sequences that might be important in destabilizing cyclin D1 mRNA.

Similar disruptions to the cyclin D1 locus have been observed in murine systems. Proviral insertions of mouse mammary tumour virus (MMTV) in mouse lymphomas at the int-2 and hst-1 loci occur within a region of mouse chromosome 7 that is syntenic with human 11q13, the region frequently found rearranged in human B cell neoplasms as described above. As a consequence of these insertions, overexpression of cyclin D1 is detectable both at the protein and RNA level in many MMTV- and Moloney murine leukaemia virus (MuLV)-induced tumours (80).

3.1.2.3 The cyclin D family

Matsushime *et al.* (38) also reported the isolation of two related genes, cyclin D2 and cyclin D3, by low stringency hybridization using a cyclin D1 probe. The three

D-type cyclins differ in their tissue distribution and periodicity of expression but little is known about their functional differences (81, 82).

By analogy, it is plausible that cyclin D2 and D3 might also be altered in human malignancies. Human cyclin D2 maps to chromosome 12p13, and cyclin D3 to chromosome band 6p21, both sites of rearrangement in certain human tumours (83). Chromosomal deletions and translocations of the short arm of 12p, where cyclin D2 has been designated to reside, are commonly rearranged in childhood acute lymphoblastic leukaemia. Abnormalities in chromosome 6 are also common with iso(p) chromosome, trisomy 6, and +del(6p), which all increase the gene dosage of the short arm of chromosome 6, are a frequent anomaly in retinoblastomas (84–86), malignant lymphomas (87), and hyperdiploid acute lymphocytic leukaemias (88). Clearly cyclin D2 and D3 are good candidates for targets of these rearrangements, and further study should reveal the role of the cyclins in the etiology of these human cancers.

Animal models provide better evidence for the involvement of cyclin D2 in tumour development. In BL/VL3 radiation leukaemia virus-induced tumours in mice and rats there is at least a 5 per cent incidence of provirus integration into a gene originally designated Vin-1 (89). The virus itself does not harbour an oncogene, and it is therefore believed that its ability to induce neoplasia may be in part through its ability to act as an insertional mutagen. The integration site is at the 5′ end of the transcriptional unit and the levels of Vin-1-specific RNA are increased in tumours which harbour a provirus in the Vin-1 region. Sequence comparisons show that this gene is cyclin D2. It would therefore seem likely that in these retrovirus-induced T cell leukaemias amplification of cyclin D2 may have played a role in carcinogenesis.

3.1.2.4 Cyclin D1 as an oncogene

The overexpression of cyclin D1 in tumours suggests that it plays a role in the oncogenic process. Although attempts at testing cyclin D1 in standard transformation assays have failed, recent work in several laboratories has suggested that cyclin D1 does indeed play a role in the control of proliferation.

Overexpression of cyclin D1 shortens the length of time cells take to progress through G1, suggesting that the loss of D1 transcriptional regulation and perhaps higher levels of cyclin D1 overcome at least one level of regulation in G1 progression (49–50). Analogous overexpression systems have been used to show that unregulated cyclin D1 expression will block differentiation of certain granulocyte cell lines *in vitro*, again arguing that cyclin D1 plays a role in key G1 decisions.

Perhaps the best example of dramatic changes induced by uncontrolled expression of D1 comes from transgenic studies. Wang *et al.* (90) have successfully developed strains of transgenic mice expressing cyclin D1 in mammary epithelium. Overexpression of cyclin D1 is associated with hyperplasia of alveoli and terminal ducts, indicating that expression of cyclin D1 can induce proliferative effects *in vivo*. A second recent example of cyclin D1's potential as an oncogene comes from transfection studies (91). While cyclin D1 will not act as an oncogene on its own in the standard NIH 3T3 transfection assay, Hinds *et al.* (91) have shown that transfection of cyclin D1 with an activated ras gene and an adenovirus E1A mutant will lead to transformation of primary rodent fibroblasts. Ras and this particular E1A mutant, which cannot bind to the retinoblastoma protein, do not induce transformation by themselves and therefore the addition of cyclin D1 to this now three-component transfection provides the necessary signals to stimulate colony formation.

In summary, studies in both solid tumours and lymphomas indicate that cyclin D1 amplification may be a common event and that an increase in cyclin D gene dosage could be an important factor in the pathogenesis of a number of cancers. It is worth noting the similarities between these alterations to cyclin D and those involved in the activation of c-myc. Many of the same characteristics that implicated myc as an important player in proliferation have now been shown for cyclin D1. These include:

- overexpression as a consequence of chromosomal translocation in human B cell lymphomas, by retroviral integration in murine tumours, and by DNA amplification in human breast cancers
- transgenic induction of breast hyperplasia
- blockage of differentiation in tissue culture
- ability to participate in colony formation in co-transfection assay.

Collectively, these data argue strongly for a role for cyclin D in certain human tumours. The requirement for cyclin D1 for proliferation is clearly demonstrated by the G1 arrest induced in fibroblasts by microinjection of cyclin D1 antibodies or antisense oligonucleotides (92). Experiments from a number of laboratories indicate that cyclin D exerts its effect at least in part by overriding the effects of the retinoblastoma gene product. An important observation was that cells arrested by overexpression of wild-type pRb can be rescued by transfection of cyclin D1 (93). There have now been several reports describing the ability of D-type cyclins to actually interact with and phosphorylate the pRb protein. These will be discussed in more detail later.

3.1.3 Cyclin A

Cyclin A accumulation commences as cells enter S phase but continues to increase through G2, followed by an abrupt degradation at the onset of mitosis (33, 94, 95). Two different views of cyclin A's roles in regulation of the cell cycle have emerged from recent studies. In vertebrate tissue culture cells, cyclin A appears to play a role in the initiation or perhaps maintenance of S phase. Microinjection of either antisense oligonucleotides or antibodies against cyclin A causes a block or delay in DNA synthesis (95, 96). However, other genetic experiments in *Drosophila* and cell cycle progression work in frog egg lysates elude to a requirement of cyclin A in G2 and mitosis decisions (97, 98). A possible explanation for these differences may be that cyclin A binds independently to both cdk2 and cdc2 kinase subunits (30, 99). The data suggest that at least one function of cyclin A is to regulate some

S phase event possibly modulating exit into G2. It remains possible, however, that cyclin A may be involved in multiple control points by virtue of its ability to bind different catalytic subunits.

3.1.3.1 Cyclin A and HBV integration

The first suggestion that cyclins may act as oncogenes came from the isolation of human cyclin A from a hepatocellular carcinoma. Hepatocellular carcinoma is often associated with chronic hepatitis B virus (HBV) infection. In chronic hepatitis, HBV frequently integrates into the genome of the infected liver cells. The significance of this integration event is unclear, but the investigation of the sequences into which the virus integrated led to the identification of one insertion into an intron of the cyclin A gene in the amino-terminus of the coding region (100). The integration event yielded a fusion protein between cyclin A and HBV, with the first 169 amino acids of the cyclin A protein being replaced by viral PreS2/S sequences. Experimental data indicate that this protein retains functional integrity with respect to kinase binding and activity, but that sequences functioning as degradation signals are lost, yielding an undegradable protein (101). Northern blot analyses showed high levels of cyclin A expression in the tumour, and coupled with the non-degradability of the resulting protein, this suggests that increased levels of cyclin A may have played a role in the onset of this tumour. Unfortunately, this tumour was not immortalized as a cell line, and other tumours containing such integrations have not been identified.

3.1.3.2 Cyclin A as a target for adenovirus E1A

Cyclins as target sites for viral insertion may be only one way in which viruses can disrupt the machinery of the cell cycle. The E1A oncoprotein of adenovirus forms complexes with a number of cellular proteins in virally infected cells (102, 103). Subsequent studies have identified some of these associated proteins, and they include negative regulators of cell growth such as the protein product of the retinoblastoma susceptibility gene (104) and a related protein, p107 (105). Also found in these complexes was cyclin A, providing the first example of interaction between a viral oncoprotein and a component of the host cell cycle machinery (33, 94). The importance of E1A's interactions with such cellular proteins is suggested by studies that show that the regions of E1A used to bind these cellular proteins correspond to the same regions required for its transforming activity (106, 107). The targeting of a component of the cell cycle machinery by E1A suggests the importance of overriding such regulatory factors to aberrantly propel the infected cells into DNA synthesis.

More recent studies have shown that E1A associates with other cell cycle proteins, namely cyclin E and cdk2 (108, 109). The virus apparently shows preference for those proteins regulating G1 and S phase events, since no interaction has been detectable with the mitotic regulators such as cyclin B/cdc2. Studies with other DNA tumour viruses suggest similar events occur within these systems. One of the viral oncogenes of the human papillomavirus type 16 (HPV16) known as E7, has been detected in stable complexes with cyclin A and cdk2 and has an associated kinase activity that peaks during S phase (110). The interaction of DNA tumour virus oncoproteins with cell cycle proteins are discussed in more detail below.

The evidence for cyclin A's participation in cancer is still limited. However, it clearly plays a role in cell cycle control, and this continues to focus attention on its possible role in tumourigenesis. Recent experiments further implicate cyclin A in promoting adhesion-independent growth, a classic feature of oncogenic transformation, when overexpressed (111).

3.1.4 Cyclin E

Cyclin E exhibits many properties characteristic of the G1 cyclins identified in yeast. Its expression and associated kinase activity are both maximal at the G1/S phase transition (112–114). Overexpression of cyclin E also shortens the G1 phase, decreases cell size, and diminishes the serum requirement for the G1 to S transition (48, 50). These experiments show that cyclin levels can be rate-limiting for G1 progression in mammalian cells. Studies have further implicated cyclin E as a downstream target for a number of external signals. Cells treated with TGF-β become arrested in G1 (115, 116). The activity of cyclin E-associated kinase is negligible in these cells, although both cyclin E and cdk2 proteins are present (112). In a similar manner, cells arrested in G1 by irradiation also contain inactive cyclin E complexes (117). In addition, cyclin E-associated kinase activity is down-regulated as cells approach senescence (118). It is now believed that signals such as TGF-β, irradiation, and senescence induce a family of inhibitory proteins which can inactivate these cyclin/cdk complexes (42, 46, 47, 117). Thus cyclin E clearly responds to a number of negative growth signals and one would predict that defects in this protein or the signals that modulate its activity would have profound effects on the state of the cell, establishing cyclin E as a prime candidate for an oncogene.

3.1.4.1 Cyclin E overexpression in tumours

Initial investigations suggest that aberrant cyclin E expression may be associated with tumours. Amplification of cyclin E at the protein level in a number of human breast tumours has been reported (119). In these studies the cyclin E gene was found to be amplified eight-fold in one breast tumour line, and there was over-expression of the protein in all (10/10) tumours examined. Overexpression of cyclins A and B and cdc2 RNAs was also noted in these tumour lines. This overexpression of cyclin E appears to be a general observation for many transformed cell lines in contrast to primary cells; however, genetic alterations in the cyclin E locus have not been identified in tumours to date. Overexpression in tumour lines may be the first example of changes to this particular cyclin, and it would seem likely that, with time, more perturbations to cyclin E expression will be revealed.

Table 2 Cyclins and cancer

Cyclin D1	11q13 translocations in parathyroid tumours (11;14)(q13;q32) translocation in BCL + CCL 11q13 amplification in 15–20% of breast, and squamous cell carcinomas of the head and neck Amplification in 25% squamous oesophageal tumours Site of proviral insertion of MMTV in mouse lymphomas
Cyclin D2	Vin-1, site of integration for BL/VL3 radiation leukaemia virus-induced tumours in rodents Chromosomal location 12p commonly deleted or translocated in childhood acute lymphoblastic leukaemia
Cyclin D3	6p21 iso(p), trisomy and del(6p) frequent anomaly in retinoblastomas, malignant lymphomas, hyperdiploid acute lymphocytic leukaemias
Cyclin A	Hepatitis B virus integration site Association with adenovirus E1A
Cyclin E	Amplification in breast tumours Association with adenovirus E1A

3.1.5 Other putative G1 cyclins

Several other cyclins have recently been cloned that may have been implicated in performing G1 functions. These include cyclin C (34), cyclin F (S. Elledge, personal communication), and cyclin G (39). Since no partners or activities have been assigned to these cyclins to date, we shall not include them in the discussion, albeit to mention their existence.

The evidence for cyclin involvement in malignancy that has been discussed is summarized in Table 2. It is of interest to note that there is an apparent selection for alterations in those cyclins regulating G2 events. At least for the time being it would appear that B-type cyclins are not subject to the perturbations that have been observed with the G1 cyclins.

3.1.6 Kinase subunits

As the other essential component in the cyclin/cdk complexes that control cell cycle progression, it would be predicted that mutations in the kinase subunits would be likely to be found in human tumours. The first possible example of this is a 20–30-fold amplification of cdk4 both at the DNA and protein level in a number of human sarcomas (120). These tumours show frequent amplification of band q13 on chromosome 12, which contains both MDM2 and cdk4. Overexpression of MDM2 has previously been described in the progression of tumours. It will clearly be important to establish which (or both) of these genes in this amplicon is important in the evolution of these tumours.

This study documents one of the few examples of amplification of kinase subunits seen in human tumours. Analysis of expression of the cdks at the RNA level in a number of different transformed cell lines revealed little variability in expression

(24) and suggested ubiquitous expression. This is true for all of the members of the cdc2-related kinase family with the exception of cdk6. A dramatic amplification of cdk6 transcripts in a myeloid leukaemia cell line, ML-1, was the only variant, and the significance of this remains to be established. Although there have not been many reports to date describing amplification of kinases in tumours, it would clearly be a possible mechanism by which to promote growth and should be explored further.

For overexpression of the kinase subunit to alter cell cycle progression, the action of the catalytic subunit must be limiting, so that the overexpression would provide a proliferative advantage to the cell. In addition to overexpression, inappropriate activation of the catalytic subunit could produce a similar phenotype. Removal of negative regulation would be the most likely step in dysregulation. In yeast we know that substitution of tyrosine 15 with phenylalanine (Y15F) causes cells to undergo mitosis prematurely, since they do not complete DNA replication before undergoing mitosis (19). In *Xenopus* egg extracts and HeLa tissue culture cells, the expression of cdc2 with a double mutation, threonine 14 to alanine (T14A) and Y15F, causes premature mitotic events (20, 22). In a similar fashion overexpression of the tyrosine 15 phosphatase, cdc25, or inactivation of the tyrosine 15 kinases, wee1 or mik1, all accelerate entry into mitosis and abolish the ability of unreplicated DNA to block entry into mitosis (54, 121, 122).

All events giving rise to a constitutively active kinase have the potential to promote proliferation. A second mechanism for loss of kinase regulation could be the loss of the inhibitory proteins that have recently been shown to bind to several of the G1 cyclin/cdk complexes and block kinase activity. Loss of this inhibition would also stimulate kinase activity inappropriately. A recent example of changes in subunit composition with transformation has been described by Xiong *et al.* (123). They have shown that in human fibroblasts, many of the kinase complexes are associated with other cellular proteins. Following transformation of these cells with DNA tumour viruses, these complexes dissociate leaving only the cyclin and cdk polypeptides in the complex. It is not clear what the functions of these higher order protein complexes are; however, their rearrangement during transformation suggests that they may indicate a change in kinase regulation. Recently, a number of these proteins have been isolated and shown to inhibit kinase activity directly (42–44). One clue as to the mechanism of these rearrangements comes from studies which show that p21, an inhibitor of kinase/cdk complexes, is regulated by p53. It is possible that loss of p53 function may lead to a reduction in inhibitor levels, therefore promoting cell growth (45).

It is likely that in the years to come, the loss of both mechanisms of negative regulation, i.e. post-translational modifications and in the form of inhibitors, will be identified as specific events in tumourigenesis. They obviously are somewhat more difficult to detect than the chromosomal changes seen with the cyclins; however, their consequences could be as dramatic in changing the regulation of the cell cycle control. Figure 2 depicts the types of alteration that could occur to produce an inappropriately active cyclin/cdk complex.

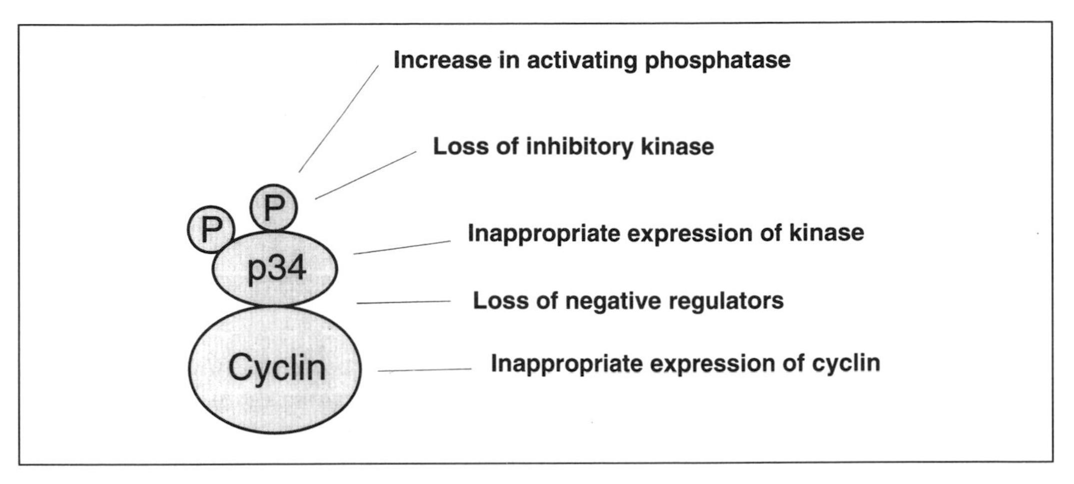

Fig. 2 Potential oncogenic alterations to cyclin/cdk complexes.

3.2 Cyclin/cdk substrates

To understand fully how inappropriate activation of cyclin/cdk complexes might promote cell proliferation, it is necessary to examine the targets of these kinases. There is a small but growing number of cdk substrates that have known roles in proliferation. These include tumour suppressor proteins, oncogenes, transcription factors, and replication factors. Below is a brief description of the key changes induced by the cyclin-dependent kinases in proteins with roles in growth control. These substrates and the speculated consequence of their phosphorylation are also summarized in Table 3.

3.2.1 Tumour suppressor genes

3.2.1.1 The retinoblastoma gene product

The retinoblastoma gene (RB) is a tumour suppressor gene whose inactivation is known to occur in a wide variety of human tumours (reviewed in ref. 124). Reintroduction of the gene into cells lacking endogenous RB results in growth

Table 3 Substrates for G1/S cyclin/cdk kinases

Substrate	Proposed role of phosphorylation
pRB	Release and activation of transcription factors
p107	Release and activation of transcription factors
p130	Release and activation of transcription factors
p53	Possible role in nuclear localization
E2F	Modulation of transcriptional activity
RPA	Initiation of DNA replication
SV40 large T antigen	Initiation of DNA replication
RNA polymerase II	Basal transcription

suppression or loss of tumourigenesis. Its loss in tumours and the reversion of the tumourigenesis phenotype by the expression of the gene suggests that the retinoblastoma protein (pRB) is a negative regulator of cell growth (125–127). pRB is expressed at reasonably constant levels through the cell cycle, but it becomes extensively modified by phosphorylation. The protein is underphosphorylated in G0/G1 and becomes highly phosphorylated in mid- to late-G1 (128–132). Several observations indicate that the active form of the protein is the hypophosphorylated form. The hypophosphorylated form of pRB is targeted by the transforming proteins of small DNA tumour viruses such as the large T antigen of SV40 (133). Similarly, only the hypophosphorylated form can interact with its physiological targets such as the transcription factor E2F (see below for details).

Thus, phosphorylation is believed to be a key regulatory event leading to the inactivation of the growth-suppressing activities of pRB. Interestingly, pRB contains a number of consensus cdc2 phosphorylation sites (134), and *in vitro* and *in vivo* mapping strongly suggest that the pRB kinase is from the cdk family (135–137). Since the timing of activation of cdc2 is a G2 event, it is unlikely that this is the initial pRB kinase. Cdk2 in association with either cyclin A or cyclin E is also able to phosphorylate these sites *in vitro* (138). Recent experiments indicate that cdk4 and cdk6 when associated with D-type cyclins can also phosphorylate physiologically relevant sites on pRB *in vitro* (26, 139, 140). The possible role of the cdk4 and cdk6 kinases in regulating pRB has been strengthened by recent data showing that several of the D-type cyclins can bind to pRB (141, 142).

Phosphorylation of pRB appears to inactivate its ability to inhibit transactivation of certain key transcription factors. In this setting, the outcome of unscheduled pRB phosphorylation will be to activate transcription. Thus, overexpression of the G1 cyclins such as the D-type cyclins would stimulate the synthesis of proteins that are important in proliferation. This could effectively be as detrimental as loss of RB alleles seen in human tumours or the sequesteration of pRB by oncoproteins of small DNA tumour viruses. The connection between the cell cycle machinery and pRB is clearly such an important one that any perturbation would have dramatic consequences.

3.2.1.2 p107 and p130

The two pRB-related proteins p107 and p130 are also substrates for cyclin E/cdk2 and cyclin A/cdk2 complexes (109). As mentioned below, p107 and p130 also associate with the transcription factor E2F. In a similar fashion to pRB/E2F complexes, we speculate that phosphorylation of p107 and p130 will promote transcription by release of factors such as E2F. As new functions for p107 and p130 are discovered, the significance of their interactions with cell cycle components will become clearer.

3.2.1.3 p53

The exact biochemical role for p53 is not fully understood, but it can act as a transactivator when p53 binding sites are found in the promoter region of certain genes. It may also act as an inhibitor for some genes that lack a p53 binding site (reviewed in ref. 143). In other experiments p53 plays an inhibitory role in controlling the initiation of DNA replication (144–147). More recent data implicate p53 as an important factor in the G1/S checkpoint monitoring and arresting growth in the event of DNA damage (this is discussed below). The p53 protein is heavily phosphorylated *in vivo*. One phosphorylation site is serine-315 in the human protein, which is a consensus phosphorylation site for the cdc2 kinase and is located close to a nuclear localization signal (148, 149). It has been noted that a peptide corresponding to this region of p53 is more heavily phosphorylated in transformed than in untransformed cells (150, 151). However, the actual role of phosphorylation for the activity of p53 remains unclear although it is speculated that serine-315 phosphorylation could modulate nuclear transport.

3.2.2 Oncogenes

The downregulation of tumour suppressor gene function is only one example of regulatory proteins that are controlled by cyclin/cdk phosphorylations. Phosphorylation can also have stimulatory actions as in the case of positive growth regulatory proto-oncogenes such as c-src.

3.2.2.1 c-src

The c-src proto-oncogene encodes a membrane-associated, tyrosine-specific protein kinase. Mutant forms of the protein bearing carboxy-terminal truncations are oncogenic and can induce morphological changes and serum- and anchorage-independent proliferation. Its role in normal cells is unclear but it appears to have diverse biological activities, many of which involve inappropriate signalling and membrane/cytoskeletal alterations. For the majority of the cell cycle the c-src protein is held in an inactive state by an inhibitory phosphorylation on tyrosine 527 (152). The kinase responsible for this, CSK (for c-terminal Src kinase) has recently been cloned (153) and appears to represent a novel tyrosine kinase. The activation of c-src at the onset of mitosis apparently requires several events, a potential contributing initial event being phosphorylation of the three amino-terminal residues by cdc2 (154–157). This phosphorylation event alone is not sufficient to stimulate its specific kinase activity. It is speculated that cdc2 phosphorylation events make c-src accessible to a phosphatase that can remove the inhibitory phosphate from tyrosine 527 or render the protein insensitive to CSK or some upstream component regulating CSK activity.

3.2.2.2 Other oncogenes

Mitotic hyperphosphorylation of a number of other proto-oncogenes has been reported including c-mos, c-abl, c-myc, and c-myb (reviewed in ref. 158). C-myb and c-mos have been shown to be phosphorylated *in vitro* by purified cdc2 (159), but it is not clear which kinase phosphorylates these sites *in vivo*. Nor is it known whether the function of these proteins is modified by these phosphorylation events.

3.2.3 Transcription factors

Transcription relies on the interplay of transcription factors which regulate the timing, magnitude, and specificity of gene expression. Evidence is accumulating that cdks can affect the activity of transcription factors either through direct phosphorylation or indirectly through phosphorylation of regulatory cofactors. Cell cycle-dependent transcription of a number of genes is expected to be mediated at least in part by such phosphorylation events. The transcription factor E2F provides a good example of such cell cycle-mediated regulation at multiple levels.

3.2.3.1 E2F

The cellular transcription factor E2F was originally identified through its role in transcription activation of the adenovirus E2 promoter (160, 161). E2F-consensus binding sites have been found upstream of cellular genes that are likely to affect proliferation in the early phases of the cell cycle. These include the early response genes c-myc, N-myc, and b-myb as well as genes encoding proteins required for DNA synthesis such as dihydrofolate reductase, thymidylate synthase, DNA polymerase α, and proliferating cell nuclear antigen (reviewed in ref. 162). In the promoters examined, the E2F sites are clearly important for transcriptional activation, with E2F playing an important role in regulating the expression of these genes (163–166).

The activity of E2F is thought to be regulated by association with other proteins, including pRB, p107, and p130. The identification of a complex between pRB and the transcription factor E2F provides the most direct link between pRB and control of cell proliferation (167–172). Recent experiments show that pRB's direct association with E2F-1 is inhibitory (173–175). These complexes with E2F appear to be cell cycle-regulated with pRB most readily detectable in association during G1 and early S phase (169, 176). These complexes are believed to be temporarily regulated by the cyclin-dependent kinases, since hypophosphorylated pRB is the only form that can bind to E2F. By phosphorylating pRB the kinases are able to alleviate transcriptional repression by preventing pRB interaction with E2F. It would appear that regulation can also occur through direct phosphorylation of E2F itself (177), which is thought to enhance DNA binding. E2F-1 protein contains at least two cdc2 consensus phosphorylation sites and has been shown to be phosphorylated by cyclin A/cdk2 *in vitro*. Figure 3 depicts how pRB allows communication between the cell cycle machinery and the transcriptional machinery. These observations imply a dual role for the cyclin/cdk complexes, both in the regulation of pRB and potentially in the modulation of E2F itself.

E2F is unlikely to be the only target for pRB regulation. Evidence would suggest that a number of transcription factors will be regulated by the hypophosphorylated forms of pRB. These include Elf-1 (178), ATF-2 (179), SP-1 (180, 181), myoD, and myogenin (182). In each of these cases it appears that pRB may exert its effects on the transcription factor through physical interaction. Again, since only the hypophosphorylated form of pRB can interact, cell cycle phosphorylation is likely to be a key regulatory event for these transcription factors.

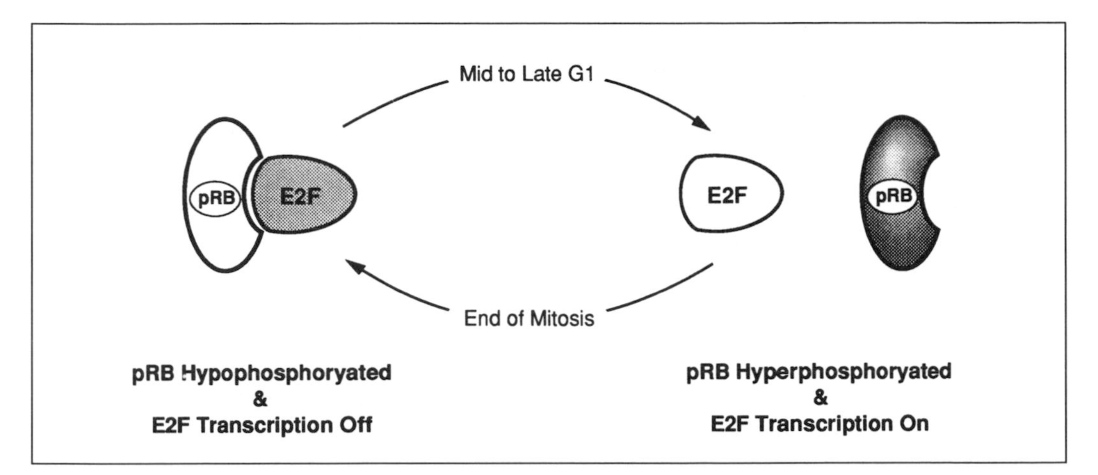

Fig. 3 pRB phosphorylation leads to the release of transcription factors.

3.2.3.2 RNA polymerase II

RNA polymerase II is a multisubunit enzyme, the largest subunit of which (IIA) is extensively modified by phosphorylation at the carboxy-terminus. This domain (CTD for carboxy-terminal domain) is comprised of tandem repeats of the consensus sequence Y–S–P–T–S–P–S, and a cdc2-like kinase has been shown to phosphorylate the CTD *in vitro* (183). It remains to be determined how phosphorylation modulates activity; however, it is believed to occur after RNA polymerase II has bound to the promoter, and is expected to promote elongation. Two likely kinases have recently been identified, but little characterization for these kinases is available (184).

3.2.4 Replication factors

The cdk complexes play a role in control of entry into S phase. In addition to regulating the synthesis of proteins needed for DNA synthesis at the transcriptional level, cdks have also been shown to modulate the activity of proteins involved in DNA replication. The replication factor RPA, an essential factor for the initiation and elongation steps of DNA replication (185, 186), is phosphorylated by cdks. Both the human and yeast forms of RPA have been purified and consist of three subunits. One of these subunits is phosphorylated by an unknown cdk, and this phosphorylation event appears to increase the activity of RPA in DNA synthesis reactions.

3.2.5 Summary

In the above section we have described a number of different proteins that are regulated by the cyclin-dependent kinase complexes. The identification of the particular kinases involved in each process remains to be established. Most studies indicate that either G1 or G2 cyclin/kinase complexes can perform such roles at least *in vitro*. However, under such circumstances these kinases are promiscuous and little specificity is displayed between the different classes. With the discovery of other cyclins and cdks whose timing of activation are most appropriate for a G1 role, it will be important to learn which kinase is the physiological regulator.

3.3 Checkpoint controls

In addition to receiving the correct environmental cues for cell division, cell cycle progression depends on the successful completion of the previous stages of the cycle (see Chapter 1). These feedback controls have come to be known as cell cycle checkpoints, and they represent intracellular signalling pathways that allow cells to monitor the correct completion of specific cell cycle events. Five cell cycle checkpoints have been identified to date:

- one that monitors for DNA damage and arrest cells in G1
- one that monitors for DNA damage and arrest cells in G2
- one that signals that DNA synthesis has been completed before mitosis begins
- one that monitors completion of mitosis before allowing entry into S phase
- one that ensures that the chromosomes are correctly aligned at the metaphase plate prior to the initiation of anaphase.

It seems likely that other checkpoints will be discovered as more sophisticated assays become available. Since checkpoints constitute the braking system for the cell cycle, enabling the cell to halt in the instance or damage of unfavourable environmental systems, any disruption could lead to cells with increased tumourigenic capacity.

3.3.1 Two different checkpoints monitor for genetic damage and arrest cells in either G1 or G2

In the preceding sections we have described the mutations that can occur to disrupt orderly growth in mammalian cells. In normal cells the frequency with which these mutations occur is minimized by monitoring systems that halt cellular division in response to DNA damage, allowing the cell time to repair its genetic material. These systems are subject to mutations, and when the checkpoint controls that monitor for DNA damage are lost, increases in the number of mutations can be expected (187, 188). The large number of mutations found in tumour cells appears to be due in part to breakdown in checkpoint controls.

An excellent example of the increased mutation rates in cancer cells is provided by colon carcinoma. In the case of colon cancer at least six separate genes are mutated during tumourigenic development (189). The probability of acquiring such a number of spontaneous mutations within the lifetime of an individual cell is so remote that it cannot account for this tumourigenesis. Even making the calculations with an exceptionally high mutation rate of 10^{-5} mutations/gene/

division, the frequency of getting six mutations in the same cell is 10^{-30}. There are only 10^{16} cell divisions needed to complete a normal adult lifespan; therefore, the frequency of recognizable tumours must be increased by factors other than spontaneous mutations. The tumour rate of mutation is obviously increased by exposure to carcinogens, but even this increase seems unlikely to account for the frequency of tumourigenesis. One possible mechanism to increase this frequency would be mutations that increased genetic instability. This is one of the characteristics of tumour cells. They are less capable of sustaining genetic integrity, as measured by frequency of mutation, chromosome amplification, or incidence of aneuploidy, than normal cells. These higher than expected mutation rates appear to result from both mutations in the DNA repair systems themselves (190) and loss of checkpoint controls (188).

In normal cells there are at least two stages in the cell cycle that can serve as regulation points where DNA damage leads to arrest. These are at the G1/S and G2/M transitions. The G1/S checkpoint is currently best understood in mammalian cells and the G2/M checkpoint in yeast. Arrest at either point is exerted on to the cell cycle machinery. The isolation and characterization of the genes responsible for these checkpoints are in progress.

3.3.1.1 G1/S DNA damage checkpoint

Recent work implicates the tumour suppressor gene p53 as a central player in the G1/S cell cycle checkpoint in mammalian cells. Levels of p53 protein transiently increase in response to DNA damage induced by several different agents (191). Tumour cells lacking p53 or having dominant mutant forms of p53 lack the G1/S delay that occurs upon exposure to ionizing irradiation. The restoration of wild-type p53 to these cells restores the delay, and addition of dominant mutant p53 to wild-type cells prevents delay (192, 193). The mechanism of p53 induction in response to damage is not well understood, but at least one of the signals that leads to increased levels of p53 is breaks in chromosomal DNA. p53 is a transcription factor, and high levels of p53 activate transcription of key genes (reviewed in ref. 143). These proteins in turn participate in the arrest of cell cycle progression. Perhaps the best understood of the genes induced by p53 is an inhibitor of G1 cyclin/cdk complexes (45). This protein, known as p21, CIP-1, or WAF-1, was originally identified through its physical association with the G1 cyclin/cdk complexes (42, 43). Through unknown mechanisms that probably include physical interaction, p21 inhibits the kinase activity of cyclin/cdk complexes. The direct induction of p21 expression by p53 through p53 responsive elements in the promoter of the p21 gene, links p53 directly to the cell cycle (45). Thus, the pathway of cell cycle arrest leads from the events of DNA damage to inhibition of the cyclin/cdk complexes that control G1/S transition. The important mediator of this arrest is p53. p53 mutations are the most common mutations found in all human tumours. Over 50 per cent of all tumours have mutations in p53 (194). This is the best understood case where mutation of a cell cycle checkpoint is known to contribute to human cancer.

A number of cancer-prone syndromes such as Li-Fraumeni syndrome, Fanconi's anaemia, Werner's syndrome, Bloom's syndrome, and ataxia-telangiectasia (AT) provide compelling evidence that genetic instability can be inherited and that such instability is associated with a predisposition to cancer. Two of these diseases point directly to roles for p53 in cancer. Li-Fraumeni syndrome is characterized by inherited predisposition to several cancers including breast carcinoma, sarcomas, and leukaemia. This cancer syndrome is due to germ line mutations in the p53 gene (195). A second disease that implicates p53 directly in checkpoint problems is AT, which is caused by a recessive mutation that inactivates the feedback control that regulates DNA replication in irradiated cells. Cells from these patients are unable to induce p53 protein levels in response to ionizing irradiation (196). These observations point to a defect in the signal transduction pathway involving p53 that controls cell cycle arrest following DNA damage. A number of other recessive mutations that are commonly found in human tumours may also inactivate functions involved in feedback controls, rather than in regulation of cell proliferation.

p53 also appears to be linked intimately with gene amplification and genetic instability. Cells with mutant p53 are genetically unstable, as these cells gain the capacity to amplify at high frequency the gene encoding CAD (a trifunctional polypeptide containing carbamoyl phosphate synthetase, aspartate transcarbamylase, and dihydroorotase), while heterozygous and wild-type cells do not (193, 197).

Analogous G1/S checkpoint controls that respond to DNA damage appear to be present in all eukaryotic organisms; however, the biochemistry is best understood in mammalian cells.

3.3.1.2 G2/M DNA damage checkpoint

The G2 checkpoint has been characterized most extensively in yeast. Work in mammalian cells has shown that a G2 DNA damage checkpoint must exist. Cells that have lost p53 function can no longer arrest in G2; however, they arrest in the next G2, arguing that a G2 checkpoint control must be active in these cells.

In both the budding and fission yeast, it has been well established that a large number of genes are needed for G2 arrest (190, 198, 199). In the budding yeast three genes, *RAD9*, *RAD17*, and *RAD24*, recognize damaged DNA and arrest cells in response to this signal (190). Deletion of *RAD9* increases chromosome loss 20-fold but is not lethal, suggesting that *RAD9* arrests cells to give them time to repair the damage. Three other genetic loci, *mec1*, *mec2*, and *mec3*, are also involved in cell cycle arrest due to damage (200). In addition to failure to arrest in the presence of damaged DNA, *mec1* and *mec2* mutants are also defective in their ability to arrest their cell cycle in response to unreplicated DNA. This suggests that although different sensors detect damaged and unreplicated DNA, both feedback controls converge to act on the same checkpoint in the cell cycle machinery. In *S. pombe* a similar, complex set of genes has been identified in this process (199).

3.3.2 DNA replication must be completed before mitosis

Mitotic checkpoints monitor completion of DNA synthesis and the presence of damaged DNA. This replication-dependent checkpoint prevents mitosis by inactivating cdc2. The proteins involved in this coupling of pathways are being sought. Strong overexpression of the tyrosine phosphatase (Cdc25), dominant activating alleles of cdc2, or inactivation of the tyrosine kinases Wee1 and Mik1 all accelerate entry into mitosis and abolish the ability of unreplicated DNA to block entry into mitosis. The induction of a cell cycle delay by inhibition of DNA synthesis requires that cdc2 is under the control of tyrosine phosphatase cdc25 (122, 199), but does not require the function of the negative cell cycle regulating protein kinase Wee1. The induction of delay by irradiation, presumably due to DNA damage, is apparently not affected by dissociating cdc2 from regulation by cdc25 nor by the absence of wee1, so the mechanism here remains to be resolved (201, 202). Such redundancy is probably a consequence of the number of ways that the activity of the cdc2 kinase can be affected. A novel protein kinase, chk1, involved in this checkpoint control has recently been isolated. Chk1 was isolated in a search for additional elements that might regulate cdc2 function, through complementation of a cold-sensitive *cdc2* mutant. It thus may represent the first direct link between the DNA damage sensing pathway and cdc2 (203).

In mammalian cells similar controls have been found to exist. In hamster cells a temperature-sensitive mutation in the RCC1 gene (repressor of chromosome condensation) causes cells at the non-permissive temperature to cease DNA synthesis and to enter into mitosis prematurely. The RCC1 protein is found in chromatin and can stimulate guanine nucleotide exchange on a small G protein called *ran*, whose function is unknown (204, 205). Relatives of RCC1 have been identified in a number of species, in *S. cerevisiae* as *pim1*, and also in *S. pombe* and *Drosophila* (206).

Cells also monitor completion of mitosis before allowing cells to re-enter DNA synthesis, demonstrated by the fact that cells blocked in G2 cannot undergo S phase. A key protein in fission yeast for regulating this step has recently been identified, and is called *rum1* (207). Overexpression of *rum1*$^+$ breaks the dependence of Start upon the completion of mitosis, initially leading to a transient G1 delay, followed by repeated rounds of DNA replication. Deletion of this gene eliminates the pre-start G1 interval indicating that *rum1*$^+$ is a major element determining the length of G1.

3.3.3 Aligning the chromosomes on the metaphase plate before anaphase

Cancer cells commonly show drastic changes in chromosome number. They are often highly aneuploid, indicating that the machinery that directs correct segregation of replicated chromosomes into each daughter cell is defective. In mammalian cells the control of chromosome segregation is not well understood; however, in yeast multiple genes have been identified that control exit from mitosis in response to the completion of spindle assembly. Mutation of these checkpoint genes in yeast would appear to be the counterpart of the loss of integrity of chromosome secretion that is so common in cancer cells.

Two classes of mutants have been isolated from budding yeast which abolish this feedback control, *bub* (budding uninhibited by benomyl) (208) and *mad* (mitotic arrest deficient) (209, 210) mutants. Both groups of mutants were identified by their ability to undergo mitosis in the presence of microtubule-depolymerizing drugs. These drugs block the formation of the mitotic spindles, and the loss-of-function mutations allow mitosis to continue even though the mitotic spindle has not been formed correctly. This feedback control probably acts by preventing the activation of the machinery that degrades the mitotic cyclins and leads to the inactivation of maturation promoting factor (MPF) (211). Very little is yet known about the biochemical events that signal the completion of correct spindle formation; however, the outcome of mutation in these signalling checkpoints is phenotypically similar to what occurs in mammalian tumour cells. Hence, improper alignment and segregation of human chromosomes is commonplace and must be permitted by some breakdown in these controls.

3.3.4 Summary

A number of checkpoints exist to ensure that proliferation occurs in a sequential fashion and to safeguard against the introduction of mutations to the genetic material of the cell. Since tumours have an increased and disorderly proliferation, and a higher rate of mutational changes, it is clear that checkpoint mutations are essential for tumour progression.

4. Discussion

In the preceding sections we have discussed a number of different ways in which perturbations to the cell cycle machinery are implicated in cancer. All stimuli that affect a cell's proliferative status must ultimately impinge on the cell cycle machinery. An incorrect response to such stimuli will give rise to aberrant growth and potentially tumourigenesis. Changes leading to genomic instability appear to be key events in the evolution of tumour cells and are likely to involve disruption of checkpoint surveillance. Alterations in oncogenes and tumour suppressor genes are likely to represent other events in malignant progression. For a tumour cell to expand, it must possess a growth advantage over its neighbours. One way to achieve this is through dysregulation of the cell cycle by inappropriate expression of cyclins and cdks or by aberrant activities of their regulators. Many substrates for these kinase complexes have been described. They include proteins that regulate key proliferative decisions in the cell and it is therefore likely that inappropriate regulation by the cell cycle machinery could have profound effects on cell growth.

The discovery of a family of inhibitory proteins that can modulate the activity of the cyclin/cdk complexes establishes another potential target for mutations that would give the cell a proliferative advantage. Figure 4 depicts how these inhibitors

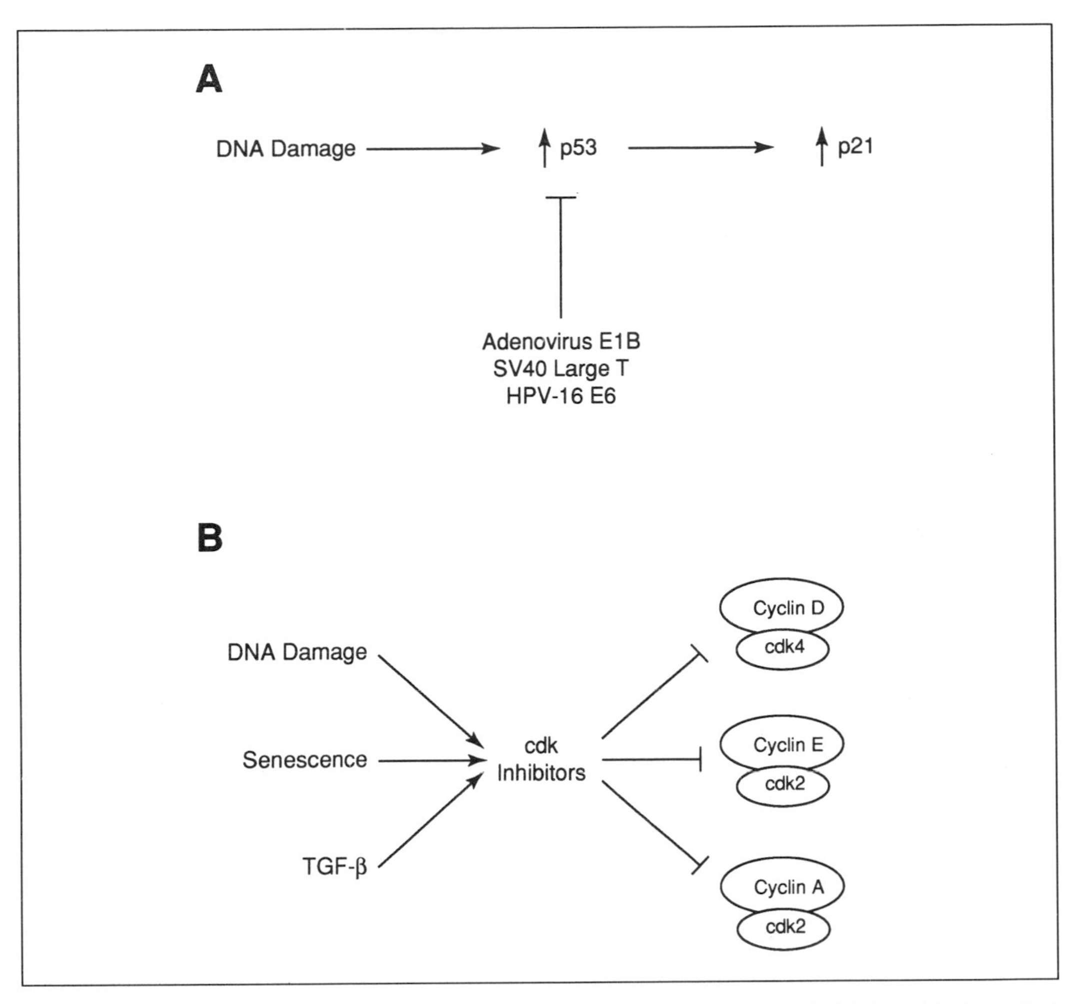

Fig. 4 Induction of cyclin/cdk inhibitors. (A) DNA damage produces elevated levels of p21 via a p53-dependent pathway. (B) Senescent and TGF-β arrested cells have elevated levels of inhibitors. In these cases the mechanism of induction of the inhibitors remains unclear.

can be induced in response to a wide variety of stimuli to execute a cell cycle arrest. Changes in the amount of inhibitors bound to cyclin/cdk complexes has already been described in a number of virally transformed cells (123). Given what we know about p21, one might predict a mechanism whereby in cells expressing it, SV40 large T antigen associates and inactivates p53 thereby reducing the transcription and thus levels of p21. Clearly any change in the expression levels of p21 could have important implications for cell cycle progression. Such changes are currently being sought in a variety of human malignancies. Maintenance of normal levels of such inhibitors will be important for controlled proliferation, and this is thus an attractive target for therapy.

Another unusual aspect of the loss of cell cycle control that is seen in tumourigenesis is the interplay between checkpoint controls and proliferation. Since the cell cycle is regulated by a series of checkpoints, each dependent on the completion of some prior event, it remains unclear how perturbation of one checkpoint can lead to deregulation of the cycle. For example, DNA damage is recognized at multiple stages of the cycle, so loss of any one control point should be compensated by others. Another unexplained conflict must follow loss of cyclin regulation. For cyclin overexpression to drive tumourigenesis, it must be capable of overriding not only proliferation signals but also checkpoint controls that rely on successful completion of one stage of the cycle before proceeding. How the interplay of these interlocking regulatory systems is lost in tumourigenesis is still not understood.

Identification of cell cycle components as potential tumourigenic agents have implications for cancer therapy. Disruption of such pathways must give the cell a proliferative advantage and allow it to grow unchecked by normal monitoring systems. Selective inhibition of such processes in inhibiting kinase activity may irradicate such cells and prevent proliferation. Understanding feedback controls should lead to better treatments for cancer. Most chemotherapeutic agents act by blocking DNA replication, inducing DNA damage, or by interfering with chromosome segregation. The lethal event for such cells is then entry into mitosis. It is possible that by additional treatment with drugs that block feedback controls, thereby minimizing the cells chance to repair the damage (212), one could increase the potency and cytotoxicity of such drugs (212).

The concept of key cell cycle regulatory proteins being directly involved in cancer is still in its infancy. Further studies will undoubtedly provide even more examples of cancers showing alterations in one or more components of the basal cell machinery.

References

1. Nurse, P. and Bissett, Y. (1981) Gene required in G1 for commitment to cell cycle and in G2 for control of mitosis in fission yeast. *Nature*, **292,** 558.
2. Piggott, J. R., Rai, R., and Carter, B. L. (1982) A bifunctional gene product involved in two phases of the yeast cell cycle. *Nature*, **298,** 391.
3. Reed, S. I. and Wittenberg, C. (1990) Mitotic role for the Cdc28 protein kinase of *Saccharomyces cerevisiae*. *Proc. Natl. Acad. Sci. USA*, **87,** 391.
4. Booher, R. N., Alfa, C. E., Hyams, J. S., and Beach, D. H. (1989) The fission yeast cdc2/cdc13/suc1 protein kinase: regulation of catalytic activity and nuclear localization. *Cell*, **58,** 485.
5. Draetta, G., Luca, F., Westendorf, J., Brizuela, L., Ruderman, J., and Beach, D. (1989) Cdc2 protein kinase is complexed with both cyclin A and B: evidence for proteolytic inactivation of MPF. *Cell*, **56,** 829.
6. Moreno, S., Hayles, J., and Nurse, P. (1989) Regulation of p34cdc2 protein kinase during mitosis. *Cell*, **58,** 361.
7. Richardson, H. E., Wittenberg, C., Cross, F., and Reed, S. I. (1989) An essential G1 function for cyclin-like proteins in yeast. *Cell*, **59,** 1127.
8. Wittenberg, C., Sugimoto, K., and Reed, S. I. (1990) G1-specific cyclins of *S. cerevisiae*:

cell cycle periodicity, regulation by mating pheromone, and association with the p34CDC28 protein kinase. *Cell*, **62,** 225.
9. Solomon, M. J., Glotzer, M., Lee, T. H., Phillipe, M., and Kirschner, M. W. (1990) Cyclin activation of p34cdc2. *Cell*, **63,** 1013
10. Cross, F. R. (1988) *DAF1*, a mutant gene affecting size control, pheromone arrest, and cell cycle kinetics of *Saccharomyces cerevisiae*. *Mol. Cell. Biol.*, **8,** 4675.
11. Hadwiger, J. A., Wittenberg, C., Richardson, H. E., de, Barros-Lopes, M., and Reed, S. I. (1989) A family of cyclin homologs that control the G1 phase in yeast. *Proc. Natl. Acad. Sci. USA*, **86,** 6255.
12. Nash, R., Tokiwa, G., Anand, S., Erickson, K., and Futcher, A. B. (1988) The *WHI1*$^+$ gene of *Saccharomyces cerevisiae* tethers cell division to cell size and is a cyclin homolog. *EMBO J.*, **7,** 4335.
13. Surana, U., Robitsch, H., Price, C., Schuster, T., Fitch, I., Futcher, A. B., and Naysmyth, K. (1991) The role of CDC28 and cyclins during mitosis in the budding yeast *S. cerevisiae*. *Cell*, **65,** 145.
14. Epstein, C. B. and Cross, F. R. (1992) CLB5: a novel B cyclin from budding yeast with a role in S phase. *Genes Dev.*, **6,** 1695.
15. Grandin, N. and Reed, S. I. (1993) Differential function and expression of *Saccharomyces cerevisiae* B-type cyclins in mitosis and meiosis. *Mol. Cell. Biol.*, **13,** 2113.
16. Richardson, H., Lew, D. H., Henze, M., Sugimoto, K., and Reed, S. I. (1990) Cyclin-B homologs in *Saccharomyces cerevisiae* function in S phase and in G2. *Mol. Cell. Biol.*, **10,** 6482.
17. Schwob, E. and Nasmyth, K. (1993) CLB5 and CLB6, a new pair of B cyclins involved in DNA replication in *Saccharomyces cerevisiae*. *Genes Dev.*, **7,** 1160.
18. Gould, K. L., Moreno, S., Owen, D. J., Sazer, S., and Nurse, P. (1991) Phosphorylation at Thr167 is required for *Schizosaccharomyces pombe* p34cdc2 function. *EMBO J.*, **10,** 3297.
19. Gould, K. L. and Nurse, P. (1989) Tyrosine phosphorylation of the fission yeast cdc2$^+$ protein kinase regulates entry into mitosis. *Nature*, **342,** 39.
20. Krek, W. and Nigg, E. A. (1991) Mutations of p34cdc2 phosphorylation sites induce premature mitotic events in HeLa cells: evidence for a double block to p34cdc2 kinase activation in vertebrates. *EMBO J.*, **10,** 3331.
21. Krek, W. and Nigg, E. A. (1991) Differential phosphorylation of vertebrate p34cdc2 kinase at the G1/S and G2/M transitions of the cell cycle: identification of major phosphorylation sites. *EMBO J.*, **10,** 305.
22. Norbury, C., Blow, J., and Nurse, P. (1991) Regulatory phosphorylation of the p34cdc2 protein kinase in vertebrates. *EMBO J.*, **10,** 3321.
23. Lee, M. G. and Nurse, P. (1987) Complementation used to clone a human homologue of the fission yeast cell cycle control gene *cdc2*. *Nature*, **327,** 31.
24. Meyerson, M., Enders, G., Wu, C.-L., Su, L.-K., Gorka, C., Nelson, C., Harlow, E., and Tsai, L.-H. (1992) A family of human cdc2-related protein kinases. *EMBO J.*, **11,** 2909.
25. Xiong, Y., Zhang, H., and Beach, D. (1992) D-type cyclins associate with multiple protein kinases and the DNA replication and repair factor PCNA. *Cell*, **71,** 505.
26. Meyerson, M. and Harlow, E. (1994) Identification of G1 kinase activity for cdk6, a novel cyclin D partner. *Mol. Cell. Biol.*, **14,** 2077.
27. Elledge, S. J. and Spottswood, M. R. (1991) A new human p34 protein kinase, CDK2, identified by complementation of a *cdc28* mutation in *Saccharomyces cerevisiae*, is a homolog of *Xenopus* Eg1. *EMBO J.*, **10,** 2653.

28. Matsushime, H., Ewen, M. E., Strom, K. K., Kato, J.-Y., Hanks, S. K., Roussel, M. F., and Sherr, C. J. (1992) Identification and properties of an atypical catalytic subunit (p34^{PSK-J3}/cdk4) for mammalian D-type cyclins. *Cell*, **71**, 323.
29. Ninomiya-Tsuji, J., Nomoto, S., Yasuda, H., Reed, S. I., and Matsumoto, K. (1991) Cloning of a human cDNA encoding a CDC2-related kinase by complementation of a budding yeast cdc28 mutation. *Proc. Natl. Acad. Sci. USA*, **88**, 9006.
30. Tsai, L.-H., Harlow, E., and Meyerson, M. (1991) Isolation of the human *cdk2* gene that encodes the cyclin A- and adenovirus E1A-associated p33 kinase. *Nature*, **353**, 174.
31. Paris, J., Le Guellec, R., Coutrier, A., Le Guellec, K., Omilli, F., Camonis, J., MacNeill, S., and Philippe, M. (1991) Cloning by differential screening of a *Xenopus* cDNA coding for a protein highly homologous to cdc2. *Proc. Natl. Acad. Sci. USA*, **88**, 1039.
32. Pines, J. and Hunter, T. (1989) Isolation of a human cyclin cDNA: evidence for cyclin mRNA and protein regulation in the cell cycle and for interaction with p34cdc2. *Cell*, **58**, 833.
33. Pines, J. and Hunter, T. (1990) Human cyclin A is adenovirus E1A-associated protein p60 and behaves differently from cyclin B. *Nature*, **346**, 760.
34. Lew, D. J., Dulic, V., and Reed, S. I. (1991) Isolation of three novel human cyclins by rescue of G1 cyclin (Cln) function in yeast. *Cell*, **66**, 1197.
35. Koff, A., Cross, F., Fisher, A., Schumacher, J., Leguellec, K., Philippe, M., and Roberts, J. M. (1991) Human cyclin E, a new cyclin that interacts with two members of the CDC2 gene family. *Cell*, **66**, 1217.
36. Motokura, T., Bloom, T., Kim, H. G., Juppner, H., Ruderman, J. V., Kronenberg, H. M., and Arnold, A. (1991) A novel cyclin encoded by a bcl1-linked candidate oncogene. *Nature*, **350**, 512.
37. Motokura, T., Keyomarsi, K., Kronenberg, H. M., and Arnold, A. (1992) Cloning and characterisation of human cyclin D3, a cDNA closely related in sequence to the PRAD-1/cyclin D1 proto-oncogene. *J. Biol. Chem.*, **267**, 20412.
38. Matasushime, H., Roussel, M. F., Ashmun, R. A., and Sherr, C. J. (1991) Colony-stimulating factor 1 regulates novel cyclins during the G1 phase of the cell cycle. *Cell*, **65**, 701.
39. Tamura, K., Kanaoka, Y., Jinno, S., Nagata, A., Ogiso, Y., Shimizu, K., Hayakawa, T., Nojima, H., and Okayama, H. (1993) Cyclin G: a new mammalian cyclin with homology to fission yeast Cig1. *Oncogene*, **8**, 2113.
40. Ducommun, B., Brambilla, P., Felix, M. A., Franza, B. J., Karsenti, E., and Draetta, G. (1991) cdc2 phosphorylation is required for its interaction with cyclin. *EMBO J.*, **10**, 3311.
41. Hunter. T. (1993) Braking the cycle. *Cell*, **75**, 839.
42. Harper, J. W., Adami, G. R., Wei, N., Keyomarsi, K., and Elledge, S. J. (1993) The p21 Cdk-interating protein Cip1 is a potent inhibitor of G1 cyclin-dependent kinases. *Cell*, **75**, 805.
43. Xiong, Y., Hannon, G. J., Zhang, H., Casso, D., Kobayashi, R., and Beach, D. (1993) p21 is a universal inhibitor of cyclin kinases. *Nature*, **366**, 701.
44. Serrano, M., Hannon, G. J., and Beach, D. (1993) A new regulatory motif in cell-cycle control causing specific inhibition of cyclin D/CDK4. *Nature*, **366**, 704.
45. El-Deiry, W. S., Tokino, T., Velculescu, V. E., Levy, D. B., Parsons, R., Trent, J. M., Lin, D., Mercer, W. E., Kinzler, K. W., and Vogelstein, B. (1993) *WAF1*, a potential mediator of p53 tumor suppression. *Cell*, **75**, 817.

46. Polyak, K., Kato, J.-Y., Solomon, M., Sherr, C. J., Massague, J., Roberts, J. M., and Koff, A. (1994) p27Kip1, a cyclin-Cdk inhibitor, links transforming growth factor-b and contact inhibition to cell cycle arrest. *Genes Dev.*, **8,** 9.
47. Hengst, L., Dulic, V., Slingerland, J. M., Lees, E., and Reed, S. I. (1994) A cell cycle regulated inhibitor of cyclin dependent kinases. *Proc. Natl. Acad. Sci. USA*, **91,** 5291.
48. Ohtsubo, M. and Roberts, J. M. (1993) Cyclin-dependent regulation of G1 in mammalian fibroblasts. *Science*, **259,** 1908.
49. Quelle, D. E., Ashmun, R. A., Shurtleff, S. A., Kato, J.-Y., Bar-Sagi, D., Roussel, M. F., and Sherr, C. J. (1993) Overexpression of mouse D-type cyclins accelerates G1 phase in rodent fibroblasts. *Genes Dev.*, **7,** 1559.
50. Resnitsky, D., Stanners, M., and Reed, S. I. (1994) Acceleration of the G1/S phase transition by conditional overexpression of cyclins D1 and E with an inducible system. *Mol. Cell. Biol.*, **14,** 1669.
51. Russell, P. and Nurse, P. (1986) cdc25^{+} functions as an inducer in the mitotic control of fission yeast. *Cell*, **45,** 145.
52. Russell, P. and Nurse, P. (1987) Negative regulation of mitosis by *wee1*$^{+}$, a gene encoding a protein kinase homolog. *Cell*, **49,** 559.
53. Russell, P. and Nurse, P. (1987) The mitotic inducer *nim1*$^{+}$ functions in a regulatory network of protein kinase homologs controlling the initiation of mitosis. *Cell*, **49,** 569.
54. Lundgren, K., Walworth, N., Booher, R., Dembski, M., Kirschner, M., and Beach, D. (1991) mik1 and wee1 cooperate in the inhibitory tyrosine phosphorylation of cdc2. *Cell*, **64,** 1111.
55. Featherstone, C. and Russell, P. (1991) Fission yeast p107wee1 mitotic inhibitor is a tyrosine/serine kinase. *Nature*, **349**, 808.
56. Parker, L. L., Atherton, F. S., Lee, M. S., Ogg, S., Falk, J. L., Swenson, K. I., and Piwnica, W. H. (1991) Cyclin promotes the tyrosine phosphorylation of p34cdc2 in a *wee1*$^{+}$ dependent manner. *EMBO J.*, **10,** 1255.
57. Chang, F. and Herskowitz, I. (1990) Identification of a gene necessary for cell cycle arrest by a negative growth factor of yeast: FAR1 is an inhibitor of a G1 cyclin, CLN2. *Cell*, **63,** 999.
58. Peter, M., Gartner, A., Horecka, J., Ammerer, G., and Herskowitz, I. (1993) FAR1 links the signal transduction pathway to the cell cycle machinery in yeast. *Cell*, **73,** 747.
59. Elion, E. A., Grisafi, P. L., and Fink, G. R. (1990) FUS3 encodes a *cdc2*$^{+}$/CDC28-related kinase required for the transition from mitosis into conjugation. *Cell*, **60,** 649.
60. Enoch, T. and Nurse, P. (1991) Coupling M phase and S phase: controls maintaining the dependence of mitosis on chromosome replication. *Cell*, **65,** 921.
61. Pardee, A. (1974) A restriction point for control of normal animal cell proliferation. *Proc. Natl. Acad. Sci. USA*, **71,** 1286.
62. Hartwell, L. H. (1974) *Saccharomyces cerevisiae* cell cycle. *Bacteriol. Rev.*, **38,** 164.
63. Sherr, C. J. (1993) Mammalian G1 Cyclins. *Cell*, **73,** 1059.
64. Murray, A. W., Solomon, M. J., and Kirschner, M. W. (1989) The role of cyclin synthesis and degradation in the control of maturation promoting factor activity. *Nature*, **339,** 280.
65. Murray, A. W. and Kirschner, M. W. (1989) Cyclin synthesis drives the early embryonic cell cycle. *Nature*, **339,** 275.
66. Xiong, Y., Connolly, T., Futcher, B., and Beach, D. (1991) Human D-type cyclin. *Cell*, **65,** 691.
67. Rosenberg, C. L., Kim, H. G., Shows, T. B., Kronenberg, H. M., and Arnold, A.

(1991) Rearrangement and overexpression of D11S-287E, a candidate oncogene on chromosome 11q13 in benign parathyroid tumors. *Oncogene*, **6**, 449.
68. Adelaide, J., Mattei, M. G., Marics, I., Raybaud, F., Planche, J., De Lapeyriere, O., and Birnbaum, D. (1988) Chromosomal localization of the hst oncogene and its co-amplification with the int-2 oncogene in a human melanoma. *Oncogene*, **2**, 413.
69. Yoshida, M. C., Wada, M., Satoh, H., Yoshida, T., Sakamoto, H., Miyagawa, K., Yokota, J., Koda, T., Kakinuma, M., Sugimura, T., and Terada, M. (1988) Human HST1 (HSTF1) gene maps to chromosome band 11q13 and coamplifies with the INT2 gene in human cancer. *Proc. Natl. Acad. Sci. USA*, **85**, 4861.
70. Varley, J. M., Walker, R. A., Casey, G., and Brammar, W. J. (1988) A common alteration to the int-2 proto-oncogene in DNA from primary breast tumors. *Oncogene*, **3**, 87.
71. Lammie, G. A., Fantl, V., Smith, R., Schuuring, E., Brookes, S., Michalides, R., Dickson, C., Arnold, A., and Peters, G. (1991) D11S287, a putative oncogene on chromosome 11q13, is amplified and expressed in squamous cell and mammary carcinomas and linked to BCL-1. *Oncogene*, **6**, 439.
72. Jiang, W., Kahan, S., Tomita, N., Zhang, Y., Lu, S., and Weinstein, B. (1992) Amplification and expression of the human cyclin D gene in esophageal cancer. *Cancer Res.*, **52**, 2980.
73. Korsmeyer, S. J. (1992) Chromosomal translocations in lymphoid malignancies reveal novel proto-oncogenes. *Annu. Rev. Immunol.* **10**, 785.
74. Erikson, J., Finan, J., Tsujimoto, Y., Nowell, P. C., and Croce, C. M. (1984) The chromosome 14 breakpoint in neoplastic B cells with the t(11;14) translocation involves the immunoglobulin heavy chain locus. *Proc. Natl. Acad. Sci. USA*, **81**, 4144.
75. Tsujimoto, Y., Jaffe, E., Cossman, J., Gorham, J., Nowell, P. C., and Croce, C. M. (1985) Clustering of breakpoints on chromosome 11 in human B-cell neoplasms with the t(11;14) chromosome translocation. *Nature*, **315**, 340.
76. Koduru, P. R., Offit, K., and Filippa, D. A., (1989) Molecular analysis of breaks in BCL-1 proto-oncogene in B-cell lymphomas with abnormalities of 11q13. *Oncogene*, **4**, 929.
77. Williams, M. E., Swerdlow, S. H., Rosenberg, C. L., and Arnold, A. (1992) Characterisation of chromosome 11 translocation breakpoints at the bcl-1 and PRAD1 loci in centrocytic lymphoma. *Cancer Res.*, **52**, 554.
78. Williams, M. E., Swerdlow, S. H., Rosenberg, C. L., and Arnold, A. (1993) Chromosome 11 translocation breakpoints at the PRAD1/cyclin D1 gene locus in centrocytic lymphoma. *Leukaemia*, **7**, 241.
79. Withers, D. A., Harvey, R. C., Faust, J. B., Melnyk, O., Carey, K., and Meeker, T. C. (1991) Characterisation of a candidate bcl-1 gene. *Mol. Cell. Biol.*, **11**, 4846.
80. Lammie, G. A., Smith, R., Silver, J., Brookes, S., Dickson, C., and Peters, G. (1992) Proviral insertions near cyclin D1 in mouse lymphomas: a parallel for BCL1 translocations in human B-cell neoplasms. *Oncogene*, **7**, 2381.
81. Ando, K., Ajchenbaum-Cymbalista, F., and Griffin, J. D. (1993) Regulation of G1/S transition by cyclins D2 and D3 in hematopoietic cells. *Proc. Natl. Acad. Sci. USA*, **90**, 9571.
82. Kato, J.-Y. and Sherr, C. J. (1993) Inhibition of granulocyte differentiation by G1 cyclins, D2 and D3, but not D1. *Proc. Natl. Acad. Sci. USA*, **90**, 11513.
83. Inaba, T., Matsushime, H., Valentine, M., Roussel, M. F., and Sherr, C. J. (1992) Genomic organization, chromosomal localization, and independent expression of human cyclin D genes. *Genomics*, **13**, 565.

84. Squire, J., Phillips, R. A., Boyce, S., Godbout, R., Rogers, B., and Gallie, B. (1984) Isochromosome 6p, a unique chromosomal abnormality in retinoblastoma: verification by standard staining techniques, new densitometric methods, and somatic cell hybridization. *Human Genet.*, **66,** 46.
85. Kusnetsova, L. E., Prigogina, E. L., Pogosianz, H. E., and Belkina, B. M. (1982) Similar chromosomal abnormalities in several retinoblastomas. *Human Genetics*, **61,** 201.
86. Benedict, W. F., Banerjee, A., Mark, C., and Murphree, A. L. (1983) Nonrandom chromosomal changes in untreated retinoblastomas. *Cancer, Genet. Cytogenet.*, **10,** 311.
87. Schouten, H. C., Sanger, W. G., Weisenburger, D. D., and Armitage, J. O. (1990) Abnormalities involving chromosome 6 in newly diagnosed patients with non-Hodgkins lymphoma. *Cancer, Genet. Cytogenet.*, **47,** 73.
88. Prigogina, E. L., Puchkova, G. P., and Mayakova, S. A. (1988) Non-random chromosomal abnormalities in acute lymphoblastic leukaemia of childhood. *Cancer, Genet. Cytogenet.*, **32,** 183.
89. Tremblay, P. J. (1992) Identification of a novel gene, Vin-1, in murine leukaemia virus-induced T-cell leukaemias by provirus insertional mutagenesis. *J. Virol.*, **66,** 1344.
90. Wang, T. C., Cardiff, R. D., Zukerberg, L., Lees, E., Arnold, A., and Schmidt, E. (1994) Mammary hyperplasia and carcinoma in MMTV-cyclin D1 transgenic mice. *Nature*, **369,** 669.
91. Hinds, P. W., Dowdy, S. F., Eaton, E. N., Arnold, A., and Weinberg, R. A. (1994) Function of a human cyclin gene as an oncogene. *Proc. Natl. Acad. Sci. USA*, **91,** 709.
92. Baldin, V., Lukas, J., Marcote, M. J., Pagano, M., and Draetta, G. (1993) Cyclin D1 is a protein required for cell cycle progression in G1. *Genes Dev.*, **7,** 812.
93. Hinds, P. W., Mittnacht, S., Dulic, V., Arnold, A., Reed, S. I., and Weinberg, R. A. (1992) Regulations of retinoblastoma protein functions by ectopic expression of human cyclins. *Cell*, **70,** 993.
94. Giordano, A., Whyte, P., Harlow, E., Franza, B. J., Beach, D., and Draetta, G. (1989) A 60 kd cdc2-associated polypeptide complexes with the E1A proteins in adenovirus-infected cells. *Cell*, **58,** 981.
95. Pagano, M., Pepperkok, R., Verde, F., Ansorge, W., and Draetta, G. (1992) Cyclin A is required at two points in the human cell cycle. *EMBO J.*, **11,** 961.
96. Girard, F., Strausfeld, U., Fernandez, A., and Lamb, N. (1991) Cyclin A is required for the onset of DNA replication in mammalian fibroblasts, *Cell*, **67,** 1169.
97. Lehner, C. F. and O'Farrell, P. H. (1990) The roles of *Drosophila* cyclins A and B in mitotic control. *Cell*, **61,** 535.
98. Walker, D. and Maller, J. L. (1991) Role for cyclin A in the dependence of mitosis on completion of DNA replication. *Nature*, **354,** 314.
99. Faha, B., Meyerson, M., and Tsai, L.-H. (1992) Adenovirus E1A targets one of two cyclin A-kinase complexes. In *Pezcoller foundation symposium*. Livingston, D. and Milich, E. (ed.) p. 107.
100. Wang, J., Chenivesse, X., Henglein, B., and Brechot, C. (1990) Hepatitis B virus integration in a cyclin A gene in a hepatocellular carcinoma. *Nature*, **343,** 555.
101. Wang, J., Zindy, F., Chenivesse, X., Lamas, E., Henglein, B., and Brechot, C. (1992) Modification of cyclin A expression by hepatitis B virus DNA integration in a hepatocellular carcinoma. *Oncogene*, **7,** 1653.
102. Harlow, E., Whyte, P., Franza, B. J., and Schley, C. (1986) Association of adenovirus early-region 1A proteins with cellular polypeptides. *Mol. Cell. Biol.*, **6,** 1579.

103. Yee, S. P. and Branton, P. E. (1985) Detection of cellular proteins associated with human adenovirus type 5 early region 1A polypeptides. *Virology*, **147**, 142.
104. Whyte, P., Buchkovich, K. J., Horowitz, J. M., Friend, S. H., Raybuck, M., Weinberg, R. A., and Harlow, E. (1988) Association between an oncogene and an anti-oncogene: the adenovirus E1A proteins bind to the retinoblastoma gene product. *Nature*, **334**, 124.
105. Ewen, M. E., Xing, Y., Lawrence, J. B., and Livingston, D. M. (1991) Molecular cloning, chromosonal mapping, and expression of the cDNA for p107, a retinoblastoma gene product-related protein. *Cell*, **66**, 1155.
106. Whyte, P., Ruley, H. E., and Harlow, E. (1988) Two regions of the adenovirus early region 1A proteins are required for transformation. *J. Virol.*, **62**, 257.
107. Giordano, A., McCall, C., Whyte, P., and Franza, B. R. (1991) Human cyclin A and the retinoblastoma protein interact with similar but distinguishable sequences in the adenovirus E1A gene product. *Oncogene*, **6**, 481.
108. Kleinberger, T. and Shenk, T. (1991) A protein kinase is present in a complex with adenovirus E1A proteins. *Proc. Natl. Acad. Sci. USA*, **88**, 11143.
109. Faha, B. F., Harlow, E., and Lees, E. M. (1993) The adenovirus E1A-associated kinase consists of cyclin E/p33cdk2 and cyclin A/p33cdk2. *J. Virol.*, **67**, 2456.
110. Tommasino, M., Adamczewski, J. P., Carlotti, F., Barth, C. F., Manetti, R., Contorni, M., Cavalieri, F., Hunt, T., and Crawford, L. V. (1993) HPV16 E7 protein associates with the protein kinase p33cdk2 and cyclin A. *Oncogene*, **8**, 195.
111. Guadagno, T. M., Ohtsubo, M., Roberts, J. M., and Assoian, R. K. (1993) A link between cyclin A expression and adhesion-dependent cell cycle progression. *Science*, **262**, 1572.
112. Koff, A., Giordano, A., Desai, D., Yamashita, K., Harper, W., Elledge, S., Nishimoto, T., Morgan, D., Franza, R., and Roberts, J. (1992) Formation and activation of a cyclin E–CDK2 complex during the G1 phase of the human cell cycle. *Science*, **257**, 1689.
113. Dulic, V., Lees, E., and Reed, S. I. (1992) Association of Human cyclin E with a period G1–S phase protein kinase. *Science*, **257**, 1958.
114. Tsai, L.-H., Lees, E., Faha, B., Harlow, E., and Riabowol, K. (1993) The cdk2 kinase is required for the G1-to-S transition in mammalian cells. *Oncogene*, **8**, 1593.
115. Howe, P. H., Draetta, G., and Leof, E. B. (1991) Transforming growth factor beta 1 inhibition of p34cdc2 phosphorylation and histone H1 kinase activity is associated with G1/S-phase growth arrest. *Mol. Cell. Biol.*, **11**, 1185.
116. Laiho, M., DeCaprio, J. A., Ludlow, J. W., Livingston, D. M., and Massague, J. (1990) Growth inhibition by TGF-beta linked to suppression of retinoblastoma protein phosphorylation. *Cell*, **62**, 175.
117. Dulic, V., Kaufmann, W. K., Wilson, S. J., Tlsty, T. D., Lees, E., Harper, W., Elledge, S., and Reed, S. I. (1994) p53-dependent inhibition of cyclin-dependent kinase activities in human fibroblasts during radiation-induced G1 arrest. *Cell*, **76**, 1013.
118. Dulic, V., Drullinger, L. F., Lees, E., Reed, S. I., and Stein, G. H. (1993) Altered regulation of G1 cyclins in senescent human diploid fibroblasts: accumulation of inactive cyclin D1/cdk2 and cyclin E/ckd2 complexes. *Proc. Natl. Acad. Sci. USA*, **90**, 11034.
119. Keyomarsi, K. and Pardee, A. B. (1993) Redundant cyclin overexpression and gene amplification in breast cancer cells. *Proc. Natl. Acad. Sci. USA*, **90**, 1112.
120. Khatib, Z. A., Matsushime, H., Valentine, M., Shapiro, D. N., Sherr, C. J., and Look, A. T. (1993) Coamplification of the CDK4 gene with MDM2 and GLI in human sarcomas. *Cancer Res.*, **53**, 5535.

121. Kumagai, A. and Dunphy, W. G. (1991) The cdc25 protein controls tyrosine dephosphorylation of the cdc2 protein in a cell-free system. *Cell*, **64,** 903.
122. Enoch, T. and Nurse, P. (1990) Mutation of fission yeast cell cycle control gene abolishes dependence of mitosis on DNA replication. *Cell*, **60,** 665.
123. Xiong, Y., Zhang, H., and Beach, D. (1993) Subunit rearrangement of the cyclin-dependent kinases is associated with cellular transformation. *Genes Dev.*, **7,** 1572.
124. Weinberg, R. A. (1992) The retinoblastoma gene and gene product. In *Tumour suppressor genes, the cell cycle and cancer*. Levine, A. J. (ed.). Cold Spring Harbor Laboratory Press, New York, p. 43.
125. Huang, H. J., Yee, J. K., Shew, J. Y., Chen, P. L., Bookstein, R., Friedmann, T., Lee, E. Y., and Lee, W. H. (1988) Suppression of the neoplastic phenotype by replacement of the RB gene in human cancer cells. *Science*, **242,** 1563.
126. Bookstein, R., Shew, J. Y., Chen, P. L., Scully, P., and Lee, W. H. (1990) Suppression of tumorigenicity of human prostate carcinoma cells by replacing a mutated RB gene. *Science*, **247,** 712.
127. Sumegi, J., Uzvolgyi, E., and Klein, G. (1990) Expression of the RB gene under the control of MuLV-LTR suppresses tumorigenicity of WERI-Rb-27 retinoblastoma cells in immunodefective mice. *Cell Growth Differ.*, **1,** 247.
128. Buchkovich, K., Duffy, L. A., and Harlow, E. (1989) The retinoblastoma protein is phosphorylated during specific phases of the cell cycle. *Cell*, **58,** 1097.
129. Chen, P.-L., Scully, P., Shew, J.-Y., Wang, J., and Lee, W.-H. (1989) Phosphorylation of the retinoblastoma gene product is modulated during the cell cycle and cellular differentiation. *Cell*, **58,** 1193.
130. DeCaprio, J. A., Ludlow, J. W., Lynch, D., Furukawa, Y., Griffin, J., Piwnica-Worms, H., Huang, C. M., and Livingston, D. M. (1989) The product of the retinoblastoma susceptibility gene has properties of a cell cycle regulatory element. *Cell*, **58,** 1085.
131. Mihara, K., Cao, X. R., Yen, A., Chandler, S., Driscoll, B., Murphree, A. L., T'Ang, A., and Fung, Y. K. (1989) Cell cycle-dependent regulation of phosphorylation of the human retinoblastoma gene product. *Science*, **246,** 1300.
132. DeCaprio, J. A., Furukawa, Y., Ajchenbaum, F., Griffin, J. D., and Livingston, D. (1992) The retinoblastoma-susceptibility gene product becomes phosphorylated in multiple stages during cell cycle entry and progression. *Proc. Natl. Acad. Sci. USA*, **89,** 1795.
133. Ludlow, J. W., DeCaprio, J. A., Huang, C. M., Lee, W. H., Paucha, E., and Livingston, D. M. (1989) SV40 large T antigens binds preferentially to an underphosphorylated member of the retinoblastoma susceptibility gene product family. *Cell*, **56,** 57.
134. Langan, T. A., Gautier, J., Lohka, M., Hollingsworth, R., Moreno, S., Nurse, P., Maller, J., and Sclafani, R. A. (1989) Mammalian growth-associated H1 histone kinase: a homolog of *cdc2*$^{+}$/CDC28 protein kinases controlling mitotic entry in yeast and frog cells. *Mol. Cell. Biol.*, **9,** 3860.
135. Taya, Y., Yasuda, H., Kamijo, M., Nakaya, K., Nakamura, Y., Ohba, Y., and Nishimura, S. (1989) *In vitro* phosphorylation of the tumor suppressor gene RB protein by mitosis-specific histone H1 kinase. *Biochem. Biophys. Res. Commun.*, **164,** 580.
136. Lees, J. A., Buchkovich, K. J., Marshak, D. R., Anderson, C. W., and Harlow, E. (1991) The retinoblastoma protein is phosphorylated on multiple sites by human cdc2. *EMBO J.*, **10,** 4279.
137. Lin, B. T., Gruenwald, S., Morla, A. O., Lee, W. H., and Wang, J. Y. (1991) Retinoblastoma cancer suppressor gene product is a substrate of the cell cycle regulator cdc2 kinase. *EMBO J.*, **10,** 857.

138. Akiyama, T., Ohuchi, T., Sumida, S., Matsumoto, K., and Toyoshima, K. (1992) Phosphorylation of the retinoblastoma protein by cdk2. *Proc. Natl. Acad. Sci. USA*, **89,** 7900.
139. Kato, J.-Y., Matsushime, H., Hiebert, S. W., Ewen, M. E., and Sherr, C. J. (1993) Direct binding of cyclin D to the retinoblastoma gene product (pRb) and pRb phosphorylation by the cyclin D dependent kinase CDK4. *Genes Dev.*, **7,** 331.
140. Matsushime, H., Quelle, D. E., Shurtleff, S. A., Shibuya, M., Sherr, C. J., and Kato, J.-Y. (1994) D-type cyclin-dependent kinase activity in mammalian cells. *Mol. Cell. Biol.*, **14,** 2066.
141. Ewen, M. E., Sluss, H. K., Sherr, C. J., Matsushime, H., Kato, J., and Livingston, D. M. (1993) Functional interactions of the retinoblastoma protein with mammalian D-type cyclins. *Cell*, **73,** 487.
142. Dowdy, S. F., Hinds, P. W., Louie, K., Reed, S. I., Arnold, A., and Weinberg, R. A. (1993) Physical interaction of the retinoblastoma protein with human D cyclins. *Cell*, **73,** 499.
143. Lane, D. P. (1992) p53, guardian of the genome. *Nature*, **358,** 15.
144. Braithwaite, A. W., Sturzbecher, H.-W., Addison, C., Palmer, C., Rudge, K., and Jenkins, J. R. (1987) Mouse p53 inhibits SV40 origin-dependent DNA replication. *Nature*, **329,** 458.
145. Sturzbecher, J.-W., Brain, R., Maimets, T., Addison, C., Rudge, K., and Jenkins, J. R. (1988) Mouse p53 blocks SV40 DNA replication *in vitro* and downregulates T antigen DNA helicase activity. *Oncogene*, **3,** 405.
146. Wang, E. H., Friedman, P. N., and Prives, C. (1989) The murine p53 protein blocks replication of SV40 DNA *in vitro* by inhibiting the initiation functions of SV40 large T antigen. *Cell*, **57,** 379.
147. Dutta, A., Ruppert, J., Aster, J., and Winchester, E. (1993) Inhibition of DNA replication factor RPA by p53. *Nature*, **365,** 79.
148. Addison, C., Jenkins, J. R., and Sturzbecher, H.-W. (1990) The p53 nuclear localisation signal is structurally linked to a p34cdc2 kinase motif. *Oncogene*, **5,** 423.
149. Bischoff, J. R., Friedman, P. N., Marshak, D. R., Prives, C., and Beach, D. (1990) Human p53 is phosphorylated by p60-cdc2 and cyclin B-cdc2. *Proc. Natl. Acad. Sci. USA*, **87,** 4766.
150. Samad, A., Anderson, C. W., and Carroll, R. B. (1986) Mapping of phosphomonoester and apparent phosphodiester bonds of the oncogene product p53 from simian virus 40-transformed 3T3 cells. *Proc. Natl. Acad. Sci. USA*, **83,** 897.
151. Meek, D. W. and Eckhart, W. (1988) Phosphorylation of p53 in normal and simian virus 40-transformed NIH 3T3 cells. *Mol. Cell. Biol.*, **8,** 461.
152. Cooper, J. A., Gould, K. L., Cartwright, O. A., and Hunter, T. (1986) Tyr 527 is phosphorylated in pp60c-src: implications for regulation. *Science*, **23,** 1431.
153. Nada, S., Okada, M., MacAuley, A., Cooper, J. A., and Nakagawa, H. (1991) Cloning of a complementary DNA for a protein-tyrosine kinase that specifically phosphorylates a negative regulatory site of p60c-src. *Nature*, **351,** 69.
154. Chackalaparampil, I. and Shalloway, D. (1988) Altered phosphorylation and activation of pp60c-src during fibroblast mitosis. *Cell*, **52,** 801.
155. Morgan, D. O., Kaplan, J. M., Bishop, J. M., and Varmus, H. E. (1989) Mitosis-specific phosphorylation of p60c-src by p34cdc2-associated protein kinase. *Cell*, **57,** 775.
156. Shenoy, S., Choi, J. K., Bagrodia, S., Copeland, T. D., Maller, J. L., and Shalloway, D. (1989) Purified maturation promoting factor phosphorylates pp60c-src at the sites phosphorylated during fibroblast mitosis. *Cell*, **57,** 763.

157. Shenoy, S., Chackalaparampil, I., Bagrodia, S., Lin, P.-H., and Shalloway, D. (1992) Role of p34cdc2-mediated phosphorylations in two-step activation of pp60c-src during mitosis. *Proc. Natl. Acad. Sci. USA*, **89,** 7237.
158. Shalloway, D. and Shenoy, S. (1991) Oncoprotein kinase in mitosis. *Adv. Cancer Res.*, **57,** 185.
159. Luscher, B. and Eisenman, R. N. (1992) Mitosis-specific phosphorylation of the nuclear oncoproteins myc and myb. *J. Cell Biol.*, **118,** 775.
160. Kovesdi, I., Reichel, R., and Nevins, J. R. (1986) Identification of a cellular transcription factor involved in E1A trans-activation. *Cell*, **45,** 219.
161. Yee, A. S., Raychaudhuri, P., Jakoi, L., and Nevins, J. R. (1989) The adenovirus-inducible factor E2F stimulates transcription after specific DNA binding. *Mol. Cell. Biol.*, **9,** 578.
162. Nevins, J. R. (1992) E2F; a link between the Rb tumor suppressor protein and viral oncoproteins. *Science*, **258,** 424.
163. Hiebert, S. W., Lipp, M., and Nevins, J. R. (1989) E1A-dependent trans-activation of the human MYC promoter is mediated by the E2F factor. *Proc. Natl. Acad. Sci. USA*, **86,** 3594.
164. Blake, M. C. and Azizkhan, J. C. (1989) Transcription factor E2F is required for efficient expression of the hamster dihydrofolate reductase gene *in vitro* and *in vivo*. *Mol. Cell. Biol.*, **9,** 4994.
165. Hiebert, S. W., Blake, M., Azizkhan, J., and Nevins, J. R. (1991) Role of E2F transcription factor in E1A-mediated trans-activation of cellular genes. *J. Virol.*, **65,** 3547.
166. Mudryj, M., Hiebert, S. W., and Nevins, J. R. (1990) A role for the adenovirus inducible E2F transcription factor in a proliferation dependent signal transduction pathway. *EMBO J.*, **9,** 2179.
167. Hiebert, S. W., Chellappan, S. P., Horowitz, J. M., and Nevins, J. R. (1992) The interaction of pRb with E2F inhibits the transcriptional activity of E2F. *Genes Dev.*, **6,** 177.
168. Weintraub, S. J., Prater, C. A., and Dean, D. C. (1992) Retinoblastoma protein switches the E2F site from positive to negative element. *Nature*, **358,** 259.
169. Chellappan, S., Hiebert, S., Mudryj, M., Horowitz, J., and Nevins, J. (1991) The E2F transcription factor is a cellular target for the RB protein. *Cell*, **65,** 1053.
170. Bandara, L. R., Adamczewski, J. P., Hunt, T., and La Thangue, N. (1991) Cyclin A and the retinoblastoma gene product complex with a common transcription factor. *Nature*, **352,** 249.
171. Chittenden, T., Livingston, D., and Kaelin, W. (1991) The T/E1A-binding domain of the retinoblastoma product can interact selectively with a sequence-specific DNA-binding protein. *Cell*, **65,** 1073.
172. Raychaudhuri, P., Bagchi, S., Devoto, S. H., Kraus, V. B., Moran, E., and Nevins, J. R. (1991) Domains of the adenovirus E1A protein that are required for oncogenic activity are also required for dissociation of E2F transcription factor complexes. *Genes Dev.*, **5,** 1200.
173. Helin, K., Lees, J. A., Vidal, M., Dyson, N., Harlow, E., and Fattaey, A. (1992) A cDNA encoding a pRB-binding protein with properties of the transcription factor E2F. *Cell*, **70,** 337.
174. Kaelin, W. G., Krek, W., Sellers, W. R., DeCaprio, J. A., Ajchanbaum, F., Fuchs, C. S., Chittenden, T., Li, Y., Farnham, P. J., Blanar, M. A., Livingston, D. M., and Flemington, E. K. (1992) Expression cloning of a cDNA encoding a retinoblastoma-binding protein with E2F-like properties. *Cell*, **70,** 351.

175. Helin, K., Harlow, E., and Fattaey, A. R. (1993) Inhibition of E2F-1 transactivation by direct binding of the retinoblastoma protein. *Mol. Cell. Biol.*, **13,** 6501.
176. Mudryj, M., Devoto, S. H., Hiebert, S. W., Hunter, T., Pines, J., and Nevins, J. R. (1991) Cell cycle regulation of the E2F transcription factor involves an interaction with cyclin A. *Cell*, **65,** 1243.
177. Bagchi, S., Raychaudhuri, P., and Nevins, J. R. (1989) Phosphorylation-dependent activation of the adenovirus-inducible E2F factor in a cell-free system. *Proc. Natl. Acad. Sci. USA*, **86,** 4352.
178. Wang, C.-J., Petryniak, B., Thompson, C. B., Kaelin, W. G., and Leiden, J. M. (1993) Regulation of the Ets-related transcription factor Elf-1 by binding to the retinoblastoma protein. *Science*, **260,** 1130.
179. Kim, S.-J., Wagner, S., Liu, F., O'Reilly, M. A., Robbins, P. D., and Green, M. R. (1992) Retinoblastoma gene product activates expression of the human TGF-β2 gene through transcription factor ATF-2. *Nature*, **358,** 331.
180. Kim, S.-J., Onwuta, U. S., Lee, Y. I., Li, R., Botchan, M. R., and Robbins, P. D. (1992) The retinoblastoma gene product regulates Sp-1 mediated transcription. *Mol. Cell. Biol.*, **12,** 2455.
181. Udvadia, A. J., Rogers, K. T., Higgins, P. D., Murata, Y., Martin, K. H., Humphrey, P. A., and Horowitz, J. M. (1993) Sp-1 binds promoter elements regulated by the Rb protein and Sp-1 mediated transcription is stimulated by RB coexpression. *Proc. Natl. Acad. Sci. USA*, **90,** 3265.
182. Gu, Y., Turck, G. W., and Morgan, D. O. (1993) Inhibition of CDK2 activity *in vivo* by an associated 20K regulatory subunit. *Nature*, **366,** 707.
183. Cisek, L. J. and Corden, J. L. (1989) Phosphorylation of RNA polymerase by the murine homologue of the cell cycle control protein cdc2. *Nature*, **339,** 679.
184. Payne, J. M. and Dahmus, M. E. (1993) Partial purification and characterization of two distinct protein kinases that differentially phosphorylate the carboxy-terminal domain of RNA polymerase subunit IIA. *J. Biol. Chem.*, **268,** 80.
185. Dutta, A. and Stillman, B. (1992) cdc2 family kinases phosphorylate a human cell DNA replication factor, RPA, and activate DNA replication. *EMBO J.*, **11,** 2189.
186. Fotedar, R. and Roberts, J. M. (1992) Cell cycle regulated phosphorylation of RPA-32 occurs within the replication initiation complex. *EMBO J.*, **11,** 2177.
187. Murray, A. (1992) Creative blocks: cell-cycle checkpoints and feedback controls. *Nature*, **359,** 599.
188. Hartwell, L. (1992) Defects in a cell cycle checkpoint may be responsible for the genetic instability of cancer cells. *Cell*, **71,** 543.
189. Vogelstein, B., Fearon, E. R., Scott, E. K., Hamilton, S. R., Preisinger, A. C., Nakamura, Y., and White, R. (1989) Allelotype of colorectal carcinomas. *Science*, **244,** 207.
190. Weinert, T. A. and Hartwell, L. H. (1988) The *RAD9* gene controls the cell cycle response to DNA damage in *Saccharomyces cerevisiae*. *Science*, **241,** 317.
191. Kastan, M. B., Onyekwere, O., Sidransky, D., Vogelstein, B., and Craig, R. W. (1991) Participation of p53 in the cellular response to DNA damage. *Cancer Res.*, **51,** 6304.
192. Kuerbitz, S. J., Plunkett, B. S., Walsh, W. V., and Kastan, M. B. (1992) Wild-type p53 is a cell cycle checkpoint determinant following irradiation. *Proc. Natl. Acad. Sci. USA*, **89,** 7491.
193. Yin, Y., Tainsky, M. A., Bischoff, F. Z., Strong, L., and Wahl, G. M. (1992) Wild-type p53 restores cell cycle control and inhibits gene amplification in cells with mutant p53 alleles. *Cell*, **70,** 937.

194. Hollstein, M., Sidransky, D., Vogelstein, B., and Harris, C. C. (1991) p53 mutations in human cancers. *Science*, **253,** 49.
195. Malkin, D., Li, F. P., Strong, L., Fraumeni, J. F., Nelson, C. E., Kim, D. H., Kassel, J., Gryka, M. A., F. Z., B., Tainsky, M. A., and Friend, S. (1990) Germ line p53 mutations in a familial syndrome of breast cancer, sarcomas, and other neoplasms. *Science*, **250,** 1233.
196. Kastan, M. B., Zhan, Q., El-Deiry, W. S., Carrier, F., Jacks, T., Walsh, W. V., Plunkett, B. S., Vogelstein, B., and Fornace, A. J. J. (1992) A mammalian cell cycle checkpoint pathway utilizing p53 and GADD45 is defective in ataxia-telangiectasia. *Cell*, **71,** 587.
197. Livingston, L. R., White, A., Sprouse, J., Livanos, E., Jacks, T., and Tlsty, T. D. (1992) Altered cell cycle arrest and gene amplification potential accompany loss of wild-type p53. *Cell*, **70,** 923.
198. Al-Khodairy, F. and Carr, A. M. (1992) DNA repair mutants defining G2 checkpoints in *Schizosaccharomyces pombe*. *EMBO J.*, **11,** 1343.
199. Enoch, T., Carr, A. M., and Nurse, P. (1992) Fission yeast genes involved in coupling mitosis to completion of DNA replication. *Genes Dev.*, **6,** 2035.
200. Weinert, T. A. (1992) Dual cell cycle checkpoints sensitive to chromosome replication and DNA damage in the budding yeast *Saccharomyces cerevisiae*. *Radiat. Res.*, **132,** 141.
201. Rowley, R., Hudson, J., and Young, P. G. (1992) The wee1 protein kinase is required for radiation-induced mitotic delay. *Nature*, **356,** 353.
202. Barbet, N. C. and Carr, A. M. (1993) Fission yeast wee1 protein kinase is not required for DNA damage-dependent mitotic arrest. *Nature*, **364,** 824.
203. Walworth, N., Davey, S., and Beach, D. (1993) Fission yeast chk1 protein kinase links the rad checkpoint to cdc2. *Nature*, **363,** 368.
204. Ohtsubo, M., Okazaki, H., and Nishimoto, T. (1989) The RCC1 protein, a regulator for the onset of chromosome condensation locates in the nucleus and binds to DNA. *J. Cell. Biol.*, **109,** 1389.
205. Bischoff, F. R. and Ponstingl, H. (1991) Catalysis of guanine nucleotide exchange on Ran by the mitotic regulator RCC1. *Nature*, **354,** 80.
206. Matsumoto, T. and Beach, D. (1991) Premature initiation of mitosis in yeast lacking RCC1 or an interacting GTPase. *Cell*, **66,** 347.
207. Moreno, S. and Nurse, P. (1994) Regulation of progression through the G1 phase of the cell cycle by the *rum1*$^{+}$ gene. *Nature*, **367,** 236.
208. Hoyt, M. A., Totis, L., and Roberts, B. T. (1991) *S. cerevisiae* genes required for cell cycle arrest in response to loss of microtubule function. *Cell*, **66,** 507.
209. Li, R. and Murray, A. W. (1991) Feedback control of mitosis in budding yeast. *Cell*, **66,** 519.
210. Li, R., Watson, J. A., and Murray, A. W. (1993) The mitotic feedback control gene MAD2 encodes the alpha-subunit of a prenyltransferase. *Nature*, **366,** 82.
211. Glotzer, M., Murray, A. W., and Kirschner, M. W. (1991) Cyclin is degraded by the ubquitin pathway. *Nature*, **349,** 132.
212. Byfield, J. E., Murnane, J., Ward, J. F., Calabro-Jones, P., Lynch, M., and Kulhanian, F. (1981) Mice, men, mustard and methylated xanthines: the potential role of caffeine and related drugs in the sensitization of human tumours to alkylating agents. *Br. J. Cancer*, **43,** 669.

10 Regulation of the cell cycle during *Drosophila* development

HELEN WHITE-COOPER and DAVID GLOVER

1. Introduction

It is perhaps surprising that relatively little is known about the regulation of the cell cycle during development, particularly as many of the organisms now used to study the cell cycle were originally the focus of studies of both cell division and development in the late nineteenth century. The classical studies that defined mitosis and its phases, together with the cyclical behaviour of chromosomes and centrosomes, originated from observations made upon developing embryos of a variety of organisms. The nineteenth century view of the cell cycle was one of alternating periods of cell division and a vegetative state. Not until the 1950s did the studies of Howard and Pelc (1) define a discrete period of DNA synthesis (S phase) within this vegetative phase, thus leading to the archetypal view of the cell cycle in which mitosis (M) is separated from S by two gap phases G1 and G2. There are, however, many variations on this theme that are particularly evident during development. Many embryos undertake a series of rapid cell cycles during their early development driven by maternally provided proteins. The *Drosophila* embryo begins its existence as a syncytium that becomes populated with several thousand nuclei as a result of 13 extremely rapid rounds of synchronous mitosis at approximately 10 min intervals. These mitoses are separated by periods of DNA synthesis without intervening gap phases. A G2 phase is introduced into the cell cycle only after individual nuclei have become incorporated into cells during division cycle 14. A G1 phase is not introduced into the *Drosophila* cell cycle until the seventeenth division, and only then do the cells of the embryo undertake 'conventional' cell cycles. These occur in the CNS and imaginal tissues, which continue to proliferate throughout larval and pupal development. In most other tissues the cell cycle is far from conventional, since *Drosophila*, in common with other dipterans, has many tissues whose cells undergo endoreduplication cycles. These comprise repeated rounds of S phase in which sister chromatids do not segregate into daughter cells but rather become laterally aggregated to form giant chromosomes. Polytenization begins during embryogenesis, but continues in such tissues as the fat body and salivary gland throughout larval development to produce cells estimated to have ploidies of up to 2^{10} (1056) in *Drosophila* and even higher in other flies. Finally, during meiosis, the ploidy of diploid cells has to be reduced to produce the haploid gametes. This is achieved in two unusual division cycles in which S phase is followed by two consecutive rounds of chromosomes segregation. The types of cell cycle seen in the life cycle of *Drosophila* are shown in Fig. 1.

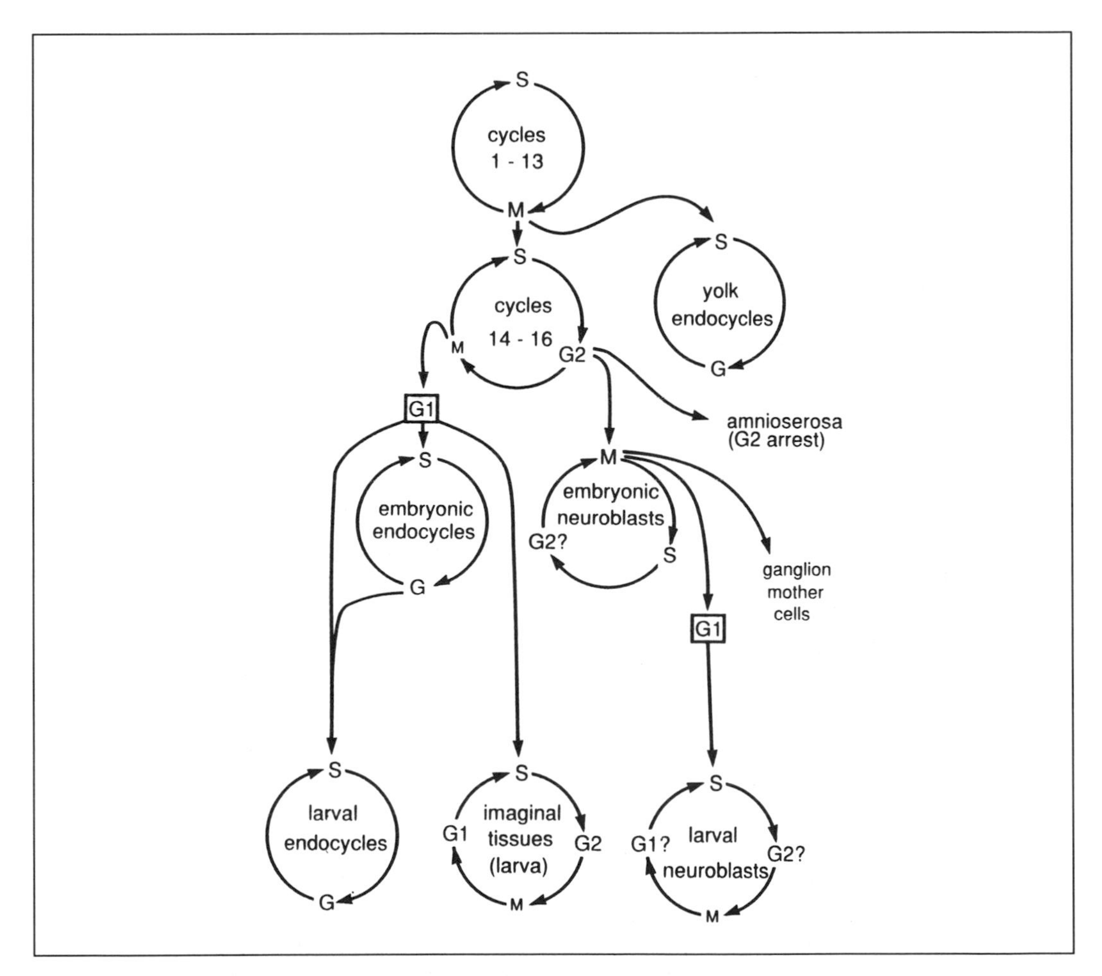

Fig. 1 Phases of the cell in *Drosophila* development. The types of cell cycle seen in different tissues during the life cycle of *Drosophila* are shown. In the early embryo the cycle consists of alternating phases of S and M, with no discernable gap phases. After cellularization a G2 phase is introduced, most lineages carry out three such divisions although the amnioserosa exit the cycle at this stage. At cycle 17 a G1 phase is introduced. Cells of the neuronal and imaginal lineages continue to divide, most other cells enter endoreduplication cycles in which S alternates with a Gap phase and there is no mitosis. Reproduced from ref. 15.

2. Regulation of the cell cycle in embryos

2.1 Fertilization and the onset of zygotic divisions in *Drosophila*

Fertilization is the trigger for the development of the *Drosophila* egg following meiosis, and if it does not occur there is no further DNA replication or mitosis. The first zygotic division is said to be gonomeric; chromosomes of the male and female pronuclei condense and undergo mitosis together but the chromosome sets do not mingle until this division is completed. A number of mutations have been identified which appear to affect this 'pronuclear fusion', but few of these are characterized. Pronuclear fusion is prevented in embryos laid by females homozygous for the mutation *fs(1)mh1184*. The male pronucleus plays no part in the subsequent mitotic divisions. This leads to an embryo populated with haploid maternally derived nuclei.

The eggs produced by females homozygous for mutations in *fs(1)Ya* also fail to develop from the very first nuclear divisions, suggesting that the product of *fs(1)Ya* is involved in initiating the first embryonic mitosis (2). Some mutant embryos leak past this stage to arrest in later stages of development, suggesting a requirement for *fs(1)Ya* in subsequent embryonic mitoses. *fs(1)Ya* encodes a 91 kDa protein that is translated during oocyte maturation independently of fertilization. It is a nuclear envelope component that disperses slightly later than lamina breakdown, and reassociates with the envelope during the next interphase period.

Before any of the mitotic divisions can get underway, the complement of DNA supplied by the pronuclei has to be replicated. Normally, a block is imposed upon DNA replication following the completion of female meiosis, and the development of the unfertilized egg is arrested. Mutations have been described at three loci, *gnu (giant nuclei), plu (plutonium),* and *png (pan gu),* that when homozygous in the mother result in the production of eggs that undergo multiple rounds of DNA replication irrespective of fertilization. This replication takes place in the absence of mitosis to produce giant nuclei (3–5). The finding of DNA replication in unfertilized *gnu*-derived eggs led to the suggestion that *gnu* regulates the onset of DNA replication following the completion of female meiosis. All of the nuclei in eggs derived from females homozygous for mutations at these loci can undergo replication. If the mutant eggs are fertilized, up to five giant nuclei can develop corresponding to the replicated products of the male and female pronuclei, and the three other products of female meiosis, the polar bodies which ordinarily would not replicate. Double mutants of *gnu* and *fs(1)mh* show the *gnu* phenotype, where the male pronucleus, which would not undertake DNA replication in the *fs(1)mh* mutant alone, participates in the formation of giant nuclei (3). Interestingly, double mutants between *gnu* and *fs(1)Ya* fail to develop giant nuclei and show the *fs(1)Ya* phenotype (6). This suggests that either the action of *gnu* is mediated through *fs(1)Ya,* or the two genes work in different pathways. In this second scenario the activity of *fs(1)Ya* establishes a precondition for the action of *gnu*.

The three genes do not function solely at the onset of zygotic development, but appear to have a continued role regulating the entry into S phase in the cleavage divisions. Although strong alleles of *png* have the phenotype described above, a weak allele allows 50 per cent of the embryos to develop between 6 and 16 nuclei. In these embryos a limited number of mitoses have taken place, suggesting that *png*, at least, has an additional role in the syncytial divisions. When females homozygous for this weak allele of *pgu* are also deficient for one wild-type copy of either *plu* or *gnu*, these limited mitoses do not occur. This indicates that the three genes share a common regulatory function.

2.2 Syncytial divisions in the early embryo; the maternal contribution

The eggs of organisms that have to undertake independent development from the time at which they are shed into the wide world are usually endowed with abundant supplies of maternal proteins to enable them to undertake many rounds of mitosis. In most cases such embryos undertake a series of extremely rapid cleavage divisions in order that they can become independent organisms as quickly as possible. This is true of the eggs of both invertebrates such as clams and sea urchins, and vertebrates such as fish and frogs. Perhaps the requirement for maternally supplied proteins is best illustrated by the large number of female sterile mutations that identify genes required for the cell cycle in *Drosophila*. Embryos derived from females homozygous for such recessive mutations undergo aberrant mitotic divisions within the syncytial stage of development. The finding of mutations with female sterile phenotypes does not necessarily imply a requirement for such genes only in the female germ line or the early embryo. Indeed the majority of genes identified in this way are also required for the cell cycle at later developmental stages. A female sterile phenotype may result from lesions in developmentally regulated enhancers, or promoters active in the germ line. Alternatively, the mutations may be weak hypomorphic alleles which survive to adulthood but fail to meet the mitotic demands of the early embryo. An embryo homozygous for a weak hypomorphic mutation can survive using wild-type gene product provided by its heterozygous mother. Complete loss of function may lead to lethality during zygotic development so that homozygous individuals fail to reach adulthood. Maternally provided proteins may survive until late embryonic development, after which point a zygotic mutant phenotype may develop. Alternatively, the protein can persist through most of larval development as this consists primarily of cell growth and polytenization. In these cases the mutant phenotype is only seen when the imaginal tissues have to proliferate to form the adult (7, 8). Indeed, many of the mutations in the *Drosophila cdc2* gene are late larval lethals in which the diploid imaginal tissues fail to proliferate. The importance of the maternal contribution is seen in the defective cell divisions in embryos derived from bringing together weaker alleles in the mother (9, 10).

It is beyond the scope of this chapter to provide a detailed list of those genes that have been identified in this way as being essential for mitosis. These have been reviewed elsewhere (11–15). Instead, we will give the single example of the gene *polo*, first identified through a female sterile allele. Embryos derived from mothers homozygous for the $polo^1$ mutation undergo highly abnormal mitoses that show disorganized arrays of condensed chromatin associated with disorganized microtubules. Homozygous $polo^1$ flies reach adulthood, despite showing defective mitoses during larval development, and can produce some sperm and eggs although they show meiotic divisions (16, 17). The defects include a high mitotic index in larval neuroblasts, which display both monopolar spindles and spindles with broad poles. In male meiosis non-disjunction occurs on multipolar spindles. *polo* encodes a 577 amino acid protein with an N-terminal kinase domain and a 300 residue C-terminal domain (17), which shows a peak of activity during late anaphase–telophase (18). The polo kinase is highly conserved. Its homologues from budding yeast, *CDC5* (19), and mouse, Plk (polo-like kinase) (20) and Snk (serum-induced kinase) (21), show respective identities of 52, 65 and 52 per cent in the kinase domain, and 8, 43 and 35 per cent in the C-terminal domain; the latter homologies lying in three instinct blocks. The murine genes are expressed only in proliferating cells, but whether they have a role in mitosis has not yet been established. In *cdc5* mutants of budding yeast, nuclear division is arrested at a late stage with an elongated spindle (22). When temperature-sensitive *cdc5* mutants are shifted to restrictive temperature after initiating the first meiotic division, two spindles form but do not elongate, resulting in the four poles being encapsulated into two diploid spores (23). A number of observations suggest that *polo/CDC5* may regulate microtubule behaviour. *cdc5* mutants show an unusual interaction with MBC, a drug that binds tubulin and depolymerizes microtubules (24); and there is a strong interaction of *polo* with mutations in *asp* which appear to affect microtubule stability (25). The mitotic stage at which polo kinase is maximally active is consistent with the enzyme having a role in orchestrating the changes in microtubule organization that occur late in anaphase, and in telophase. The availability of mutations in the genes, and antibodies to their proteins, now offers the means of exploring some of these potential roles for the enzyme.

2.3 Oscillators in early embryonic cell cycles

The rapidity of the cleavage divisions in the absence of transcription in organisms such as *Xenopus* originally led to the feeling that these cycles might be fundamentally different from those in organisms such as yeast in which a sequential pathway of gene activity could be demonstrated. The idea of an oscillating clock underlying these cycles was substantiated by the experiments of Hara *et al.* (26) who showed that even following enucleation of the *Xenopus* embryo, cyclical contractions of the cortical cytoplasm still occurred at regular intervals. The independent function of components of the cell cycle has also been dramatically demonstrated from experiments with *Drosophila* embryos.

The first few nuclear divisions take place at approximately 10 min intervals in the interior of the embryo towards the anterior, clustered around the region in which pronuclear fusion has taken place. During division cycles 4–6, nuclei become distributed posteriorly, and then in cycles 7–10 migrate out to the cortex of the embryo. These migrations are the result of two distinct processes both requiring functional microtubules (27, 28). Once the nuclei have reached the surface, a stage known as syncytial blastoderm, the division cycle lengthens such that divisions 10–13 undergo a gradual lengthening of interphase to 15–20 min that allows zygotic gene expression to begin in response to positional information laid down by maternally active genes. Figure 2 demonstrates the distribution of the nuclei in the embryo throughout the syncytial stages of development. A hierarchical set of interactions between sets of transcriptional regulators establishes the key features of segmentation. The independence of major cell cycle events can be seen in experiments in which DNA synthesis is inhibited by the microinjection of aphidicolin into syncytial embryos (29). Chromosomes containing incompletely replicated DNA initially associated with the mitotic spindles, but their chromatids cannot resolve at metaphase. Consequently, centrosomes dissociate from nuclei and then undergo repeated cycles of replication. Moreover, other mitotic events also continue to cycle, including nuclear envelope breakdown and reformation, chromosome condensation and decondensation, and the actin filament-mediated cortical budding cycles. These division cycles are therefore unusual in both their rapidity and their apparent lack of feedback control mechanisms. We saw in earlier chapters of this book that checkpoint mechanisms operate in yeasts and mammalian cells to ensure that mitosis does not take place until the genome is completely replicated. Indeed, if cellularized embryos are treated with aphidicolin, this will arrest the cell cycle (see Section 5.1). Thus on cellularization the embryo acquires a replication checkpoint. It is at this time that the cycle comes under the control of *string* (see below), perhaps indicating a role of this mitotic activator in sensing the completion of DNA synthesis as in fission yeast. The mechanisms that autoregulate these early syncytial cycles are, however, poorly understood, and are difficult to explain solely in terms of known modifications to $p34^{cdc2}$, as we will discuss below.

2.4 Rate-limiting components

The phosphorylation state of $p34^{cdc2}$ has been examined in each of the first 14 divisions of the *Drosophila* embryo (30). Figure 3 shows the levels of the string, cdc2, and cyclin B proteins during the syncytial cycles. Active string protein appears to be absent from mature oocytes, and is produced in embryos by translation of maternal mRNA during ovulation. $p34^{cdc2}$/cyclin complexes are dephosphorylated on tyrosine 15 and so activated, allowing the mitotic cycles to begin. The level of $p34^{cdc2}$ remains constant through these cycles. $p34^{cdc2}$ remains dephosphorylated on tyrosine 15, phosphorylated on threonine 161, and associated with high levels of cyclins throughout cycles 2–7, hence its activity remains high. As

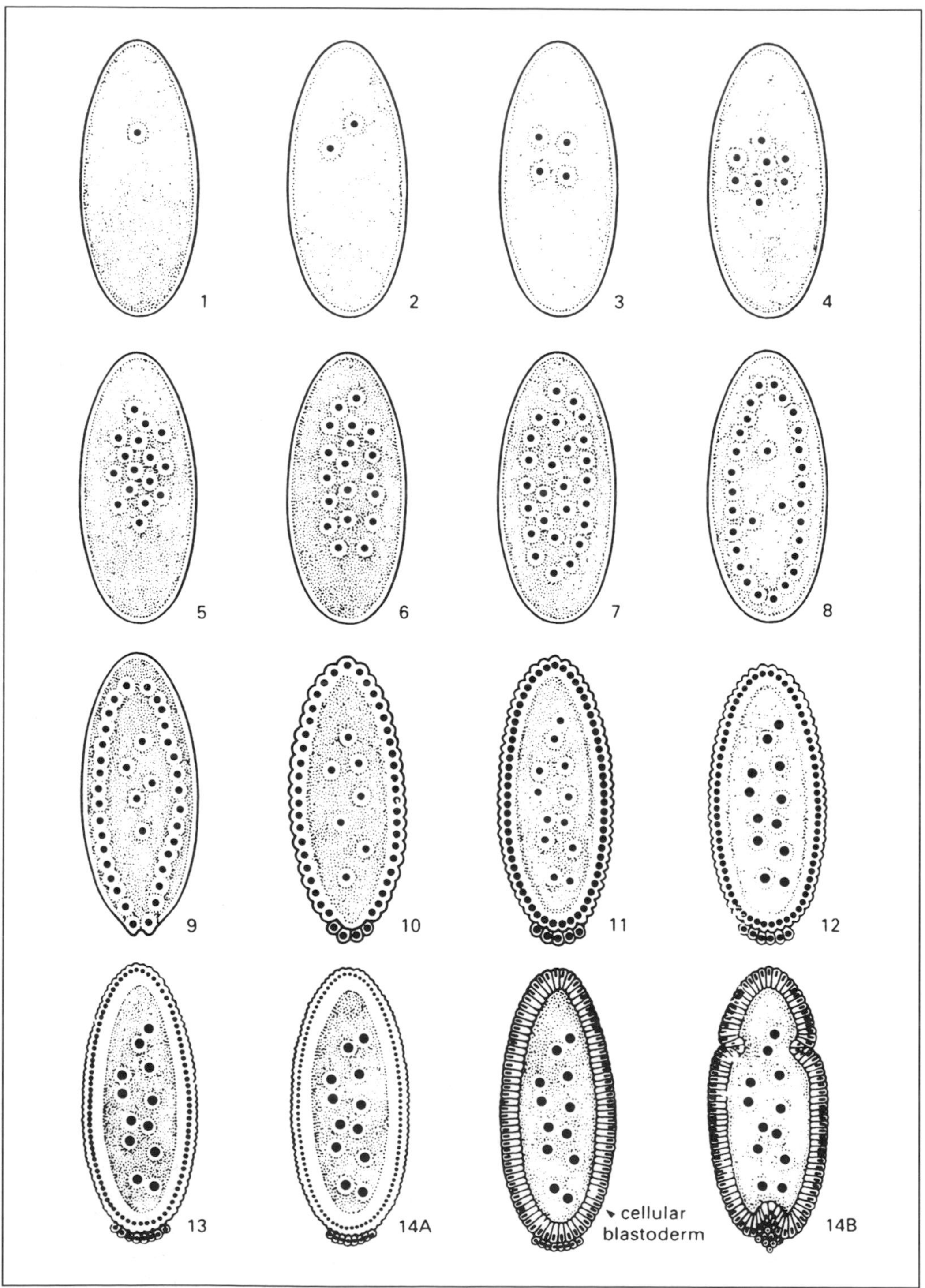
1
2
3
4
5
6
7
8
9
10
11
12
13
14A
cellular blastoderm
14B

Fig. 2 Distribution of the nuclei in the *Drosophila* syncytial embryo. The distribution of nuclei within the syncytium during mitotic cycles 1–14. Pronuclear fusion occurs towards the anterior of the egg. Throughout the first 5 or 6 divisions the nuclei remain concentrated in the anterior, and interior region of the development embryo. By cycle 7 the nuclei have become spaced regularly along the length of the embryo. At cycles 8 and 9 the nuclei migrate towards the cortex, leaving behind those destined to exit the mitotic cycles and undergo endoreduplication in the interior of the egg as the yolk nuclei. Those nuclei that reach the posterior pole first cellularize during cycles 9 and 10, and take up the germ line fate. After they have cellularized they loose synchrony with the somatic nuclei. In cycles 10–13, the nuclei divide with near synchrony within the cortical layer, a stage known as the syncytial blastoderm. After mitosis 14 the cells pass through S phase and enter the first G2 phase, during which time they cellularize. The first post-cellularization mitotic division, mitosis 14, then occurs in 27 discrete domains. Reproduced from ref. 31.

the cycle begins to lengthen during cycles 8–13, there is a progressive increase in the destruction of the cyclins at anaphase. In cycle 8, 30 per cent of the cyclin B is degraded at anaphase; this proportion increases to more than 80 per cent at cycle 13. The phosphorylation of threonine 161 also correlates with maximal $p34^{cdc2}$ kinase activity at metaphase during these divisions. The progressive lengthening of interphases during cycle 10–13 seems to correlate with the extent of cyclin

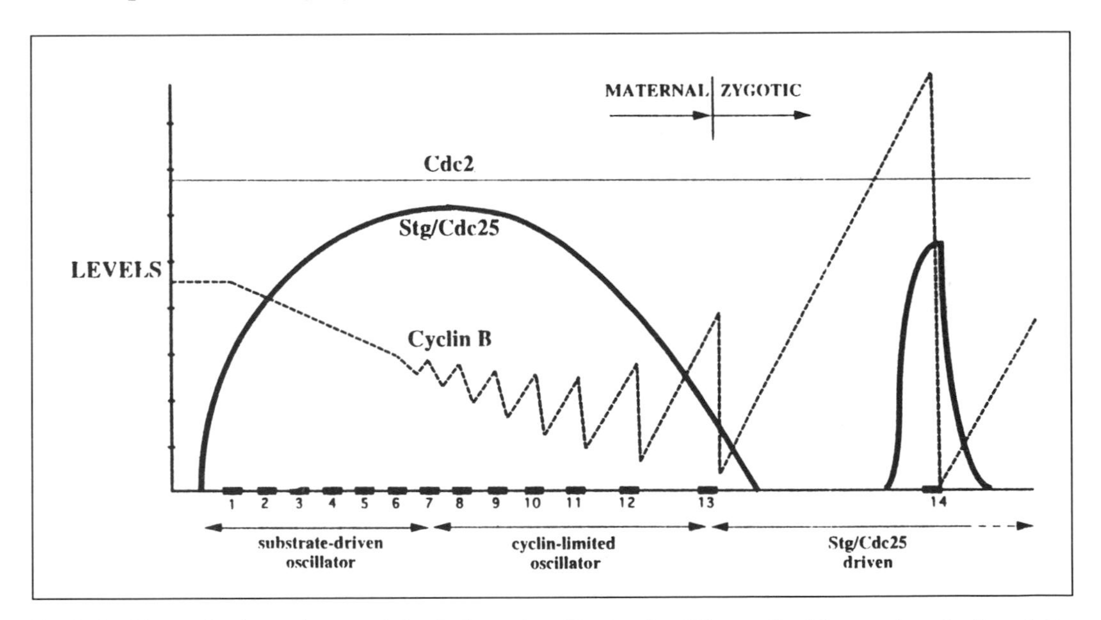

Fig. 3 Oscillators in the early syncytial mitotic cycles. The levels of the cdc2, string, and cyclin B proteins during the syncytial cycles. The levels of the proteins are shown on the Y axis, arbitrary units. The X axis shows the time since egg deposition, the time in which mitosis is occurring is shown by the dark boxes on this axis. The level of the cdc2 protein remains constant throughout these cycles, although its phosphorylation state changes from cycle 8. The transcription of the maternally provided string transcripts occurs at ovulation, the level of this protein peaks around cycle 8 and then begins to fall. All of the string protein is degraded after mitosis 13 so that the first post-cellularization mitosis is reliant on zygotically produced transcript. Cyclin B protein levels show little variation during nuclear cycles 1–7, from cycle 8 there is a partial degradation of the protein at each metaphase to anaphase transition. This coincides with the beginning of the gradual lengthening of the cycle and the changes in the phosphorylation state of cdc2. Thus from cycle 8 it appears that the synthesis of cyclins is rate-limiting for the entry into mitosis. Reproduced from ref. 30.

breakdown, suggesting that cyclin synthesis could be rate-limiting at this time. This is supported by the finding that the embryos of mothers heterozygous for mutations in both cyclin genes take some 30 per cent longer to undertake these blastoderm cycles. The nature of the oscillator that derives cycles 2–7 is unclear. Edgar *et al.* (30) have suggested that the availability of $p34^{cdc2}$ substrate might regulate the timing of these cycles. If cyclin degradation is dependent upon the formation and function of the mitotic apparatus, then the progressive increase in numbers of mitotic spindles might lead to the increased levels of cyclin of at least cyclin B. Cyclin B is closely associated with the mitotic apparatus. This has been shown in many systems and would be consistent with a role for cyclin B in targeting $p34^{cdc2}$ to potential substrates. It can be dramatically visualized in *Drosophila* when the nuclei have migrated out to the cortex during division cycles 8–10 (31). At this stage, the chromosomes become oriented with their centromeres on the apical (surface) side of the cortical nuclei. Nuclear envelope breakdown begins on the apical side of the nucleus and then spreads basally. The spindle is asymmetrically positioned on the apical side, and it is within this region that chromosome congression begins. The centromeric regions of the chromosomes are the first to align on the metaphase plate followed by the chromosome arms (32). Cyclin B is distributed within a layer of approximately ten microns predominantly on the apical side of the cortical nuclei, within which it is most abundant in a cloud that surrounds the centrosome within the region occupied by the astral microtubules (33). This association of at least some cyclin B with the spindle poles does however persist during anaphase in these syncytial embryos.

3. Cell cycle transitions during embryonic development

3.1 Introduction of G2 phase in the *Drosophila* embryo

3.1.1 Cellularization

Cellularization is triggered by the ratio of nuclei to cytoplasm. This was most elegantly shown by Edgar *et al.* (34) who applied compression to the embryo to divide it into anterior and posterior halves at an early nuclear cycle when most of the nuclei are in the anterior part of the embryo. Nuclei, initially present at lower density in the posterior half, have to undertake compensatory rounds of division in order to achieve the correct ratio of nuclei to cytoplasm to enable cellularization to take place. Thus, as with the midblastula transition in *Xenopus*, cellularization is dependent upon the nuclear to cytoplasm ratio. The subsequent lengthening of the division cycles permits transcription of a larger number of zygotic genes for the first time. The relative lengths of the pre- and post-cellularization mitotic cycles are shown in Fig. 4.

Cellularization is immediately followed by gastrulation. The cephalic furrow forms to separate the head region of the embryo from the thoracic and abdominal regions. The ventral furrow will define ectoderm and mesodermal tissues along the anterior–posterior axis. Towards the end of gastrulation the embryo begins to

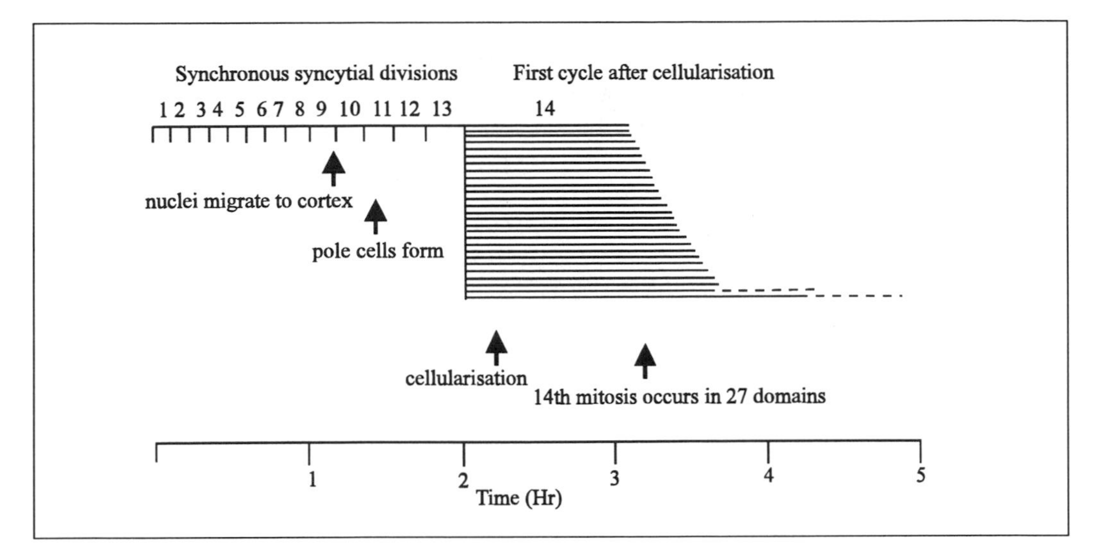

Fig. 4 The introduction of a G2 phase at cycle 14. The timing of the first 14 mitotic cycles. The length of each cycle is indicated by the length of the lines towards the top of the figure. At the onset of nuclear migration the length of the cycle begins to increase. The first G2 phase at cycle 14 varies in length between 69 and 170 min in 27 domains. Mitosis in 25 of these domains is relatively synchronous, in the other 2 domains it occurs over a wide time window, indicated by the dashed lines.

undergo elongation and, as it is confined within the vitelline membrane, this has the effect of pushing the posterior-most part of the embryo dorsally and anteriorly. Mitosis no longer takes place uniformly throughout the embryo at this stage, but within 27 discrete domains in a spatially and temporally regulated pattern (35). These domains differ in the length of time that their cells spend in G2 before undertaking mitosis. The time-course of the entry into the fourteenth mitotic division is shown in Fig. 5. This figure shows the DNA and cyclin B distributions in embryos at various stages of cycle 14. The cyclin B protein gradually accumulates in interphase and is degraded at metaphase, so the gaps in the cyclin B staining indicate the mitotic domains.

3.1.2 *String*

The introduction of a G2 phase and the need to undertake true cell division, rather than the more simple nuclear divisions of the syncytium, is reflected in the profound changes in cell cycle regulation upon cellularization. In this transition, maternal mRNAs encoding proteins essential for the cell cycle are degraded. Those encoding unstable proteins such as cyclin A, cyclin B, and string are replaced by zygotic transcripts. Three lines of evidence point to the importance of the onset of the zygotic transcription of *string* in regulating the fourteenth cycle of mitosis. First of all, its pattern of expression anticipates each of the mitotic domains by 25–35 min (36). Secondly, in a *string* mutant embryo the fourteenth and subsequent cycles of division do not occur, although the embryo goes on to undertake further

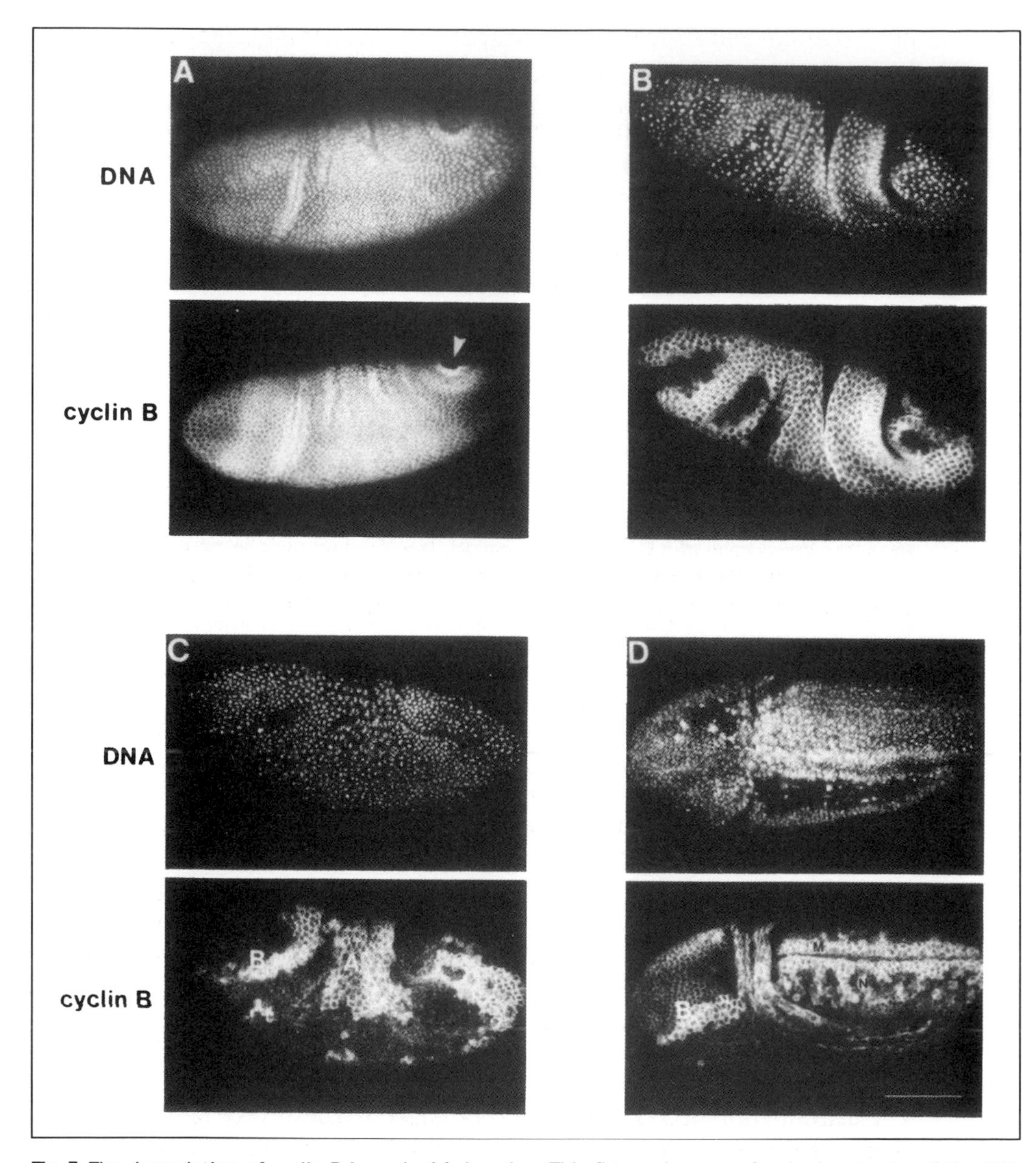

Fig. 5 The degradation of cyclin B in cycle 14 domains. This figure shows confocal micrographs of the DNA and cyclin B staining at four stages of cycle 14. The embryo in panel A is approximately 20 min into G2. All of the cells remain in interphase and cyclin B has accumulated to a high level. The embryo in B is at a later stage, the cells in domains 1–5 have entered mitosis and degraded their cyclin B. In panel C the early domains have finished this division and some of the later domains have entered mitosis 14. A and B indicate the non-dividing regions of the amnioserosa and the cephalic region. In panel D the late dividing domains are entering mitosis 14, the two relatively asynchronous domains M and N are yet to enter M-phase, as can be seen by the high cyclin B in these cells. Photographs courtesy of G. Maldonado-Codina.

development with a reduced number of cells. Finally, Edgar and O'Farrell (37) have constructed transgenic lines of flies in which the *string* gene has been placed downstream of the temperature inducible promoter of the gene for the 70 kDa heat-shock protein (hsp70). A temperature shift induces *string* expression throughout the embryos of such transgenic flies, and consequently cells enter mitosis synchronously. The most likely interpretation of this experiment is that the *cdc25* homologue is causing the premature activation of $p34^{cdc2}$ kinase throughout the embryo rather than in individual mitotic domains, thus driving cells into the mitotic cycle. This effect would appear to be over and above the effect of heat shock alone to delay mitosis and partially synchronize the mitotic domains (38). The expression pattern of *stg* in a gastrulating embryo is shown in Fig. 6.

3.1.3 The mitotic cyclins

The maternal transcripts encoding cyclin A and B also undergo extensive degradation at cellularization. Furthermore, these proteins also begin to show the characteristic pattern of degradation at the metaphase–anaphase transition during mitosis 14. Lehner and O'Farrell (39) have described recessive embryonic lethal mutations in the cyclin A gene, demonstrating that it has essential functions that cannot be supplied by any of the other cyclins. They suggest that maternal supply of cyclin A RNA is sufficient to permit the first 15 rounds of mitotic divisions, and S phase 16 leading to arrest in G2. The rate of accumulation of cyclin A protein in wild-type cycle 14 embryos appears to be uniform, even though divisions no longer occur synchronously throughout the embryo, suggesting that cyclin A is unlikely to be rate-limiting. This is further supported by the normal progression of cycle 14 divisions in the *cyclin A* mutant embryos, despite the low levels of cyclin A protein relative to wild-type (39). Mutations in the *cyclin B* gene have yet to be isolated, although Knoblich and Lehner (40) have generated a small deficiency of the chromosomal region in which the gene lies. Embryos homozygous for this deficiency die late in their development, although it is not clear whether or not this is a result of lack of *cyclin B* or some other essential gene. These embryos appear to complete mitosis 16 normally. In those regions in which divisions occur in such *cyclin B*-deficient embryos, the mitotic index is higher than usual suggesting an aberrant metaphase–anaphase progression. Some spindle defects are described, but these are not dramatic and so the nature of this aspect of the defect is difficult to evaluate. In a *cyclin A* and *cyclin B* double mutant embryo, however, mitosis 15 does not occur, indicating that the two mitotic cyclins can act synergistically to activate $p34^{cdc2}$ kinase, and mediate entry into mitosis.

In vertebrate cells, cyclin A has been clearly demonstrated to have a role in DNA replication, consistent with its localization to the nucleus. The incorporation of BrdU into S phase cells in cycle 16 of *cyclin A* mutant embryos, and the apparent G2 block in cycle 15 embryos in the double mutant argues against an S phase role for cyclin A in *Drosophila,* although a role in the completion of S phase cannot be ruled out. Moreover, in the wild-type embryo, cyclin A is degraded after mitosis in cycles 14–16, to be followed immediately by S phase. Of course it is possible

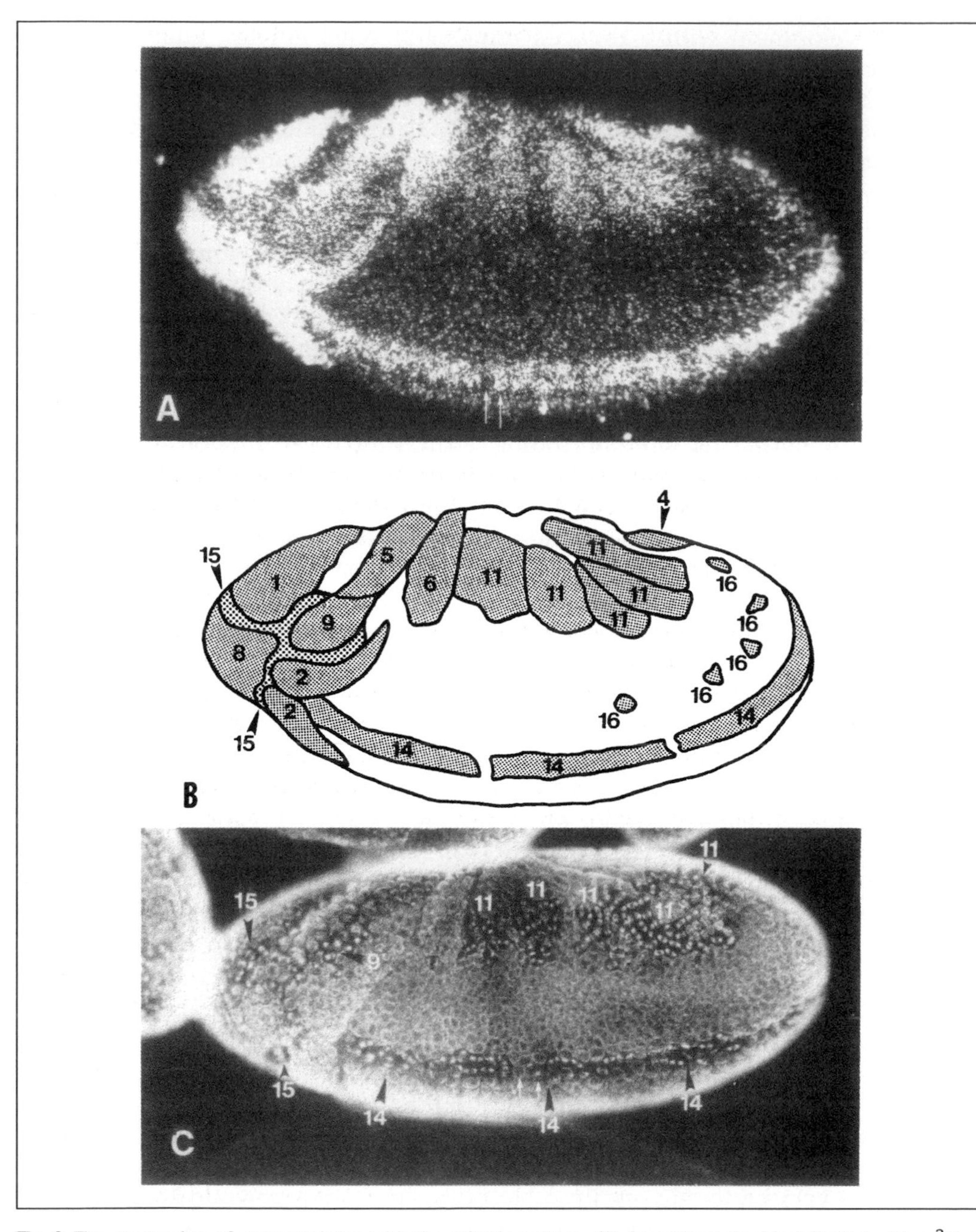

Fig. 6 The expression of *stg* correlates with the mitotic pattern. (A) An embryo double labelled with a ^{3}H *stg* cDNA probe (white) and hoechst (grey). The arrows indicate adjacent cells with different levels of *stg* message. (B) A tracing of the embryo in A showing the correlation between the expression of the *stg* transcript and the mitotic domain map. The levels of *stg* message are highest in the regions due to divide earliest. (C) A slightly older embryo stained with antibodies to tubulin to reveal the mitotic domains. In this embryo mitosis in domains 1–8 has already occurred, divisions in domains 9, 11, 14 and 15 can be seen. The arrows indicate pairs of cells in different mitotic states. Reproduced from ref. 36.

that either sufficient cyclin A persists following mitosis to permit entry into S, or a cyclin A-dependent kinase acts before anaphase of one cycle to allow immediate entry into the S phase of the next cycle at this developmental stage.

Cyclin A degradation precedes that of cyclin B by about 1–2 min (41, 42). The points at which the two cyclins are degraded can be clearly separated by blocking the mitotic cycle with the microtubule destabilizing drug, colchicine. Larval neuroblasts treated in this way accumulate and degrade cyclin A normally, whereas cyclin B protein accumulates to a very high level (41). This implies that the cue for cyclin B degradation is dependent on the correct functioning of the spindle. Indeed, although the introduction of non-degradable cyclins into a *Xenopus* cell-free system appeared to lead to metaphase arrest (43), when similar experiments are carried out with extracts capable of spindle function, the arrest is in late anaphase (44). This implies that cyclin B degradation is not necessarily the signal for the commencement of anaphase. It is supported by the finding that destruction of the CDC28/CLB mitotic kinase in budding yeast is not required in the metaphase–anaphase transition (45). Moreover, *in vivo* experiments in *Drosophila* demonstrate when non-degradable cyclin B is expressed in a system driven by the yeast GAL4 transcriptional activator; it also results in delay at late anaphase (46).

It should be possible to dissect the events of the metaphase–anaphase transition by using mutations that affect mitosis at this point. Two genes have been described that lead to a delay in the progression through metaphase in newly cellularized embryos. One of these, *fizzy*, was identified because development of the embryonic nervous system is severely disrupted (47). A high mitotic index with a striking predominance of metaphase figures is seen in *fizzy* embryos in cycle 15. *fizzy* appears to act earlier in the cycle than microtubule depolymerizing drugs; both A- and B-type cyclins are present at high levels in the metaphase figures (48). Mutations in the gene *three rows* have a similar phenotype, but do not result in a cell cycle arrest, but rather a metaphase-like delay in mitosis 15 in which chromosomes are aligned on a normal-looking spindle, with both cyclins A and B having been degraded (49, 50). The *thr* mutation prevents any anaphase chromatid movement, but telophase does eventually occur. The chromosomes decondense in the mid-region of the spindle, and this is later followed by DNA replication. As the regulation of the segregation of chromatids into daughter cells at anaphase is crucial in cell division, the failure of the cell cycle to arrest at this point in *thr* embryos is unexpected. The *thr* protein contains a small region of similarity with the fission yeast gene *nuc2*. Mutations in this gene block chromosome segregation and lead to metaphase arrest in fission yeast. However, as other characteristic features of *nuc2* are not present in *thr*, the significance of this remains unclear.

3.1.4 Cytokinesis

As cytokinesis has to take place for the first time during division cycle 14, it is perhaps not unexpected that a class of genes would be specifically required for this process at this developmental stage. A number of maternally active genes, *grapes*, *scrambled*, and *nuclear fall-out*, have been described whose products are required for the actin-based formation of pseudo-cleavage furrows in the syncytial embryo and the cellularization furrows (51). The embryos derived from females homozygous for any of these loci exhibit extensive nuclear division abnormalities, but only after nuclei have migrated to the surface of the embryo. Additional genes are required for complete cytokinesis, including *pebble*. Mutations in this gene result in embryos that contain fewer and larger cells than wild-type embryos as mitoses 14, 15, and 16 take place without cytokinesis (52, 53). Other gene products are required for cytokinesis, but some of these appear to be supplied maternally and persist until later development. Mutations in *spaghetti squash*, for example, a gene encoding the regulatory light chain of non-muscle myosin, do not show defects until larval development (54).

3.2 Introduction of G1 phase

Three approaches have been used to identify other cyclin genes in *Drosophila*. Two groups have independently identified a *Drosophila* gene able to complement mutant *S. cerevisiae* that are defective for *CLN1*, *2*, and *3* (55, 56). This new cyclin is closely related to a human cyclin identified by a similar approach, and is given the name cyclin C. *Drosophila* cyclin C maps to cytological position 88E on chromosome 3R. A single transcript of about 1.2 kb is found in Kc tissue culture cells and all developmental stages examined (55). Its tissue distribution during development is broadly consistent with a role in the cell cycle. However, as no mutants in cyclin C have yet been identified, its role is as enigmatic as those of cyclin C-like molecules in vertebrates.

Another successful approach has been to use the interaction trap, in which the gene for a protein of interest, in this case *Dmcdc2* or *Dmcdc2c*, is expressed as a LexA fusion in a yeast strain containing LexA binding sites upstream of a selectable marker gene, *LEU2*. A cDNA library is then expressed in this strain with the cDNA encoded proteins fused to a transcription activation domain such that those interacting with the LexA fusion will activate transcription of *LEU2*. Using this system, Finley and Brent (57) have identified four unique cDNAs encoding proteins that interact with *Dmcdc2* termed CDIs for cdc2 interactors. Of these, one (CDI2) is a homologue of the *S. pombe* gene *suc1*, and another (CDI3) is a homologue of the vertebrate cyclin D. These two genes plus the gene for one other protein interacting with *Dmcdc2* were also picked out in a search for interactors with *Dmcdc2c*. This second screen identified three further genes, one of which (CDI5) identifies a new type of cyclin sharing some identity with cyclins A and B, but missing amino acids that would place it into either of these classes. Both CDI3 and CDI5 are able to complement an *S. cerevisiae CLN1 CLN2 CLN3* strain, indicating that they are bona fide cyclins. The proteins identified by the remaining CDI genes are of unknown function, but could conceivably be substrates for cyclin-associated kinases.

Finally, a *Drosophila* cyclin E gene has been identified through its homology with its vertebrate counterpart (58). By analogy with the vertebrate kinases, it would seem likely that this cyclin associates with a kinase having a role in S phase.

Maternal transcripts are supplied to the syncytial embryo, and in cell cycles 14–16 of the cellularized embryo, zygotic transcripts appear to be constitutively expressed. Concomitant with the introduction of a G1 phase into the cells of the central and peripheral nervous system in cycle 17, transcription of the gene appears to become cell cycle regulated, and is expressed around the time of the G1–S transition. When the gene is ectopically expressed in cells which would normally exit the cell cycle, an extra mitosis is induced. Mutations of the cyclin E gene have now been described; in embryos homozygous for this mutation the cell cycle is arrested after mitosis 16. The arrest occurs prior to S phase in cells destined to undergo endoreduplication as well as those continuing in the mitotic cycle. This supports a role for cyclin E in the G1–S transition (59). A thorough understanding of the functions of the *Drosophila* cyclins C and D awaits the finding of mutations in their genes.

4. Cell cycle regulation in later development

4.1 Endoreduplication and cell division in imaginal development of *Drosophila*

We alluded to the late larval phenotype characteristic of cell proliferation mutants earlier in this chapter. It arises because many proteins essential for cell division that are supplied maternally are stable, and suffice until this developmental stage. This is largely because larval tissues undergo extensive polytenization and so have no need for such proteins. Endoreduplication cycles begin early in embryonic development (60), but continue in tissues such as the salivary gland and fat body throughout the whole of larval development, as well as in several adult tissues. Very little is known about the biology of the onset of the endoreduplication cycles and of the genes required for the switch between mitotic division and endoreduplication cycles. The major groups of cells within the larva that remain diploid include the cells of the central nervous system (CNS) which continue to proliferate until adulthood, cells of the imaginal discs which proliferate throughout larval development and form the cuticular structures of the adult head, thorax, and genitalia, and the abdominal nests of histoblasts that only start to divide after pupation to form the epidermis of the adult abdomen. Gatti and Baker (8) have described many mutations with a late larval or early pupal phenotype that identify genes whose products are essential for mitosis. It is not the purpose of this chapter to fully document all of such genes characterized by these authors and others, so the reader is referred to other reviews for details (11–14).

Mutations in the gene *escargot (esg)* lead to a variety of defects in adult structures, including the loss of abdominal cuticle (61). In this mutant the histoblast nests fail to proliferate, and instead replicate their DNA to estimated ploidies of up to 64C, becoming similar in appearance to the polytene larval epidermal cells. This response is not seen in diploid imaginal disc cells unless a double mutant is made between *esg* and a *D-raf* gene that blocks imaginal cell division (ref. 62, and see below). This suggests that if imaginal cells enter a resting state of the cell cycle, as is normal for histoblast nests during larval development and which can be imposed upon disc cells through a *D-raf* mutation, then the *escargot* gene product is required to prevent them from entering an endoreduplication cycle.

Another interaction between genes regulating mitosis and endoreduplication is suggested by the phenotypes of mutations in *morula* (63). In hypomorphic alleles of *morula*, the nurse cells of the developing ovarian follicles only undertake the first one or two of their ten endoreduplication cycles. They then revert to a mitotic-like state in which chromosomes condense and associate with spindles. In contrast to the female sterility resulting from such mutations, total loss of *morula* function leads to larval lethality as a consequence of a metaphase-like blocking diploid cell proliferation. Such phenotypes could be explained if the *morula* gene product were required for a common process in the exit of cells from metaphase, and in keeping nurse cells out of the mitotic cycle. An understanding of the true role of this gene awaits further genetic and molecular studies. It seems likely that additional genes, yet to be identified, will be involved in regulating the endoreduplication cycles.

4.2 The coordination of cell division with larval development

Mutations that lead to the arrest of the diploid cell cycle affect the development of the *Drosophila* CNS, which undergoes a programmed pattern of cell proliferation that begins at the time these cells are set aside during embryogenesis, continuing throughout larval and pupal development to form the adult nervous system. One characteristic feature of the developing brain is a sort of segmentally arranged giant neuroblasts that essentially serve as stem cells to populate the CNS with neurones. These cells undertake repeated asymmetrical divisions that generate a new giant cell and a smaller ganglion mother cell. The ganglion mother cells undergo symmetrical divisions to produce neurones. In the developing optic lobes of the brain, a distinct population of neuroblasts initially undergoes symmetric divisions, and once this population of cells has been expanded, a series of asymmetrical divisions is undertaken. The proliferation of these cell types is under precise developmental control, although the mechanisms that allow these cells to remain in what is probably a G0 state for long periods are unknown. In the mutation *anachronism*, quiescent post-embryonic neuroblasts enter S phase precociously. The glycoprotein encoded by this gene is not expressed in the neuroblasts themselves, but is secreted from neighbouring glial cells (64). This protein joins a growing list of secreted or cell surface proteins that can regulate cell proliferation (see below) and suggests a role for the glia as negative regulators of cell proliferation in the developing brain.

CNS cells have to respond to a number of external signals that regulate their proliferation, one of the most critical occurring in the process by which neurones from the developing eye disc establish their connections with the optic lobes. This occurs at a stage when cells of both the eye disc and the optic lobes are undergoing a coordinated pattern of proliferation. Each ommatidium in the adult eye contains a set of eight photoreceptor cells that make precise connections with corresponding arrays of neurones in two regions of the brain, the lamina and the medulla. The development of cells within the lamina of the larval brain requires the ingrowth of axons from photoreceptor cells R1–R6. Similarly, development of the medulla requires R7 and R8. The patterns of cell division in lamina development have been studied in some detail. There are two major regions of cell proliferation within the optic lobe, the so-called outer and inner proliferative centres. Cells destined to become part of the lamina migrate as an epithelial sheet from the outer proliferative centre across a furrow that forms a large arc across each optic lobe. As they do so, they pass through two rounds of cell division. The ingrowing axons from R1–R6 selectively approach cells in the G1 phase preceding the second division cycle. In the mutant *sine oculis (so)*, axon ingrowth does not occur and cells do not undertake their final S phase, arresting in the preceding G1. Consequently, the lamina does not develop. It seems therefore that cell contact in G1 is needed to trigger this second round of division, but the nature of the contacting molecules is not known (65).

A parallel set of events also takes place as the eye disc differentiates. During the development of the eye disc, a groove, known as the morphogenetic furrow, forms at its posterior edge and then moves anteriorly. Immediately posterior to the furrow, reiterated patterns of cells differentiate into clusters of photoreceptor neurones in a process that takes about two days to complete. Photoreceptor cell 8 (R8) is the first cell to differentiate followed by R2 and R5, R3 and R4, R1 and R6, and finally R7. The induction of R7 development involves a signal transduction pathway that has been dissected in considerable molecular detail (66, 67). Cells R2, R3, R4, R5, and R8 arise from a wave of mitosis anterior to the morphogenetic furrow. The remaining three R3 cells arise from a second wave of divisions in proximity to the furrow. It is not known how these waves of cell division are regulated, but *string* transcripts are expressed in a broad arc across the eye disc preceding the mitoses (68).

The morphogenetic furrow divides the eye disc into two separate parts. The cells ahead of the furrow divide more or less at random. As the furrow approaches, the cells arrest in G1, so that in the region immediately in front of the furrow all of the cells are synchronized. The cells within the furrow are therefore all in G1. The cells are then faced with the choice of differentiating as neurones (R2–5 and R8) or of entering another division cycle to form R1, 6, and 7, and the non-neuronal cells of the eye. These divisions occur in a band just to the posterior of the morphogenetic furrow. Several genes have been implicated in the synchronization of the cells in G1, and their subsequent development. Mutations at the *roughex (rux)* locus prevent the accumulation of cells in G1 so that the region of synchronous cell division is moved anteriorly. In addition, cells which would normally differentiate without undertaking this division are seen to divide (69). The phenotype of *rux* can be suppressed by a reduction in the dose of either *cyclin A* or *string*, but not *cyclin B* or *twine*. It is suggested that *rux* may be required for the negative regulation of these genes to establish the G1 synchronization. The overexpression of the *Drosophila cyclin E* gene in the eye disc using a heat shock promoter produces a phenotype similar to that described for mutation in *rux*. The band of S phase is moved anteriorly so that cells just in front of the furrow are in S rather than G1 (70).

5. Cell division in the germ line

5.1 Setting aside the germ line in *Drosophila* embryogenesis

The germ cells are the first to undergo cellularization during the syncytial embryo and do so some three rounds of cell division before cellularization of the cells in the main part of the embryo. As nuclei migrate out from the interior to the cortex during cycles 8 and 9, those at the very posterior of the embryo encounter the polar granules, ribonuclear protein bodies within the posterior that contain maternal determinants that direct the future development of posterior structures. Many of the maternally active genes responsible for the assembly and for encoding components of the polar granules have been identified (reviewed in ref. 71). As nuclei migrate to the cortex of the embryo, those that encounter the polar granules undergo cellularization to form the pole cells, precursors to the germ line cells of the organism (Fig. 7). These cells then undergo one or two more rounds of cell division before cellularization of the nuclei in the soma of the embryo. It is not an interaction between the nuclei *per se* and the polar granules that results in the formation of the pole cells, but rather one between the polar granules and centrosomes (72). If aphidicolin is microinjected into embryos in the early division cycles when the nuclei are still in the interior, then centrosomes dissociate from the nuclei, continue their replication cycles, and migrate out towards the cortex. Those centrosomes that interact with polar granules trigger the formation of pole cells lacking nuclei (Fig. 8). Centrioles have been observed within polar granules by electron microscopy, and when protein or RNA components of the polar granules are followed either by immunostaining or by *in situ* hybridization respectively, then these molecules appear to distribute into two bodies concomitantly with cell division being clustered around the polar regions of the spindles of the pole cells. The mechanisms whereby pole cells are formed around the centrosomes are not understood, but present an intriguing problem for cell biology. The selection for a mechanism that enables a group of nuclei to be sequestered away to form the germ line as soon as possible makes good sense, since the syncytial nuclear divisions are likely to be error-prone with their lack of the replication checkpoint. This is not to say that all errors in these syncytial cycles are tolerated. If abnormal divisions occur resulting in a disruption of the cytoskeletal network of the syncytium in the later cycles, as happens for example in embryos derived from *daughterless abo-like* mothers, then affected nuclei can fall into the interior of the egg and do not participate in further development (73).

Both the cyclin A and cyclin B genes are abundantly transcribed in the nurse cells during oogenesis, the transcripts becoming deposited in the developing oocyte (74). The two forms of cyclin B transcript that are synthesized in the developing egg chamber differ as a result of a splicing event in the 3′ non-

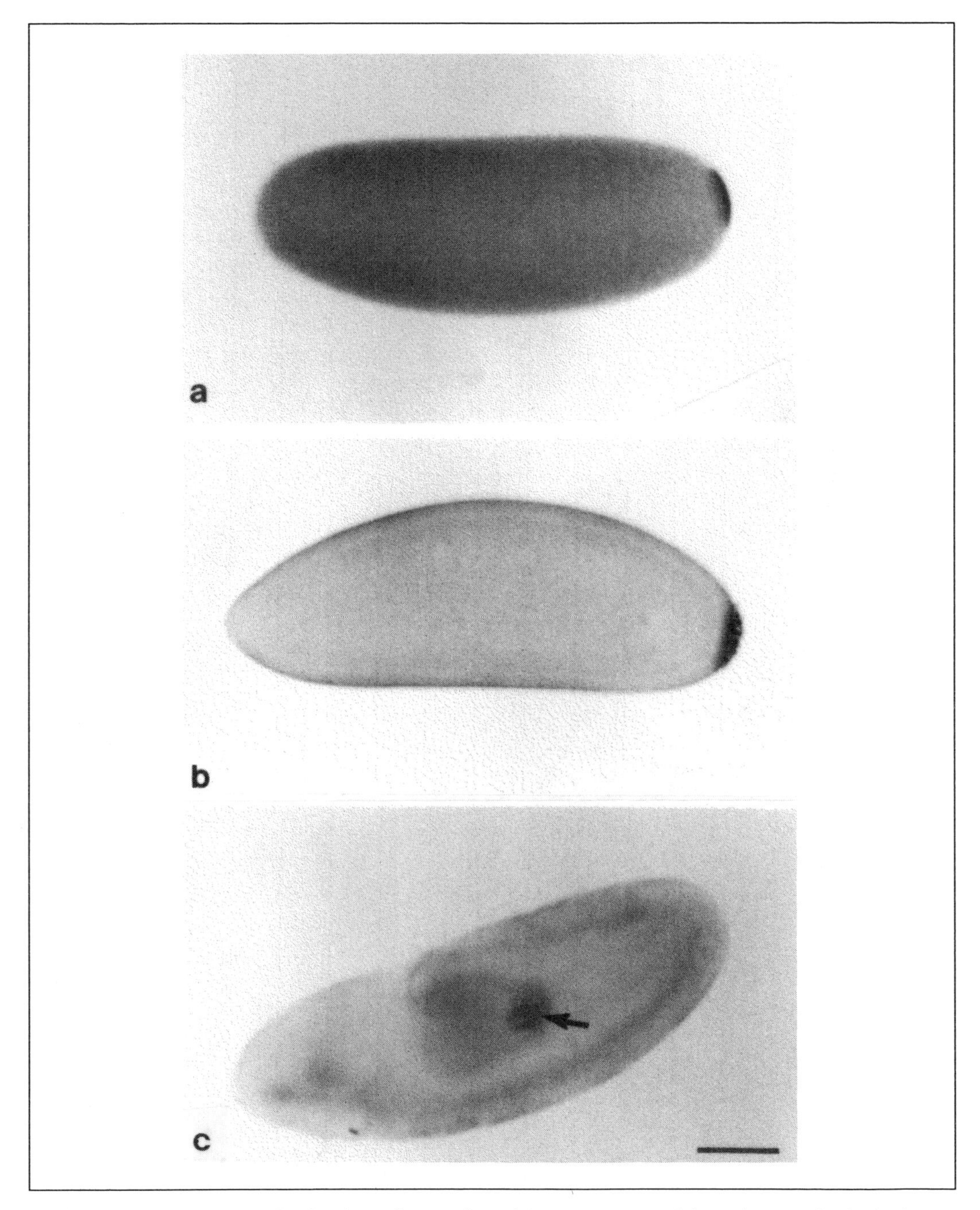

Fig. 7 Pole cell formation and polar determinants. One of the components of the polar granules is the longer cyclin B mRNA. This figure shows RNA *in situ* hybridizations to this message at different stages of embryogenesis. The message is already localized in unfertilized eggs (a). When the pole cells form they incorporate the polar granules (b). The RNA remains in the pole cells even after the extension of the germ band and the migration of the germ cells (arrowed in c). Micrographs courtesy of B. Dalby.

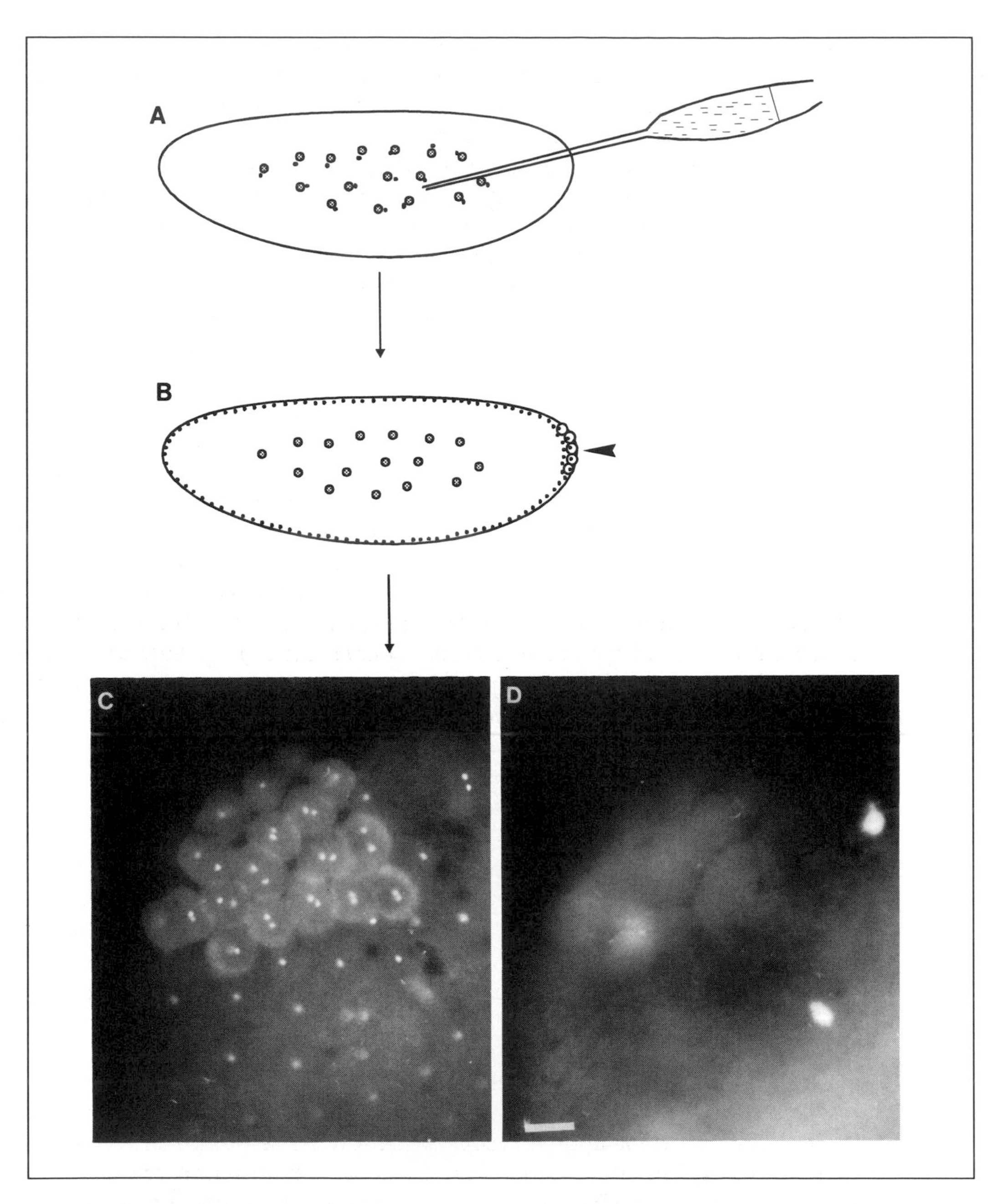

Fig. 8 Pole cells form as a result of the interaction of the polar granules and the centrosomes in embryos treated with aphidicolin. When embryos are injected with aphidicolin the replication of the DNA is inhibited so the mitotic cycles are arrested (A). The replication of the centrosomes continues in these embryos. The centrosomes (black) dissociate from the nuclei (grey) and migrate to the cortex of the embryo (B). Those that reach the posterior pole trigger the formation of pole cells even in the absence of nuclei (arrow). Panels C and D show the presence of centrosomes, labelled with an antibody to a centrosomal antigen (B×63) in pole cells formed in this way. The hoechst staining of these cells reveals the lack of DNA.

translated region. The spliced form of the transcript is found in the developing oocyte at earlier stages of oogenesis, whereas the unspliced form is synthesized in the nurse cells. The transcripts synthesized in the nurse cells are deposited in the developing oocyte. Cyclin A transcripts are uniformly distributed throughout the early syncytial embryo, whereas cyclin B transcripts are in addition concentrated at the posterior pole of the embryo. This concentration begins at stages 13 and 14 of oogenesis onwards (74). The unspliced maternal messenger for cyclin B is one of the components of the polar granules that appears to be added at a late stage in the assembly of these bodies. The accumulation of cyclin B maternal RNA at the posterior pole is disrupted in the eggs of females homozygous for mutations in *cappuccino, spire, staufen,* and *oskar* (75), functions required for the formation of the polar granules themselves. The posterior pole accumulation seem to be a late event of oogenesis, and the transcripts would appear to be interacting with some component of the polar granules. In addition, the cyclin B transcripts come to have a perinuclear distribution that can be disrupted by microtubule-destabilizing drugs (75). Dalby and Glover (74) showed that a sequence element within the 3′ untranslated sequence of the unspliced transcript is required both for the posterior accumulation of the RNA and also for the perinuclear localization in the somatic part of the embryo, and that this sequence could be used to localize foreign mRNAs to these regions in germ-line transformation experiments. The region required for posterior retention of cyclin B mRNA has been more tightly defined by experiments in which biotinylated RNAs tagged with segments of this sequence were injected into early syncytial embryos (76). Injected RNAs having 3′ sequence elements become stabilized within the pole cells. The minimum sequence requirement for this localization/stabilization has been defined as lying within a 181 nucleotide region. It is possible that the mechanisms that localize maternal RNAs in the perinuclear and posterior polar regions could share a common intermediary, since a microtubular transport system is thought to localize other RNAs to posterior pole.

Once the pole cells have formed, they undergo two divisions that are completed shortly before the onset of gastrulation. The pole cells then migrate from the posterior pole of the cellularized embryo to become incorporated into the gonads of the stage 14 embryo, over a period of about 9 h. During this time the pole cells cease dividing, to resume in the last 2–3 h of embryogenesis (77). Dalby and Glover (76) have found that the pole cell cyclin B transcripts, but not the shorter, spliced RNA, contain a control element that represses the translation of this RNA until late stage 14 of embryogenesis, when the divisions resume. The region required for this translational repression contains sequence motifs similar to the *nanos*-response element that mediates the *nanos*-dependent repression of *bicoid* and *hunchback* translation (78), although the functional significance of this is not yet clear. It is possible that repression of translation of maternal cyclin B RNA results in cyclin B becoming rate-limiting for mitosis until the gonads form. Translation of cyclin B could be the trigger for the resumption of cell division in this lineage. Certainly, zygotic transcription of cyclin B in the pole cells is not seen until the first larval instar (76).

5.2 Meiosis in *Drosophila*

The two meiotic divisions are profoundly different. Meiosis I is usually preceded by recombination. This requires the pairing of the maternally and paternally derived homologues, each comprising replicated chromatids. This is mediated through the synaptonemal complex. The paired homologues become aligned on the meiosis I spindle, and following the resolution of any recombination events, each homologue segregates to a spindle pole, the sister chromatids remaining attached at their centromeric regions. Most, but not all, chromosomes undergo recombination before meiosis I, and so this process is intimately linked to the first meiotic division. However, mechanisms do exist that enable non-exchange chromosomes to segregate correctly at meiosis I. In *Drosophila* females, for example, a system for the segregation of non-exchange chromosomes has been well defined at the genetic level, although molecular understanding of this process is still rudimentary (reviewed in ref. 79). Male *Drosophila* do not undertake recombination, however, and so a separate system must exist to permit homologues to pair and then segregate from each other at meiosis I. The second meiotic division is a modified form of mitosis in that the pairs of chromatids, linked at their centromeres, align on the spindle and separate at anaphase.

In the females of most species, meiosis becomes arrested, although the precise stage at which this happens varies from one species to another. In *Drosophila* it is the passage of the egg down the oviduct that releases the block at meiosis I. Fertilization normally occurs internally from sperm retained in storage glands in the female. In the absence of fertilization, development does not proceed further, and so unfertilized females produce eggs that have undergone the two meiotic divisions.

The difficulties of obtaining sufficient quantities of germ-cells at specific stages has hindered any biochemical analysis of p34 function during meiosis in *Drosophila*. Nevertheless, genetic studies offer a powerful means for gaining understanding of this process. Of particular interest is the role of one of the two *Drosophila* homologues of the mitotic activator *cdc25*. One of these genes, *string*, functions to activate $p34^{cdc2}$ primarily in somatic tissues (Section 3.1), whereas the second homologue, *twine*, seems to function only in the germ line. *twine* was first identified by complementation of *cdc25-22* in *S. pombe* (80, 68). Its protein shows a high degree of sequence conservation with other *cdc25* tyrosine phosphatases in the C-terminal catalytic domain, enabling Courtot *et al.* (81) to isolate the gene utilizing a PCR method.

The apex of the testes contains a population of dividing stem cells that produce the gonial precursor cells. *string* but not *twine* transcripts are seen in these cells. The precursor cells undergo four rounds of mitotic division to produce a cyst of 16 cells which remain linked by cytoplasmic bridges or ring canals. *twine*, but not *string*, is expressed in these cells over a period of cell growth lasting about 90 h which immediately precedes meiosis. The two meiotic divisions generate 64 cells within the cyst, which elongate and differentiate into a cluster of 64 mature sperm.

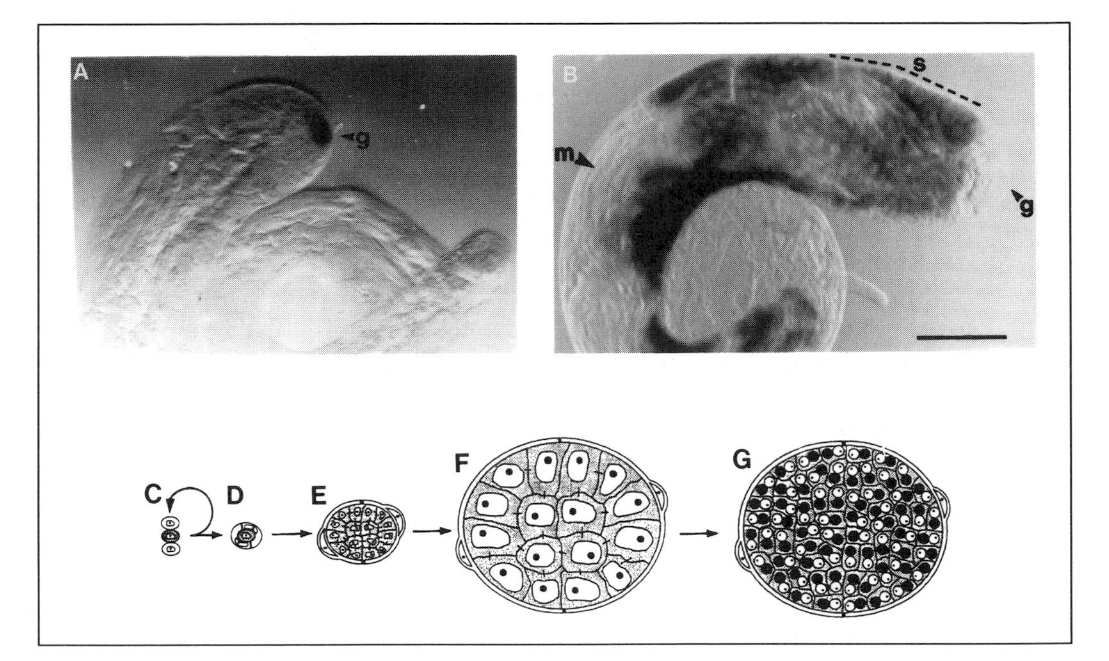

Fig. 9 Spermatogenesis in *Drosophila*. The expression patterns of the *string* and *twine* genes in the adult testes, determined by *in situ* hybridization. (A) *string* expression is limited to a region at the apical tip of the testis in which the mitotic divisions are occurring, the germinal proliferative centre (g). (b) *twine* is not expressed in these cells, the transcript is first detected in the growing stage, during which time S phase occurs (s) and reaches a peak just before meiosis. The transcript is degraded before or during meiosis (m). C–G depict the early stages of spermatogenesis. (C) A stem cell undergoes an asymmetric division to produce another stem cell and a cyst progenitor cell. (D) The cyst progenitor cell undergoes four rounds of mitosis to produce a cyst of 16 cells. (E) The cells in this cyst grow for a period of 90 h, and express all of the transcripts required for post meiotic development. (F) The cells then undergo the two meiotic divisions to produce a 64 cell onion stage cyst (G).

The early stages of spermatogenesis, and the distribution of the *stg* and *twine* transcripts in the adult testis is shown in Fig. 9. A single mutant allele of *twine* has been identified that appears to be a null. Spermatogenesis appears normal up to the end of the growing stages in *twine* males but the cysts do not go through meiosis (68). The $twine^{HB5}$ allele has mis-sense mutation that changes a conserved proline residue in the tyrosine phosphatase domain of the protein to a leucine (81). It would therefore seem that $p34^{cdc2}$ kinase is not correctly activated at the point of entry into meiosis in $twine^{HB5}$ mutants. However, although a meiotic spindle never forms, chromosome condensation does occur and moreover it is accompanied by nuclear envelope breakdown (82). This could be explained either if enzymes other than $p34^{cdc2}$ could mediate some aspects of the G2–M transition for the entry into male meiosis, or if some forms of the $p34^{cdc2}$ complex were not regulated by the phosphorylation/dephosphorylation of tyrosine 15. In extracts of activated *Xenopus* eggs, $p34^{cdc2}$ complexed to cyclin A is not subject to inhibitory phosphorylation of tyrosine 15, in contrast to the $p34^{cdc2}$/cyclin B complex (83,

84). If this were the case in the *Drosophila* spermatocyte, *twine* function would only be required to activate the $p34^{cdc2}$/cyclin B complex, which may be specifically required to modify microtubule behaviour for spindle formation.

The gene *rux* appears to be involved in the coordination of the two meiotic divisions in male meiosis. In males deficient for *rux* function the two meiotic divisions occur normally. Instead of continuing through post-meiotic development, these cell then attempt a further meiosis II-like division. This phenotype is suppressed when the dose of *twine* or *cyclin A* is reduced. Overexpression of *rux* in the testes results in the failure of the second meiotic division. In wild-type males the cyclin A protein is present during the first, but not the second, meiotic division. In both *rux* and *twine* the breakdown of this protein is delayed so that it is present in the nucleus for longer than normal (85, 86).

Oogenesis in *Drosophila* shows some similarities with spermatogenesis in that the gonial precursor cells undergo four mitotic divisions in the germarium of the ovary to produce a cyst of 16 cells interconnected by cytoplasmic bridges. However, as the egg chamber develops only one cell becomes the oocyte while the other 15 cells develop into nurse cells. The nurse cells undertake up to ten rounds of endoreduplication during which time *twine* and *string* are coexpressed together with many other maternally provided mRNAs that are required for early embryonic development (Fig. 10). Prophase of meiosis is initiated in the presumptive oocyte nucleus in the germarium and continues during oogenesis. The meiotic spindle is not formed until the late stages of oogenesis, by the end of which, the oocyte is arrested in metaphase of meiosis I, with the exchange chromosomes in a bundle at the metaphase plate and the non-exchange fourth chromosomes displaced from the plate towards the poles. The meiotic divisions are completed only after activation of the oocyte on passage through the oviduct.

In contrast to the meiotic block seen in *twine* males, meiosis continues in females, but does so abnormally. Repeated abnormal meiotic divisions are seen in *twine* oocytes which appear to resemble the reductional division rather than the equational division. Repeated attempts at the reductional division would explain the dramatic

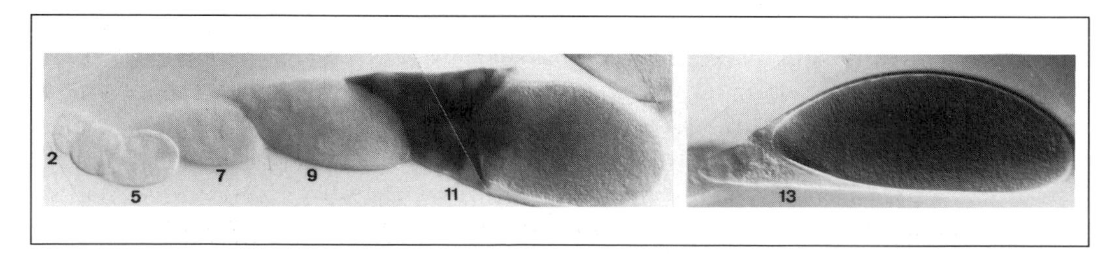

Fig. 10 Oogenesis in *Drosophila*. The pattern of expression of the *stg* transcript during oogenesis was determined by *in situ* hybridization. Oocytes at various stages of oogenesis from the same follicle are shown, the stage is indicated by the number below each egg chamber. The expression of *stg* message can be seen weakly in the nurse cells of the stage 9 egg chamber, by stage 11 the transcript is being expressed to high levels. By stage 13 all of the transcript has been translocated to the oocyte from the nurse cells, ready for transcription on oviposition. The expression of the *twin* transcript is essentially identical to that shown for *stg*.

non-disjunction that occurs during *twine* meiosis. This suggests that *twine* function is required to maintain meiotic arrest by keeping $p34^{cdc2}$ dephosphorylated at tyrosine 15 and thereby active. In *Drosophila*, meiotic recombination only occurs in the female. Thus it might be expected that the mechanisms regulating entry into the first meiotic division might differ between the sexes. Furthermore, the mechanisms whereby the meiotic spindle is established in female meiosis is also quite characteristic, and could also help explain the differing requirement for twine (and $p34^{cdc2}$) function between male and female meiosis. A cytological study of spindle assembly in female meiosis led Theurkauf and Hawley (87) to propose that the major microtubule nucleating activity is provided by paired centromeres of the major chromosomes rather than the centrosomes. Such a different mechanism of spindle formation might be expected to be under different regulation and so not blocked by the twine mutation as in male meiosis. The bundling of microtubules emanating from the chromosomal nucleation points requires the activity of a kinesin-like molecule encoded by the *ncd* gene (88–90). In *ncd* mutants, this bundling is not complete, leading to spindles with broad poles that are often distorted around the metaphase plate and which resemble the abnormal twine spindles. Normally, an equilibrium exists at metaphase I in which the chromosomes that have undergone recombination remain at the equator still connected through chiasmata that will eventually ensure their correct segregation. The separation of non-exchange chromosomes is controlled in part through the kinesin-like protein encoded by *nod* (91). This imparts a force upon these chromosomes in the direction of the metaphase plate, and is counteracted by a pole-ward directed force that allows non-exchange chromosomes to move toward the poles in a size-dependent manner. In this way the tiny fourth chromosomes becomes positioned between the poles and the equator. This type of arrangement is not seen in the second meiotic metaphase in which all chromosomes align on the metaphase plate before undertaking the equational division. Premature separation of the fourth chromosome is seen in the multiple meiotic-like divisions that occur in *twine* oocytes. Mutation in *nod* leads to the dissociation of non-exchange chromosomes from the spindle or their premature movement to the pole. Similar events can also be seen in *twine* mutants.

Arrest at metaphase I in female meiosis is normally also dependent upon recombination having taken place to produce chiasmate bivalents (92). Thus in mutants that prevent recombination, the meiotic arrest at metaphase I does not occur. However, the absence of any significant zygotic lethality indicates that meiosis is otherwise normal and relies entirely upon the mechanisms for segregating non-exchange chromosomes. Thus the failure to arrest in *twine* mutants differs profoundly from the effects of mutations preventing meiotic recombination. It has been suggested that the formation of chiasmata leads to the establishment of mechanical tension at the metaphase plate that signals a meiotic block (92). The gross abnormalities observed in meiosis in *twine* females suggests that its function is likely to be a prerequisite for the block imposed through the mechanisms that senses the presence of chiasmata.

6. *Drosophila* genes regulating cell proliferation

Mutations that disrupt mitosis can affect all of the proliferating larval tissues resulting in small or missing imaginal discs, and underdeveloped brains. Weak mutant alleles of many mitotic genes will occasionally give rise to the production of adults which often have nicked wings, reduced numbers of bristles, and roughened eyes, reflecting abnormal cell division in the production of these structures. The gene *rough-deal* (93), for example, acquired its name not only because mutations in the gene result in lagging chromatids at anaphase leading to the unequal distribution of chromatids between daughter cells, but also because of the rough-eyed phenotype of mutant adults. Scores of genes have now been described that lead to this general phenotype.

Equally exciting are a set of mutations in over 50 genes that leads to overgrowth of the brain, imaginal discs, hematopoietic system, or gonads. Several of these may be formally equivalent to the tumour suppresser genes in humans that have been discovered through the identification of the genetic basis for familial disposition to cancer. The functions of these genes were considered in Chapter 9. To reiterate, the retinoblastoma gene encodes a nuclear protein that appears to control entry into S phase, the Wilms' tumour gene a Zn finger protein that is probably a transcription factor, the neurofibromatosis gene NF1, a GTPase activating protein probably involved in signal transduction, and the DDC gene, one copy of which is lost during the progression of colorectal carcinoma, encodes a putative cell adhesion molecule. The normal functions of these proteins during developing is poorly understood (see ref. 94 for review).

Seven tumour suppresser loci have been identified in *Drosophila* as recessive lethal mutations that lead to excessive cell proliferation. Mutations at *lethal(2)giant larvae (lgl)* (95, 96) and *lethal(1)discs large (dlg)* (97) result in cells losing their characteristic shape together with their ability to differentiate upon transplantation into wild-type hosts. The *dlg* gene product is localized at septate junctions of epithelial cells. These are thought to be the equivalent of tight junctions in vertebrate cells, functioning to restrict mobile proteins to apical or basal domains of polarized cells. The *dlg* protein contains an SH3 domain, a motif present in several cell surface-associated proteins with roles in signal transduction, and a catalytic domain with homology to guanylate kinase, suggesting a role in guanine nucleotide-mediated signal transduction. The *lgl* gene product is a 127 kDa protein that is localized to the inner surface of the cell membrane where it appears to be associated with protein kinase activity (98).

In the other group of *Drosophila* tumour suppresser gene mutants, there is hyperplastic overgrowth, although the cells retain their morphology and ability to differentiate. The *fat* gene is a member of this group that encodes a very large protein of the cadherin super-family. Cadherins are a family of calcium-dependent cell adhesion molecules that can mediate homophilic cell aggregation. Subclasses of these molecules may enable groups of cells to remain together during development (see ref. 99 for review). The 5000 amino acid *fat* protein contains no less than

34 contiguous cadherin domains, four EGF-like repeats, a transmembrane domain, and a 500 amino acid cytoplasmic domain. The role of this protein in regulating cell proliferation is not clear, but it seems possible that molecules such as this and the human DCC protein could have a signalling role beyond a cell adhesion role.

In recent years it has become clear that central pathways of signal transduction have been conserved. The activation of cell surface tyrosine kinase receptors, for example, transduces a cytoplasmic signal sequentially through ras and a guanine nucleotide exchange factor, raf kinase, MAP kinase kinase, MAP kinase, and ultimately to the nucleus whereupon transcription factors can be activated. This pathway is used in all eukaryotes. In mammalian cells, the signal transduced through this pathway can mediate either cell differentiation or bring about cell proliferation. The response appears to depend upon the nature of the signal received by the cell, but whether this represents a response to different ligand–receptor interactions or is more of a graded response to the intensity of an incoming signal is not yet clear. Studies in *Drosophila* have given a clear picture of the roles of this pathway in developmental processes. In the development of the terminal structures of the egg, for example, the uniformly distributed tyrosine kinase receptor encoded by *torso* becomes locally activated at the termini to trigger the above phosphorylation cascade to activate the transcription factors *tailless* and *huckebein*. The signal transduction pathway also serves in the differentiation of photoreceptor cell R7 in the eye, in response to the activation of the tyrosine kinase receptor *sevenless*. There are undoubtedly many more processes that use this signalling pathway yet to be discovered. The role of this pathway in regulating cell proliferation in *Drosophila* is less well understood, although it is clear that in addition to its role in the *torso* and *sevenless* signalling pathways, the mutant phenotype of *D-raf* indicates a role in the regulation of disc cell proliferation (62, 100). One of the major challenges that lies ahead is to understand the interplay between the signals that regulate cell differentiation and cell proliferation, and to determine how this interleaves with the mechanics of cell cycle control.

Acknowledgements

We thank Bruce Edgar and Victoria Foe for allowing the reproduction of figures. We acknowledge the support of the Cancer Research Campaign.

References

1. Howard, A. and Pelc, S. R. (1951) Synthesis of nucleoprotein in bean root cells. *Nature*, **167**, 599.
2. Lin, H. and Wolfner, M. (1989) Cloning and analysis of *fs(1)Ya*, a maternal effect gene required for the initiation of *Drosophila* embryogenesis. *Mol. Gen. Genet.*, **215**, 257.
3. Freeman, M., Nusslein-Volhard, C., and Glover, D. M. (1986). The dissociation of nuclear and centrosomal division in *gnu*, a nuclear replication mutant of *Drosophila*. *Cell*, **46**, 457.

4. Freeman, M. and Glover, D. M. (1987) The *gnu* mutation of *Drosophila* causes inappropriate DNA synthesis in unfertilised and fertilised eggs. *Genes Dev.*, **1,** 924.
5. Shamanski, F. and Orr-Weaver, T. (1991) The Drosophila *plutonium* and *pan gu* genes regulate entry into S phase at fertilisation. *Cell*, **66,** 1289.
6. Lin, H. and Wolfner, M. (1991) The *Drosophila* maternal effect gene *fs(1)Ya* encodes a cell cycle dependent nuclear envelope component required for embryonic mitoses. *Cell*, **64,** 49.
7. Szabad, J. and Bryant, P. J. (1982) The mode of action of the discless mutations in *Drosophila melanogaster*. *Dev. Biol.*, **93,** 240.
8. Gatti, M. and Baker, B. S. (1989) Genes controlling essential cell cycle functions in *Drosophila melanogaster*. *Genes Dev.*, **3,** 438.
9. Clegg, N. J., Whitehead, I. P., Brock, J. K., Sinclair, D. A., Mottus, R., Stromotich, G., Harrington, M. J., and Grigliatti, T. A. (1993). A cytogenetic analysis of chromosomal region 31 of *Drosophila melanogaster*. *Genetics*, **134,** 221.
10. Stern, B., Reid, G., Clegg, N. J., Grigliatti, T. A., and Lehner, C. F. (1993) Genetic analysis of the *Drosophila cdc2* homologue. *Development*, **117,** 219.
11. Glover, D. M. (1989) Mitosis in *Drosophila*. *J. Cell Sci.*, **92,** 137.
12. Glover, D. M. (1991) Mitosis in the *Drosophila* embryo—in and out of control. *Trends Genet.*, **7,** 125.
13. Gatti, M. and Goldberg, M. L. (1991) Mutations affecting cell division in *Drosophila*. *Methods Cell Biol.*, **35,** 543.
14. Gonzalez, C., Alphey, L., and Glover, D. M. (1994) Cell cycle genes of *Drosophila*. *Adv. Genet.*, **31,** 79.
15. Foe, V. E., Odell, G. M., and Edgar, B. A. (1993) Mitosis and morphogenesis in the *Drosophila* embryo: point and counterpoint. In *The development of Drosophila melanogaster*. M. Bate and A. Martinez-Arias (ed.), p. 149. Cold Spring Harbor Press, New York.
16. Sunkel, C. and Glover, D. (1988) *polo*, a mitotic mutant of *Drosophila* displaying abnormal spindle poles. *J. Cell Sci.*, **89,** 25.
17. Llamazares, S., Moreira, A., Tavares, A., Girdham, C., Spruce, B., Gonzalez, C., Karess, R., Glover, D., and Sunkel, C. (1991) *polo* encodes a protein kinase homologue required for mitosis in *Drosophila*. *Genes Dev.*, **5,** 2153.
18. Fenton, B. and Glover, D. (1993) A conserved mitotic kinase active at late anaphase–telophase in syncytial *Drosophila* embryos. *Nature*, **363,** 637.
19. Kitada, K., Johnson, A. L., Johnston, L. H., and Sugino, A. (1993) A multicopy suppressor gene of the *Saccharomyces cerevisiae* G1 cell cycle mutant gene *DBF4* encodes a protein kinase and is identified as *CDC5*. *Mol. Cell. Biol.*, **13,** 4445.
20. Clay, F., McEwen, S. J., Bertoncello, I., Wilks, A. F., and Dunn, A. (1993) Identification and cloning of a protein kinase-encoding mouse gene, *plk*, related to the *polo* gene of *Drosophila*. *Proc. Natl. Acad. Sci. USA*, **90,** 4882.
21. Simmons, D. L., Neel, B. G., Stevens, R., Evett, G., and Erikson, R. (1992) Identification of an early growth response gene encoding a novel putative protein kinase. *Mol. Cell. Biol.*, **12,** 4164.
22. Byers, B. and Goetsch, L. (1974) Duplication of spindle plaques and integration of the yeast cell cycle. *Cold Spring Harbor Symp. Quant. Biol.*, **38,** 123.
23. Schild, D. and Byers, B. (1980) Diploid spore formation and other meiotic effects of 2 cell-division-cycle mutations of *Saccharomyces cerevisiae*. *Genetics*, **96,** 859.
24. Wood, J. S. and Hartwell, L. H. (1982) A dependent pathway of gene functions leading to chromosome segregation in *Saccharomyces cerevisiae*. *J. Cell Biol.*, **94,** 718.

25. Gonzalez, C., Sunkel, C., and Glover, D. M. (1991) unpublished observations.
26. Hara, K., Tydeman, P., and Kirschner, M. (1980) A cytoplasmic clock with the same period as the division cycle in *Xenopus* eggs. *Proc. Natl. Acad. Sci. USA*, **77,** 462.
27. Zalokar, M. and Erk, I. (1976) Division and migration of nuclei during early embryogenesis of *Drosophila melanogaster*. *J. Microscopie Biol. Cell*, **25,** 97–106.
28. Baker, J., Theurkauf, W. E., and Schubiger, G. (1993). Dynamic changes in microtubule configuration correlate with nuclear migration in the pre-blastoderm *Drosophila* embryo. *J. Cell. Biol.*, **122,** 113.
29. Raff, J. and Glover, D. M. (1988) Nuclear and cytoplasmic mitotic cycles continue in *Drosophila* embryos in which DNA synthesis is inhibited with aphidicolin. *J. Cell Biol.*, **107,** 2009.
30. Edgar, B. A., Sprenger, F., Duronio, R. J., Leopold, P., and O'Farrell, P. (1994) Distinct molecular mechanisms time mitosis at four successive stages of *Drosophila* embryogenesis. *Genes Dev.*, **8,** 440.
31. Foe, V. E. and Alberts, B. M. (1983) Studies of nuclear and cytoplasmic behaviour during the five mitotic cycles that precede gastrulation in *Drosophila* embryogenesis. *J. Cell Sci.*, **61,** 31.
32. Hiraoka, Y., Minden, J. S., Swedlow, J. R., Sedat, J. W., and Agard, D. A (1990) Focal points for chromosome condensation and decondensation revealed by three-dimensional *in vivo* time-lapse microscopy. *Nature*, **342,** 293.
33. Maldonado-Codina, G. and Glover, D. M. (1992) Cyclins A and B associate with chromatin and the polar regions of the spindle, respectively, and do not undergo complete breakdown at anaphase in syncytial *Drosophila* embryos. *J. Cell Biol.*, **116,** 967.
34. Edgar, B. A., Kiehle, C. P., and Schubiger, G. (1986) Cell cycle control by the nucleo-cytoplasmic ratio in early *Drosophila* development. *Cell*, **44,** 365.
35. Foe, V. (1989). Mitotic domains reveal early commitment of cells in *Drosophila* development. *Development*, **107,** 1.
36. Edgar, B. and O'Farrell, P. (1989) Genetic control of the cell division patterns in the *Drosophila* embryo. *Cell*, **57,** 177.
37. Edgar, B. and O'Farrell, P. (1990) The three post-blastoderm cell cycles of *Drosophila* are regulated by *string*. *Cell*, **62,** 469.
38. Maldonado-Codina, G., Llamazares, S., and Glover, D. M. (1993) Heat shock results in cell cycle delay and synchronisation of mitotic domains in cellularised *Drosophila melanogaster* embryos. *J. Cell Sci.*, **105,** 711.
39. Lehner, C. F. and O'Farrell, P. H. (1989) Expression and function of *Drosophila* cyclin A during embryonic cell cycle progression. *Cell*, **56,** 957.
40. Knoblich, J. A. and Lehner, C. F. (1993) Synergistic action of *Drosophila* cyclins A and B during the G2–M transition. *EMBO J.*, **12,** 66.
41. Whitfield, W. G., Gonzalez, C., Maldonado-Codina, G., and Glover, D M. (1990) The A- and B-type cyclins of *Drosophila* are accumulated and destroyed in temporally distinct events that define separable phases of the G2–M transition. *EMBO J.*, **9,** 2563.
42. Lehner, C. G. and O'Farrell, P. H. (1990) The roles of cyclins A and B in mitotic control. *Cell*, **61,** 535.
43. Murray, A. W., Solomon, M. J., and Kirschner, M. W. (1989) The role of cyclin synthesis and degradation in the control of maturation promoting factor activity. *Nature*, **339,** 280.
44. Holloway, S. L., Glotzer, M., King, R. W., and Murray, A. W. (1993) Anaphase is

initiated by proteolysis rather than by the inactivation of maturation promoting factor. *Cell*, **73,** 1393.
45. Surana, U. H., Amon, A., Dowzer, C., McGrew, J., Byers, B., and Nasmyth, K. (1993) Destruction of the *CDC28/CLB* mitotic kinase is not required for the metaphase to anaphase transition. *EMBO J.*, **12,** 1969.
46. Rimmington, G. A., Dalby, B., and Glover, D. M. (1994) Expression of N-terminally truncated cyclin B in the *Drosophila* larval brain leads to mitotic delay in late anaphase. *J. Cell. Sci.*, **107,** 2729.
47. Dawson, I. A., Roth, S., Akam, M., and Artavanis-Tsakonas, S. (1993) Mutations of the *fizzy* locus cause metaphase arrest in *Drosophila melanogaster* embryos. *Development*, **117,** 359.
48. Philp, A. and Dawson, I. (1994) Personal communication.
49. Philp, A. V., Axton, J. M., Saunders, R. D. C., and Glover, D. M. (1993) Mutations in the *Drosophila melanogaster* gene *three rows* permit aspects of mitosis to continue in the absence of chromatid segregation. *J. Cell Sci.*, **106,** 87.
50. D'Andrea, R. J., Stratmann, R., Lehner, C. F., John, U. P., and Saint, R. (1993) The *three rows* gene of *Drosophila melanogaster* encodes a novel protein that is required for chromosome disjunction during mitosis. *Mol. Biol. Cell*, **4,** 1161.
51. Sullivan, W., Fogarty, P., and Theurkauf, W. (1993) Mutations affecting the cytoskeletal organisation of syncytial *Drosophila* embryos. *Development*, **118,** 1245.
52. Hime, G. and Saint, R. (1992) Zygotic expression of the *pebble* locus is required for cytokinesis during the postblastoderm mitoses of *Drosophila*. *Development*, **114,** 165.
53. Lehner, C. F. (1992) The *pebble* gene is required for cytokinesis in *Drosophila*. *J. Cell Sci.*, **103,** 1021.
54. Karess, R. E., Chang, X. J., Edwards, K. A., Kulkarni, S., Aguilera, I., and Kielhart, D. P. (1991). The regulatory light chain of non-muscle myosin is encoded by *spaghetti-squash*, a gene required for cytokinesis in *Drosophila*. *Cell*, **65,** 1177.
55. Lahue, E. E., Smith, A. V., and Orr-Weaver, T. L. (1991) A novel cyclin gene from *Drosophila* complements CLN function in yeast. *Genes Dev.*, **5,** 2166.
56. Leopold, P. and O'Farrell, P. H. (1991) An evolutionary conserved cyclin homologue from *Drosophila* rescues yeast deficient in G1 cyclins. *Cell*, **66,** 1207.
57. Finlay, R. and Brent, R. (1992) Personal communication.
58. Richardson, H. E., O'Keefe, L. V., Reed, S. I., and Saint, R. (1993) A *Drosophila* G1-specific cyclin E homologue exhibits different modes of expression during embryogenesis. *Development*, **119,** 673.
59. Knoblich, J. A., Sauer, K., Jones, L., Richardson, H., Saint, R., and Lehner, C. F. (1994) Cyclin E controls S phase progression and its down-regulation during *Drosophila* embryogenesis is required for the arrest of cell proliferation. *Cell*, **77,** 1.
60. Smith, A. V. and Orr-Weaver, T. L. (1991) The regulation of the cell cycle during *Drosophila* embryogenesis: the transition to polyteny. *Development*, **112,** 997.
61. Hayashi, S., Hirose, S., Metcalfe, T., and Shirras, A. D. (1993) Control of imaginal cell development by the *escargot* gene of *Drosophila*. *Development*, **118,** 105.
62. Nishida, Y., Hata, M., Ayaki, T., Ryo, H., Yamagata, M., Shimizo, K., and Nishizuka, Y. (1988) Proliferation of both somatic and germ cells is affected in the *Drosophila* mutants of *raf* proto-oncogene. *EMBO J.*, **7,** 775.
63. Reed, B. (1992) The genetic analysis of endoreduplication in *Drosophila melanogaster*. Ph.D. thesis, University of Cambridge, England.
64. Ebens, A. J., Garren, H., Cheyette, B. N. R., and Zipursky, S. L. (1993) The *Drosophila*

anachronism locus—a glycoprotein secreted by glia inhibits neuroblast proliferation. *Cell*, **74,** 15.

65. Selleck, S. B., Gonzalez, C., Glover, D. M., and White, K. (1992) Regulation of the G1–S transition in post-embryonic neuronal precursors by axon ingrowth. *Nature*, **355,** 253.
66. Wolff, T. and Ready, D. F. (1993) Pattern formation in the *Drosophila* retina. In *The development of Drosophila melanogaster*. M. Bate and A. Martinez-Arias (ed.), p. 1277. Cold Spring Harbor Press, New York.
67. Dickson, B. and Hafan, E. (1993) Genetic dissection of eye development in *Drosophila*. In *The development of Drosophila melanogaster*. M. Bate and A. Martinez-Arias (ed.), p. 1327. Cold Spring Harbor Press, New York.
68. Alphey, L., Jimenez, J., White-Cooper, H., Dawson, I., Nurse, P., and Glover, D. M. (1992) *twine*, a *cdc25* homologue that functions in the male and female germlines of *Drosophila*. *Cell*, **69,** 977.
69. Thomas, B. J., Gunning, D. A., Cho, J., and Zipursky, S. L. (1994) Cell cycle progression in the developing *Drosophila* eye: *roughex* encodes a novel protein required for the establishment of G1. *Cell*, **77,** 1003.
70. Saint, R. (1994) Personal communication.
71. St. Johnston, D. (1993) Pole plasm and the posterior group genes. In *The development of Drosophila melanogaster*. M. Bate and A. Martinez-Arias (ed.), p. 325. Cold Spring Harbor Press, New York.
72. Raff, J. W. and Glover, D. M. (1989) Centrosomes, and not nuclei, initiate pole cell formation in *Drosophila* embryos. *Cell*, **57,** 611.
73. Sullivan, W., Minden, J. S., and Alberts, B. M. (1990) *daughterless-abo-like*, a *Drosophila* maternal-effect mutation that exhibits abnormal centrosome separation during the late blastoderm divisions. *Development*, **110,** 311.
74. Dalby, B. and Glover, D. (1992) 3′ non-translated sequences in *Drosophila* cyclin B transcripts direct posterior pole accumulation late in oogenesis and peri-nuclear association in syncytial embryos. *Development*, **115,** 989.
75. Raff, J. W., Whitfield, W. G. F., and Glover, D. M. (1990) Two distinct mechanisms localise cyclin B transcripts in syncytial *Drosophila* embryos. *Development*, **110,** 1249.
76. Dalby, B. and Glover, D. (1993) Discrete sequence elements control the posterior pole accumulation and translational repression of maternal cyclin B RNA in *Drosophila*. *EMBO J.*, **12,** 1219.
77. Sonnenblick, B. (1950) The early embryology of *Drosophila melanogaster*. In *Biology of Drosophila*. Demerec, M. (ed.). Hafner Publishing Company, New York.
78. Wharton, R. and Struhl, G. (1991) RNA regulatory elements mediate control of *Drosophila* body pattern by the posterior morphogen *nanos*. *Cell*, **67,** 955.
79. Hawley, R. S. and Theurkauf, W. E. (1993) Requiem for distributive segregation—achiasmate segregation in *Drosophila* females. *Trends Genet.*, **9,** 310.
80. Jimenez, J., Alphey, L., Nurse, P., and Glover, D. M. (1990) Complementation of fission yeast *cdc2*ts and *cdc25*ts mutants identifies two cell cycle genes from *Drosophila*: a *cdc2* homologue and *string*. *EMBO J.*, **9,** 3565.
81. Courtot, C., Frankhauser, C., Simanis, V., and Lehner, C. F. (1992) The *Drosophila cdc25* homologue *twine* is required for meiosis. *Development*, **116,** 405.
82. White-Cooper, H., Alphey, L., and Glover, D. M. (1993) The *cdc25* homologue *twine* is required for only some aspects of the entry into meiosis in *Drosophila*. *J. Cell Sci.*, **106,** 1035.

83. Clarke, P., Leiss, D., Pagano, M., and Karsenti, E. (1992) Cyclin A- and cyclin B-dependent protein kinases are regulated by different methods in *Xenopus* egg extracts. *EMBO J.*, **11,** 1757.
84. Devault, A., Fesquet, D., Cavadores, J.-C., Garrigues, A.-M., Labbé, J.-C., Lorca, T., Picard, A., Philippe, M., and Dorée, M. (1992) Cyclin A potentiates maturation-promoting factor in the early *Xenopus* embryo via inhibition of the tyrosine kinase that phosphorylates cdc2. *J. Cell Biol.*, **188,** 1109.
85. Gonczy, P., Thomas, B. T., and DiNardo, S. (1994). *roughex* is a dose-dependent regulator of the second meiotic division during *Drosophila* spermatogenesis. *Cell*, **77,** 1015.
86. White-Cooper, H. (1993) Unpublished observations.
87. Theurkauf, W. E. and Hawley, R. S. (1992) Role of the nod protein, a kinesin homologue, in meiotic spindle assembly. *J. Cell Biol.*, **116,** 1167.
88. Walker, R. A., Salmon, E. D., and Endow, S. A. (1990) The *Drosophila claret* segregation protein is a minus-end directed motor molecule. *Nature*, **347,** 780.
89. Hatsumi, M. and Endow, S. (1992) Mutants of the microtubule protein, *non claret dysjunctional*, affect spindle structure and chromosome movement in meiosis and mitosis. *J. Cell Sci.*, **101,** 547.
90. Sequeria, W., Nelson, C. R., and Szauter, P. (1989) Genetic analysis of the *claret* locus of *Drosophila melanogaster*. *Genetics*, **123,** 511.
91. Zhang, P. and Hawley, R. S. (1990) The genetic analysis of distributive segregation in *Drosophila melanogaster*. II. Further genetic analysis of the *nod* locus. *Genetics*, **125,** 115.
92. McKim, K. S., Jang, J. K., Therkauf, W. E., and Hawley, R. S. (1993) The mechanical basis of meiotic metaphase arrest. *Nature*, **362,** 346.
93. Karess, R. E. and Glover, D. M. (1989) *rough deal:* a gene required for proper mitotic segregation in *Drosophila*. *J. Cell Biol.*, **109,** 2951.
94. Marshall, C. J. (1991) Tumor suppressor genes. *Cell*, **64,** 313.
95. Mechler, B. M., McGinnis, W., and Gehring, W. J. (1985) Molecular cloning of *lethal(2)giant larvae*, a recessive oncogene of *Drosophila melanogaster*. *EMBO J.*, **4,** 1551.
96. Klaembt, C., Lutzelschwab, R., Muller, S., Rossa, R., Schmidt, O., and Totzke, F. (1989) The *Drosophila melanogaster 1(2)gl* gene encodes a protein homologous to the cadherin cell-adhesion molecular family. *Dev. Biol.*, **133,** 425.
97. Woods, D. F., and Bryant, P. J. (1991) The *discs-large* tumor suppressor gene of *Drosophila* encodes a guanylate kinase homolog localised at septate junctions. *Cell*, **66,** 451.
98. Mechler, B. (1994) Personal communication.
99. Takeichi, M. (1990) Cadherins—a molecular family important in selective cell–cell adhesion. *Annu. Rev. Biochem.*, **59,** 237.
100. Ambrosio, L., Mahowald, A. P., and Perrimon, N. (1989) Requirement of the *Drosophila raf* homologue for torso function. *Nature*, **342,** 288.

Index